PRIMES OF THE FORM $x^2 + ny^2$

PRIMES OF THE FORM $x^2 + ny^2$

Fermat, Class Field Theory, and Complex Multiplication

David A. Cox
Department of Mathematics
Amherst College
Amherst, Massachusetts

A WILEY-INTERSCIENCE PUBLICATION

JOHN WILEY & SONS

New York / Chichester / Brisbane / Toronto / Singapore

Library of Congress Cataloging in Publication Data:

Cox, David A.

Primes of the form $x^2 + ny^2$: Fermat, class field theory, and complex multiplication / David A. Cox

p. cm.

Bibliography: p.

Includes index.

1. Numbers, Prime. I. Title. II. Title: Primes of the form x^2 plus ny^2.

QA246.C69 1989

512'.72—dc19 89-5555

ISBN 0-471-50654-0

CIP

Printed in the United States of America

10 9 8 7 6 5 4 3 2 1

CONTENTS

CHAPTER TWO: CLASS FIELD THEORY

CHAPTER THREE: COMPLEX MULTIPLICATION

PREFACE

Several years ago while reading Weil's *Number Theory: An Approach Through History*, I noticed a conjecture of Euler concerning primes of the form $x^2 + 14y^2$. That same week I picked up Cohn's *A Classical Invitation to Algebraic Numbers and Class Fields* and saw the same example treated from the point of view of the Hilbert class field. The coincidence made it clear that something interesting was going on, and this book is my attempt to tell the story of this wonderful part of mathematics.

I am an algebraic geometer by training, and number theory has always been more of an avocation than a profession for me. This will help explain some of the curious omissions in the book. There may also be errors of history or attribution (for which I take full responsibility), and doubtless some of the proofs can be improved. Corrections and comments are welcome!

I would like to thank my colleagues in the number theory seminars of Oklahoma State University and the Five Colleges (Amherst College, Hampshire College, Mount Holyoke College, Smith College and the University of Massachusetts) for the opportunity to present material from this book in preliminary form. Special thanks go to Dan Flath and Peter Norman for their comments on earlier versions of the manuscript. I am also grateful to the reference librarians at Amherst College and Oklahoma State University for their help in obtaining books through interlibrary loan.

DAVID A. COX

Amherst, Massachusetts
August 1989

NOTATION

The following standard notation will be used throughout the book.

$\mathbb{Z}$	The integers.
$\mathbb{Q}$	The rational numbers.
$\mathbb{R}$	The real numbers.
$\mathbb{C}$	The complex numbers.
$\mathfrak{h}$	The upper half plane $\{x + iy \in \mathbb{C} : y > 0\}$.
$\mathbb{Z}/n\mathbb{Z}$	The ring of integers modulo n.
$[a] \in A/B$	The coset of $a \in A$ in the quotient A/B.
R^*	The group of units in a commutative ring R with identity.
$\mathrm{GL}(2, R)$	The group of invertible matrices $\left(\begin{smallmatrix} a & b \\ c & d \end{smallmatrix}\right)$, $a, b, c, d \in R$.
$\mathrm{SL}(2, R)$	The subgroup of $\mathrm{GL}(2, R)$ of matrices with determinant 1.
$\mathrm{Gal}(L/K)$	The Galois group of the field extension $K \subset L$.
$\mathcal{O}_K$	The ring of algebraic integers in a finite extension K of $\mathbb{Q}$.
$\zeta_n = e^{2\pi i/n}$	The standard primitive nth root of unity.
$[a, b]$	The set $\{ma + nb : m, n \in \mathbb{Z}\}$.
$\gcd(a, b)$	The greatest common divisor of the integers a and b.
$\lvert S \rvert$	The number of elements in a finite set S.
Q.E.D.	The end of a proof or the absence of a proof.

PRIMES OF THE FORM $x^2 + ny^2$

INTRODUCTION

Most first courses in number theory or abstract algebra prove a theorem of Fermat which states that for an odd prime p,

$$p = x^2 + y^2, \ x, y \in \mathbb{Z} \iff p \equiv 1 \bmod 4.$$

This is only the first of many related results that appear in Fermat's works. For example, Fermat also states that if p is an odd prime, then

$$p = x^2 + 2y^2, \ x, y \in \mathbb{Z} \iff p \equiv 1, 3 \bmod 8$$

$$p = x^2 + 3y^2, \ x, y \in \mathbb{Z} \iff p = 3 \text{ or } p \equiv 1 \bmod 3.$$

These facts are lovely in their own right, but they also make one curious to know what happens for primes of the form $x^2 + 5y^2$, $x^2 + 6y^2$, etc. This leads to the basic question of the whole book, which we formulate as follows:

Basic Question 0.1. *Given a positive integer n, which primes p can be expressed in the form*

$$p = x^2 + ny^2$$

where x and y are integers?

We will answer this question completely, and along the way we will encounter some remarkably rich areas of number theory. The first steps will be easy, involving only quadratic reciprocity and the elementary theory of quadratic forms in two variables over $\mathbf{Z}$. These methods work nicely in the special cases considered above by Fermat. Using genus theory and cubic and biquadratic reciprocity, we can treat some more cases, but elementary methods fail to solve the problem in general. To proceed further, we need class field theory. This provides an abstract solution to the problem, but doesn't give explicit criteria for a particular choice of n in $x^2 + ny^2$. The final step uses modular functions and complex multiplication to show that for a given n, there is an algorithm for answering our question of when $p = x^2 + ny^2$.

This book has several goals. The first, to answer the basic question, has already been stated. A second goal is to bridge the gap between elementary number theory and class field theory. Although our basic question is simple enough to be stated in any beginning course in number theory, we will see that its solution is intimately bound up with higher reciprocity laws and class field theory. A related goal is to provide a well-motivated introduction to the classical formulation of class field theory. This will be done by carefully stating the basic theorems and illustrating their power in various concrete situations.

Let us summarize the contents of the book in more detail. We begin in Chapter One with the more elementary approaches to the problem, using the works of Fermat, Euler, Lagrange, Legendre and Gauss as a guide. In §1, we will give Euler's proofs of the above theorems of Fermat for primes of the form $x^2 + y^2$, $x^2 + 2y^2$ and $x^2 + 3y^2$, and we will see what led Euler to discover quadratic reciprocity. We will also discuss the conjectures Euler made concerning $p = x^2 + ny^2$ for $n > 3$. Some of these conjectures, such as

$$p = x^2 + 5y^2 \iff p \equiv 1, 9 \bmod 20, \tag{0.2}$$

are similar to Fermat's theorems, while others, like

$$p = x^2 + 27y^2 \iff \begin{cases} p \equiv 1 \bmod 3 \text{ and } 2 \text{ is a} \\ \text{cubic residue modulo } p, \end{cases}$$

are quite unexpected. For later purposes, note that this conjecture can be written in the following form:

$$p = x^2 + 27y^2 \iff \begin{cases} p \equiv 1 \bmod 3 \text{ and } x^3 \equiv 2 \bmod p \\ \text{has an integer solution.} \end{cases} \tag{0.3}$$

In §2, we will study Lagrange's theory of positive definite quadratic forms. After introducing the basic concepts of reduced form and class number, we will develop an elementary form of genus theory which will enable us to prove (0.2) and similar theorems. Unfortunately, for cases like (0.3), genus theory can only prove the partial result that

$$(0.4) \qquad p = \begin{Bmatrix} x^2 + 27y^2 \\ \text{or} \\ 4x^2 + 2xy + 7y^2 \end{Bmatrix} \Longleftrightarrow p \equiv 1 \bmod 3.$$

The problem is that $x^2 + 27y^2$ and $4x^2 + 2xy + 7y^2$ lie in the same genus and hence can't be separated by simple congruences. We will also discuss Legendre's tentative attempts at a theory of composition.

While the ideas of genus theory and composition were already present in the works of Lagrange and Legendre, the real depth of these theories wasn't revealed until Gauss came along. In §3 we will present some basic results in Gauss' *Disquisitiones Arithmeticae*, and in particular we will study the remarkable relationship between genus theory and composition. But for our purposes, the real breakthrough came when Gauss used cubic reciprocity to prove Euler's conjecture (0.3) concerning $p = x^2 + 27y^2$. In §4 we will give a careful statement of cubic reciprocity, and we will explain how it can be used to prove (0.3). Similarly, biquadratic reciprocity can be used to answer our question for $x^2 + 64y^2$. We will see that Gauss clearly recognized the role of higher reciprocity laws in separating forms of the same genus. This section will also begin our study of algebraic integers, for in order to state cubic and biquadratic reciprocity, we must first understand the arithmetic of the rings $\mathbb{Z}[e^{2\pi i/3}]$ and $\mathbb{Z}[i]$.

To go further requires class field theory, which is the topic of Chapter Two. We will begin in §5 with the Hilbert class field, which is the maximal unramified Abelian extension of a given number field. This will enable us to prove the following general result:

Theorem 0.5. *Let $n \equiv 1, 2 \bmod 4$ be a positive squarefree integer. Then there is an irreducible polynomial $f_n(x) \in \mathbb{Z}[x]$ such that for a prime p dividing neither n nor the discriminant of $f_n(x)$,*

$$p = x^2 + ny^2 \Longleftrightarrow \begin{cases} (-n/p) = 1 \text{ and } f_n(x) \equiv 0 \bmod p \\ \text{has an integer solution.} \end{cases}$$

While the statement of Theorem 0.5 is elementary, the polynomial $f_n(x)$ is quite sophisticated: it is the minimal polynomial of a primitive element of the Hilbert class field L of $K = \mathbb{Q}(\sqrt{-n})$.

As an example of this theorem, we will study the case $n = 14$. We will show that the Hilbert class field of $K = \mathbb{Q}(\sqrt{-14})$ is $L = K(\alpha)$, where $\alpha = \sqrt{2\sqrt{2} - 1}$. By Theorem 0.5, this will show that for an odd prime p,

$$(0.6) \qquad p = x^2 + 14y^2 \Longleftrightarrow \begin{cases} (-14/p) = 1 \text{ and } (x^2 + 1)^2 \equiv 8 \bmod p \\ \text{has an integer solution,} \end{cases}$$

which answers our basic question for $x^2 + 14y^2$. The Hilbert class field will also enable us in §6 to give new proofs of the main theorems of genus theory.

The theory sketched so far is very nice, but there are some gaps in it. The most obvious is that the above results for $x^2 + 27y^2$ and $x^2 + 14y^2$ ((0.3) and (0.6) respectively) both follow the same format, but (0.3) does *not* follow from Theorem 0.5, for $n = 27$ is *not* squarefree. There should be a unified theorem that works for *all* positive n, yet the proof of Theorem 0.5 breaks down for general n because $\mathbb{Z}[\sqrt{-n}]$ is not in general the full ring of integers in $\mathbb{Q}(\sqrt{-n})$.

The goal of §§7–9 is to show that Theorem 0.5 holds for *all* positive integers n. This, in fact, is the main theorem of the whole book. In §7 we will study the rings $\mathbb{Z}[\sqrt{-n}]$ for general n, which leads to the concept of an *order* in an imaginary quadratic field. In §8 we will summarize the main theorems of class field theory and the Čebotarev Density Theorem, and in §9 we will introduce a generalization of the Hilbert class field called the ring class field, which is a certain (possibly ramified) Abelian extension of $\mathbb{Q}(\sqrt{-n})$ determined by the order $\mathbb{Z}[\sqrt{-n}]$. Then, in Theorem 9.2, we will use the Artin Reciprocity Theorem to show that Theorem 0.5 holds for *all* $n > 0$, where the polynomial $f_n(x)$ is now the minimal polynomial of a primitive element of the above ring class field. To give a concrete example of what this means, we will apply Theorem 9.2 to the case $x^2 + 27y^2$, which will give us a class field theory proof of (0.3). In §§8 and 9 we will also discuss how class field theory is related to higher reciprocity theorems.

The major drawback to the theory presented in §9 is that it is not constructive: for a given $n > 0$, we have no idea how to find the polynomial $f_n(x)$. From (0.3) and (0.6), we know $f_{27}(x)$ and $f_{14}(x)$, but the methods used in these examples hardly generalize. Chapter Three will use the theory of complex multiplication to remedy this situation. In §10 we will study elliptic functions and introduce the idea of complex multiplication, and then in §11 we will discuss modular functions and show that the j-function can be used to generate ring class fields. As an example of the wonderful formulas that can be proved, in §12 we will give Weber's computation that

$$j(\sqrt{-14}) = 2^3\left(323 + 228\sqrt{2} + \left(231 + 161\sqrt{2}\right)\sqrt{2\sqrt{2}-1}\right)^3.$$

These methods will also enable us to prove the Baker–Heegner–Stark Theorem on imaginary quadratic fields of class number 1. The final section of the book will discuss the class equation, which is the minimal polynomial of $j(\sqrt{-n})$. We will learn how to compute the class equation, and this in turn will lead to a constructive solution of $p = x^2 + ny^2$. We will then describe some more recent work by Deuring and by Gross and Zagier. In 1946 Deuring proved a result about the difference of singular j-invariants, which implies an especially elegant version of our main theorem, and drawing on Deuring's work, Gross and Zagier discovered yet more remarkable properties of the class equation. The book will end with a discussion of elliptic curves and an application of the class equation to primality testing.

Number theory is usually taught at three levels, as an undergraduate course, a beginning graduate course, or a more advanced graduate course. These levels correspond roughly to the three chapters of the book. Chapter One requires only beginning number theory (up to quadratic reciprocity) and a semester of abstract algebra. Since the proofs of quadratic, cubic and biquadratic reciprocity are omitted, this book would be best suited as a supplementary text in a beginning course. For Chapter Two, the reader should know Galois theory and some basic facts about algebraic number theory (these are reviewed in §5), but no previous exposure to class field theory is assumed. The theorems of class field theory are stated without proof, so that this book would be most useful as a supplement to the topics covered in a first graduate course. Chapter Three requires a knowledge of complex analysis, but otherwise it is self-contained. (Brief but complete accounts of the Weierstrass $\wp$-function and modular functions are included in §§10 and 11.) This portion of the book should be suitable for use in a graduate seminar.

There are exercises at the end of each section, many of which consist of working out the details of arguments sketched in the text. Readers learning this material for the first time should find the exercises to be useful, while more sophisticated readers may skip them without loss of continuity.

Many important (and relevant) topics are not covered in the book. An obvious omission in Chapter One concerns forms such as $x^2 - 2y^2$, which were certainly considered by Fermat and Euler. Questions of this sort lead to Pell's equation and the class field theory of real quadratic fields. We have also ignored the problem of representing arbitrary integers, not just primes, by quadratic forms, and there are interesting questions to ask about the *number* of such representations (this material is covered in Grosswald's recent book [47]). In Chapter Two we do not discuss adeles or ideles—we give only a classical formulation of class field theory. For a more modern treatment, see either Neukirch [80] or Weil [104]. We also do not do justice to the use of analytic methods in number theory. For a nice introduction in the case of quadratic fields, see Zagier [111]. Our treatment of elliptic curves is rather incomplete. See Husemöller [58] or Silverman [93] for the basic theory, while more advanced topics are covered by Lang [73] and Shimura [90].

There are many books which touch on the number theory encountered in studying the problem of representing primes by $x^2 + ny^2$. Four books that we particularly recommend are Cohn's *A Classical Invitation to Algebraic Numbers and Class Fields* [19], Lang's *Elliptic Functions* [73], Scharlau and Opolka's *From Fermat to Minkowski* [86], and Weil's *Number Theory: An Approach Through History* [106]. These books, as well as others to be found in the bibliography, open up an extraordinarily rich area of mathematics. The purpose of this book is to reveal some of this richness and to encourage the reader to learn more about it.

CHAPTER ONE

FROM FERMAT TO GAUSS

§1. FERMAT, EULER AND QUADRATIC RECIPROCITY

In this section we will discuss primes of the form $x^2 + ny^2$, where n is a fixed positive integer. Our starting point will be the three theorems of Fermat

$$
\begin{aligned}
& p = x^2 + y^2,\ x,y \in \mathbb{Z} \iff p \equiv 1 \bmod 4 \\
(1.1) \qquad & p = x^2 + 2y^2,\ x,y \in \mathbb{Z} \iff p \equiv 1 \text{ or } 3 \bmod 8 \\
& p = x^2 + 3y^2,\ x,y \in \mathbb{Z} \iff p = 3 \text{ or } p \equiv 1 \bmod 3
\end{aligned}
$$

mentioned in the introduction. The goals of §1 are to prove (1.1) and, more importantly, to get a sense of what's involved in studying the equation $p = x^2 + ny^2$ when $n > 0$ is arbitrary. This last question was best answered by Euler, who spent 40 years proving Fermat's theorems and thinking about how they can be generalized. Our exposition will follow some of Euler's papers closely, both in the theorems proved and in the examples studied. We will see that Euler's strategy for proving (1.1) was one of the primary things that led him to discover quadratic reciprocity, and we will also discuss some of his conjectures concerning $p = x^2 + ny^2$ for $n > 3$. These remarkable conjectures touch on quadratic forms, composition, genus theory, cubic and biquadratic reciprocity, and will keep us busy for the rest of the chapter.

A. Fermat

Fermat's first mention of $p = x^2 + y^2$ occurs in a 1640 letter to Mersenne [35, Vol. II, p. 212], while $p = x^2 + 2y^2$ and $p = x^2 + 3y^2$ come later, first appearing in a 1654 letter to Pascal [35, Vol. II, pp. 310–314]. Although no proofs are given in these letters, Fermat states the results as theorems. Writing to Digby in 1658, he repeats these assertions in the following form:

> Every prime number which surpasses by one a multiple of four is composed of two squares. Examples are 5, 13, 17, 29, 37, 41, etc.
>
> Every prime number which surpasses by one a multiple of three is composed of a square and the triple of another square. Examples are 7, 13, 19, 31, 37, 43, etc.
>
> Every prime number which surpasses by one or three a multiple of eight is composed of a square and the double of another square. Examples are 3, 11, 17, 19, 41, 43, etc.

Fermat adds that he has solid proofs—"firmissimis demonstratibus" [35, Vol. II, pp. 402–408 (Latin), Vol. III, pp. 314–319 (French)].

The theorems (1.1) are only part of the work that Fermat did with $x^2 + ny^2$. For example, concerning $x^2 + y^2$, Fermat knew that a positive integer N is the sum of two squares if and only if the quotient of N by its largest square factor is a product of primes congruent to 1 modulo 4 [35, Vol. III, Obs. 26, pp. 256–257], and he knew the number of different ways N can be so represented [35, Vol. III, Obs. 7, pp. 243–246]. Fermat also studied forms beyond $x^2 + y^2$, $x^2 + 2y^2$ and $x^2 + 3y^2$. For example, in the 1658 letter to Digby quoted above, Fermat makes the following conjecture about $x^2 + 5y^2$, which he admits he can't prove:

> If two primes, which end in 3 or 7 and surpass by three a multiple of four, are multiplied, then their product will be composed of a square and the quintuple of another square.
>
> Examples are the numbers 3, 7, 23, 43, 47, 67, etc. Take two of them, for example 7 and 23; their product 161 is composed of a square and the quintuple of another square. Namely 81, a square, and the quintuple of 16 equal 161.

Fermat's condition on the primes is simply that they be congruent to 3 or 7 modulo 20. In §2 we will present Lagrange's proof of this conjecture, which uses ideas from genus theory and the composition of forms.

Fermat's proofs used the method of infinite descent, but that's often all he said. As an example, here is Fermat's description of his proof for $p = x^2 + y^2$ [35, Vol. II, p. 432]:

> If an arbitrarily chosen prime number, which surpasses by one a multiple of four, is not a sum of two squares, then there is a prime number of the same form, less than the given one, and then yet a third still less, etc., descending infinitely until you arrive at the number 5, which is the least of all of this nature, from which it would follow was not the sum of two squares. From this one must infer, by deduction of the impossible, that all numbers of this form are consequently composed of two squares.

This explains the philosophy of infinite descent, but doesn't tell us how to produce the required lesser prime. In fact, we have only one complete proof by Fermat. It occurs in one of his marginal notes (the area of a right triangle with integral sides cannot be an integral square [35, Vol. III, Obs. 45, pp. 271–272]—for once the margin was big enough!). The methods of this proof (see Weil [106, p. 77] or Edwards [31, pp. 10–14] for modern expositions) do not apply to our case, so that we are still in the dark. In his recent book [106], Weil makes a careful study of Fermat's letters and marginal notes, and with some hints from Euler, he reconstructs some of Fermat's proofs. Weil's arguments are quite convincing, but we won't go into them here. For the present, we prefer to leave things as Euler found them, i.e., wonderful theorems but no proofs.

B. Euler

Euler first heard of Fermat's results through his correspondence with Goldbach. In fact, Goldbach's first letter to Euler, written in December 1729, mentions Fermat's conjecture that $2^{2^n}+1$ is always prime [40, p. 10]. Shortly thereafter, Euler read some of Fermat's letters that had been printed in Wallis' *Opera* [100] (which included the one to Digby quoted above). Euler was intrigued by what he found. For example, writing to Goldbach in June 1730, Euler comments that Fermat's four-square theorem (every positive integer is a sum of four or fewer squares) is a "non inelegans theorema" [40, p. 24]. For Euler, Fermat's assertions were serious theorems deserving of proof, and finding the proofs became a life-long project. Euler's first paper on number theory, written in 1732 at age 25, disproves Fermat's claim about $2^{2^n}+1$ by showing that 641 is a factor of $2^{32}+1$ [33, Vol. II, pp. 1–5]. Euler's interest in number theory continued unabated for the next 51 years—there was a steady stream of papers introducing many of the fundamental concepts of number theory, and even after his death in 1783, his papers continued to appear until 1830 (see [33, Vol. IV–V]). Weil's book [106] gives a detailed survey of Euler's work on number theory (other references are Burkhardt [14], Edwards [31, Chapter 2], Scharlau and Opolka [86, Chapter 3], and the introductions to Volumes II–V of Euler's collected works [33]).

We can now present Euler's proof of the first of Fermat's theorems from (1.1):

Theorem 1.2. *An odd prime p can be written as $x^2 + y^2$ if and only if $p \equiv 1 \bmod 4$.*

Proof. If $p = x^2 + y^2$, then congruences modulo 4 easily imply that $p \equiv 1 \bmod 4$. The hard work is proving the converse. We will give a modern version of Euler's proof. Given an odd prime p, there are two basic steps to be proved:

Descent Step: If $p \mid x^2 + y^2$, $\gcd(x, y) = 1$, then p can be written as $x^2 + y^2$.

Reciprocity Step: If $p \equiv 1 \bmod 4$, then $p \mid x^2 + y^2$, $\gcd(x, y) = 1$.

It will soon become clear why we use the names "Descent" and "Reciprocity."

We'll do the Descent Step first since that's what happened historically. The argument below is taken from a 1747 letter to Goldbach [40, pp. 416–419] (see also [33, Vol. II, pp. 295–327]). We begin with the classical identity

$$(x^2 + y^2)(z^2 + w^2) = (xz \pm yw)^2 + (xw \mp yz)^2 \tag{1.3}$$

(see Exercise 1.1) which enables one to express composite numbers as sums of squares. The key observation is the following lemma:

Lemma 1.4. *Suppose that N is a sum of two relatively prime squares, and that $q = x^2 + y^2$ is a prime divisor of N. Then N/q is also a sum of two relatively prime squares.*

Proof. Write $N = a^2 + b^2$, where a and b are relatively prime. We also have $q = x^2 + y^2$, and thus q divides

$$\begin{aligned} x^2N - a^2q &= x^2(a^2 + b^2) - a^2(x^2 + y^2) \\ &= x^2b^2 - a^2y^2 = (xb - ay)(xb + ay). \end{aligned}$$

Since q is prime, it divides one of these two factors, and changing the sign of a if necessary, we can assume that $q \mid xb - ay$. Thus $xb - ay = dq$ for some integer d.

We claim that $x \mid a + dy$. Since x and y are relatively prime, this is equivalent to $x \mid (a + dy)y$. However,

$$\begin{aligned} (a + dy)y &= ay + dy^2 = xb - dq + dy^2 \\ &= xb - d(x^2 + y^2) + dy^2 = xb - dx^2, \end{aligned}$$

which is obviously divisible by x. Furthermore, if we set $a + dy = cx$, then the above equation implies that $b = dx + cy$. Thus we have

$$\begin{aligned} a &= cx - dy \\ b &= dx + cy. \end{aligned} \tag{1.5}$$

Then, using (1.3), we obtain

$$\begin{aligned} N = a^2 + b^2 &= (cx - dy)^2 + (dx + cy)^2 \\ &= (x^2 + y^2)(c^2 + d^2) = q(c^2 + d^2). \end{aligned}$$

Thus $N/q = c^2 + d^2$ is a sum of squares, and (1.5) shows that c and d must be relatively prime since a and b are. This proves the lemma. Q.E.D.

To complete the proof of the Descent Step, let p be an odd prime dividing $N = a^2 + b^2$, where a and b are relatively prime. If a and b are changed by multiples of p, we still have $p \mid a^2 + b^2$. We may thus assume that $|a| < p/2$ and $|b| < p/2$, which in turn implies that $N < p^2/2$. The new a and b may have a greatest common divisor $d > 1$, but p doesn't divide d, so that dividing a and b by d, we may assume that $p \mid N$, $N < p^2/2$, and $N = a^2 + b^2$ where $\gcd(a,b) = 1$. Then all prime divisors $q \neq p$ of N are less than p. If q were a sum of two squares, then Lemma 1.4 would show that N/q would be a multiple of p, which is also a sum of two squares. If all such q's were sums of two squares, then repeatedly applying Lemma 1.4 would imply that p itself was of the same form. So if p is not a sum of two squares, there must be a smaller prime q with the same property. Since there is nothing to prevent us from repeating this process indefinitely, we get an infinite decreasing sequence of prime numbers. This contradiction finishes the Descent Step.

This is a classical descent argument, and as Weil argues [106, pp. 68–69], it is probably similar to what Fermat did. In §2 we will take another approach to the Descent Step, using the reduction theory of positive definite quadratic forms.

The Reciprocity Step caused Euler a lot more trouble, taking him until 1749. Euler was clearly relieved when he could write to Goldbach "Now have I finally found a valid proof" [40, pp. 493–495]. The basic idea is quite simple: since $p \equiv 1 \bmod 4$, we can write $p = 4k + 1$. Then Fermat's Little Theorem implies that

$$(x^{2k} - 1)(x^{2k} + 1) \equiv x^{4k} - 1 \equiv 0 \bmod p$$

for all $x \not\equiv 0 \bmod p$. If $x^{2k} - 1 \not\equiv 0 \bmod p$ for *one* such x, then $p \mid x^{2k} + 1$, so that p divides a sum of relatively prime squares, as desired. For us, the required x is easy to find, since $x^{2k} - 1$ is a polynomial over the field $\mathbb{Z}/p\mathbb{Z}$ and hence has at most $2k < p - 1$ roots. Euler's first proof is quite different, for it uses the calculus of finite differences—see Exercise 1.2 for details. This proves Fermat's claim (1.1) for primes of the form x^2+y^2. Q.E.D.

Euler used the same two-step strategy in his proofs for x^2+2y^2 and x^2+3y^2. The Descent Steps are

$$\text{If } p \mid x^2+2y^2,\ \gcd(x,y)=1, \text{ then } p \text{ is of the form } x^2+2y^2$$

$$\text{If } p \mid x^2+3y^2,\ \gcd(x,y)=1, \text{ then } p \text{ is of the form } x^2+3y^2,$$

and the Reciprocity Steps are

$$\text{If } p \equiv 1,3 \bmod 8, \text{ then } p \mid x^2+2y^2,\ \gcd(x,y)=1$$

$$\text{If } p \equiv 1 \bmod 3, \text{ then } p \mid x^2+3y^2,\ \gcd(x,y)=1,$$

where p is always an odd prime. In each case, the Reciprocity Step was harder to prove than the Descent Step, and Euler didn't succeed in giving complete proofs of Fermat's theorems (1.1) until 1772, 40 years after he first read about them. Weil discusses the proofs for x^2+2y^2 and x^2+3y^2 in [106, pp. 178–179, 191, and 210–212], and in Exercises 1.4 and 1.5 we will present a version of Euler's argument for x^2+3y^2.

C. $p = x^2+ny^2$ and Quadratic Reciprocity

Let's turn to the general case of $p = x^2+ny^2$, where n is now any positive integer. To study this problem, it makes sense to start with Euler's two-step strategy. This won't lead to a proof, but the Descent and Reciprocity Steps will both suggest some very interesting questions for us to pursue.

The Descent Step for arbitrary $n>0$ begins with the identity

$$(1.6) \qquad (x^2+ny^2)(z^2+nw^2) = (xz \pm nyw)^2 + n(xw \mp yz)^2$$

(see Exercise 1.1), and Lemma 1.4 generalizes easily for $n>0$ (see Exercise 1.3). Then suppose that $p \mid x^2+ny^2$. As in the proof of the Descent Step in Theorem 1.2, we can assume that $|x|,|y| \le p/2$. For $n \le 3$, it follows that $x^2+ny^2 < p^2$ when p is odd, and then the argument from Theorem 1.2 shows that p is of the form x^2+ny^2 (see Exercise 1.4). One might conjecture that this holds in general, i.e., that $p \mid x^2+ny^2$ always implies $p = x^2+ny^2$. Unfortunately this fails even for $n=5$: for example, $3 \mid 21 = 1^2+5\cdot 2^2$ but $3 \ne x^2+5y^2$. Euler knew this, and most likely so did Fermat (remember his speculations about x^2+5y^2). So the question becomes: how are prime divisors of x^2+ny^2 to be represented? As we will see in §2, the proper language for this is Lagrange's theory of quadratic forms, and in particular a complete solution to the Descent Step will follow from the properties of reduced forms.

Turning to the Reciprocity Step for $n > 0$, the general case asks for congruence conditions on a prime p which will guarantee $p \mid x^2 + ny^2$. To see what kind of congruences we need, note that the conditions of (1.1) can be unified by working modulo $4n$. Thus, given $n > 0$, we're looking for a congruence of the form $p \equiv \alpha, \beta, \ldots \bmod 4n$ which implies $p \mid x^2 + ny^2$, $\gcd(x, y) = 1$. To give a modern formulation of this last condition, we first define the Legendre symbol (a/p). If a is an integer and p an odd prime, then

$$\left(\frac{a}{p}\right) = \begin{cases} 0 & p \mid a \\ 1 & p \nmid a \text{ and } a \text{ is a quadratic residue modulo } p \\ -1 & p \nmid a \text{ and } a \text{ is a quadratic nonresidue modulo } p. \end{cases}$$

We can now restate $p \mid x^2 + ny^2$ as follows:

Lemma 1.7. *Let n be a nonzero integer, and let p be an odd prime not dividing n. Then*

$$p \mid x^2 + ny^2,\ \gcd(x, y) = 1 \iff \left(\frac{-n}{p}\right) = 1.$$

Proof. The basic idea is that if $x^2 + ny^2 \equiv 0 \bmod p$ and $\gcd(x, y) = 1$, then y must be relatively prime to p and consequently has a multiplicative inverse modulo p. The details are left to the reader (see Exercise 1.6).

Q.E.D.

The arguments of the above lemma are quite elementary, but for Euler they were not so easy—he first had to realize that quadratic residues were at the heart of the matter. This took several years, and it's fun to watch his terminology evolve: in 1744, he writes "prime divisors of numbers of the form $aa - Nbb$" [33, Vol. II, p. 216]; by 1747 this changes to "residues arising from the division of squares by the prime p" [33, Vol. II, p. 313]; and by 1751 the transition is complete—Euler now uses the terms "residua" and "non-residua" freely, with the "quadratic" being understood [33, Vol. II, p. 343].

Using Lemma 1.7, the Reciprocity Step can be restated as the following question: is there a congruence $p \equiv \alpha, \beta, \ldots \bmod 4n$ which implies $(-n/p) = 1$ when p is prime? This question also makes sense when $n < 0$, and in the following discussion n will thus be allowed to be positive or negative. We will see in Corollary 1.19 that the full answer is intimately related to the law of quadratic reciprocity, and in fact the Reciprocity Step was one of the primary things that led Euler to discover quadratic reciprocity.

Euler became intensely interested in this question in the early 1740s, and he mentions numerous examples in his letters to Goldbach. In 1744 Euler collected together his examples and conjectures in the paper *Theoremata circa divisores numerorum in hac forma* $paa \pm qbb$ *contentorum* [33, Vol. II, pp. 194–222]. He labels his examples as "theorems," but they are really "theorems found by induction," which is eighteenth-century parlance for conjectures based on working out some particular cases. Here are of some of Euler's conjectures, stated in modern notation:

$$
\begin{aligned}
\left(\frac{-3}{p}\right) &= 1 \iff p \equiv 1, 7 \bmod 12 \\
\left(\frac{-5}{p}\right) &= 1 \iff p \equiv 1, 3, 7, 9 \bmod 20 \\
\left(\frac{-7}{p}\right) &= 1 \iff p \equiv 1, 9, 11, 15, 23, 25 \bmod 28 \\
\left(\frac{3}{p}\right) &= 1 \iff p \equiv \pm 1 \bmod 12 \\
\left(\frac{5}{p}\right) &= 1 \iff p \equiv \pm 1, \pm 11 \bmod 20 \\
\left(\frac{7}{p}\right) &= 1 \iff p \equiv \pm 1, \pm 3, \pm 9 \bmod 28,
\end{aligned}
\tag{1.8}
$$

where p is an odd prime not dividing n. In looking for a unifying pattern, the bottom three look more promising because of the $\pm$'s. If we rewrite the bottom half of (1.8) using $11 \equiv -9 \bmod 20$ and $3 \equiv -25 \bmod 28$, we obtain

$$
\begin{aligned}
\left(\frac{3}{p}\right) &= 1 \iff p \equiv \pm 1 \bmod 12 \\
\left(\frac{5}{p}\right) &= 1 \iff p \equiv \pm 1, \pm 9 \bmod 20 \\
\left(\frac{7}{p}\right) &= 1 \iff p \equiv \pm 1, \pm 25, \pm 9 \bmod 28.
\end{aligned}
$$

All of the numbers that appear are odd squares!

Before getting carried away, we should note another of Euler's conjectures:

$$\left(\frac{6}{p}\right) = 1 \iff p \equiv \pm 1, \pm 5 \bmod 24.$$

Unfortunately, ± 5 is not a square modulo 24, and the same thing happens for $(10/p)$ and $(14/p)$. But 3, 5 and 7 are prime, while 6, 10 and 14 are composite. Thus it makes sense to make the following conjecture for the prime case:

Conjecture 1.9. *If p and q are distinct odd primes, then*

$$\left(\frac{q}{p}\right) = 1 \iff p \equiv \pm\beta^2 \bmod 4q \text{ for some odd integer } \beta\,.$$

The remarkable fact is that this conjecture is equivalent to the usual statement of quadratic reciprocity:

Proposition 1.10. *If p and q are distinct odd primes, then Conjecture 1.9 is equivalent to*

$$\left(\frac{p}{q}\right)\left(\frac{q}{p}\right) = (-1)^{(p-1)(q-1)/4}.$$

Proof. Let $p^* = (-1)^{(p-1)/2}$. Then the standard properties

$$\text{(1.11)} \qquad \begin{aligned} \left(\frac{-1}{p}\right) &= (-1)^{(p-1)/2} \\ \left(\frac{ab}{p}\right) &= \left(\frac{a}{p}\right)\left(\frac{b}{p}\right) \end{aligned}$$

of the Legendre symbol easily imply that quadratic reciprocity is equivalent to

$$\text{(1.12)} \qquad \left(\frac{p^*}{q}\right) = \left(\frac{q}{p}\right)$$

(see Exercise 1.7). Since both sides are ± 1, it follows that quadratic reciprocity can be stated as

$$\left(\frac{q}{p}\right) = 1 \iff \left(\frac{p^*}{q}\right) = 1.$$

Comparing this to Conjecture 1.9, we see that it suffices to show

$$\text{(1.13)} \qquad \left(\frac{p^*}{q}\right) = 1 \iff p \equiv \pm\beta^2 \bmod 4q,\ \beta \text{ odd}.$$

The proof of (1.13) is straightforward and is left to the reader (see Exercise 1.8). Q.E.D.

With hindsight, we can see why Euler had trouble with the Reciprocity Steps for $x^2 + 2y^2$ and $x^2 + 3y^2$: he was working out special cases of quadratic reciprocity! Exercise 1.9 will discuss which special cases were involved. We will not prove quadratic reciprocity in this section, but later in §8 we will give a proof using class field theory. Proofs of a more elementary nature can be found in most number theory texts.

The discussion leading up to Conjecture 1.9 is pretty exciting, but was it what Euler did? The answer is yes and no. To explain this, we must look more closely at Euler's 1744 paper. In addition to conjectures like (1.8), the paper also contained a series of Annotations where Euler speculated on what was happening in general. For simplicity, we will concentrate on the case of (N/p), where $N > 0$. Euler notes in Annotation 13 [33, Vol. II, p. 216] that for such N's, all of the conjectures have the form

$$\left(\frac{N}{p}\right) = 1 \iff p \equiv \pm\alpha \bmod 4N$$

for certain odd values of α. Then in Annotation 16 [33, Vol. II, pp. 216–217], Euler states that "while 1 is among the values [of the α's], yet likewise any square number, which is prime to $4N$, furnishes a suitable value for α." This is close to what we want, but it doesn't say that the odd squares fill up all possible α's when N is prime. To see this, we turn to Annotation 14 [33, Vol. II, p. 216], where Euler notes that the number of α's that occur is $(1/2)\phi(N)$. When N is prime, this equals $(N-1)/2$, exactly the number of incongruent squares modulo $4N$. Thus what Euler states is fully equivalent to Conjecture 1.9. In 1875, Kronecker identified these Annotations as the first complete statement of quadratic reciprocity [68, Vol. II, pp. 3–4].

The problem is that we have to read between the lines to get quadratic reciprocity—why didn't Euler state it more explicitly? He knew that the prime case was special, for why else would he list the prime cases before the composite ones? The answer to this puzzle, as Weil points out [106, pp. 207–209], is that Euler's real goal was to characterize the α's for *all* N, not just primes. To explain this, we need to give a modern description of the $\pm\alpha$'s. The following lemma is at the heart of the matter:

Lemma 1.14. *If $D \equiv 0, 1 \bmod 4$ is a nonzero integer, then there is a unique homomorphism $\chi : (\mathbb{Z}/D\mathbb{Z})^* \to \{\pm 1\}$ such that $\chi([p]) = (D/p)$ for odd primes p not dividing N. Furthermore,*

$$\chi([-1]) = \begin{cases} 1 & \text{when } D > 0 \\ -1 & \text{when } D < 0. \end{cases}$$

Proof. The proof will make extensive use of the Jacobi symbol. Given $m > 0$ odd and relatively prime to M, recall that the Jacobi symbol (M/m) is defined to be the product

$$\left(\frac{M}{m}\right) = \prod_{i=1}^{r} \left(\frac{M}{p_i}\right)$$

where $m = p_1 \cdots p_r$ is the prime factorization of m. Note that $(M/m) = (N/m)$ when $M \equiv N \bmod m$, and there are the multiplicative identities

$$\text{(1.15)} \qquad \begin{aligned} \left(\frac{MN}{m}\right) &= \left(\frac{M}{m}\right)\left(\frac{N}{n}\right) \\ \left(\frac{M}{mn}\right) &= \left(\frac{M}{m}\right)\left(\frac{M}{n}\right) \end{aligned}$$

(see Exercise 1.10). The Jacobi symbol also satisfies the following version of quadratic reciprocity:

$$\text{(1.16)} \qquad \begin{aligned} \left(\frac{-1}{m}\right) &= (-1)^{(m-1)/2} \\ \left(\frac{2}{m}\right) &= (-1)^{(m^2-1)/8} \\ \left(\frac{M}{m}\right) &= (-1)^{(M-1)(m-1)/4}\left(\frac{m}{M}\right) \end{aligned}$$

(see Exercise 1.10).

For this lemma, the crucial property of the Jacobi symbol is one usually not mentioned in elementary texts: if $m \equiv n \bmod D$, where m and n are odd and positive and $D \equiv 0, 1 \bmod 4$, then

$$\text{(1.17)} \qquad \left(\frac{D}{m}\right) = \left(\frac{D}{n}\right).$$

The proof is quite easy when $D \equiv 1 \bmod 4$ and $D > 0$: using quadratic reciprocity (1.16), the two sides of (1.17) become

$$\text{(1.18)} \qquad \begin{aligned} &(-1)^{(D-1)(m-1)/4}\left(\frac{m}{D}\right) \\ &(-1)^{(D-1)(n-1)/4}\left(\frac{n}{D}\right). \end{aligned}$$

To compare these, first note that the two Jacobi symbols are equal since $m \equiv n \bmod D$. From $D \equiv 1 \bmod 4$ we see that

$$(D-1)(m-1)/4 \equiv (D-1)(n-1)/4 \equiv 0 \bmod 2$$

since m and n are odd. Thus the signs in front of (1.18) are both $+1$, and (1.17) follows. When D is even or negative, a similar argument using the supplementary laws from (1.16) shows that (1.17) still holds (see Exercise 1.11).

It follows from (1.17) that $\chi([m]) = (D/m)$ gives a well-defined homomorphism from $(\mathbb{Z}/D\mathbb{Z})^*$ to $\{\pm 1\}$ (see Exercise 1.12), and the statement concerning $\chi([-1])$ follows from the above properties of the Jacobi symbol (see Exercise 1.12). Finally, the condition that $\chi([p]) = (D/p)$ for p prime determines χ uniquely follows because every class in $(\mathbb{Z}/D\mathbb{Z})^*$ contains a prime—this is a consequence of Dirichlet's theorem on primes in arithmetic progressions (to be proved in §8). Q.E.D.

The above proof made heavy use of quadratic reciprocity, which is no accident: Lemma 1.14 is in fact equivalent to quadratic reciprocity and the supplementary laws (see Exercise 1.13). For us, however, the main feature of Lemma 1.14 is that it gives a complete solution of the Reciprocity Step of Euler's strategy:

Corollary 1.19. *Let n be a nonzero integer, and let $\chi : (\mathbb{Z}/4n\mathbb{Z})^* \to \{\pm 1\}$ be the homomorphism from Lemma 1.14 when $D = -4n$. If p is an odd prime not dividing n, then the following are equivalent:*

(i) $p \mid x^2 + ny^2$, $\gcd(x,y) = 1$.

(ii) $(-n/p) = 1$.

(iii) $[p] \in \ker(\chi) \subset (\mathbb{Z}/4n\mathbb{Z})^*$.

Proof. (i) and (ii) are equivalent by Lemma 1.7, and since $(-4n/p) = (-n/p)$, (ii) and (iii) are equivalent by Lemma 1.14. Q.E.D.

To see how this solves the Reciprocity Step, note that if $\ker(\chi) = \{[\alpha],[\beta],[\gamma],\ldots\}$, then $[p] \in \ker(\chi)$ is equivalent to the congruence $p \equiv \alpha,\beta,\gamma,\ldots \bmod 4n$, which is exactly the kind of condition we were looking for. Actually, Lemma 1.14 allows us to refine this a bit: when $n \equiv 3 \bmod 4$, then congruence can be taken to be of the form $p \equiv \alpha,\beta,\gamma,\ldots \bmod n$ (see Exercise 1.14). We should also note that in all cases, the usual statement of quadratic reciprocity makes it easy to compute the classes in question (see Exercise 1.15 for an example).

To see how this relates to what Euler did in 1744, let N be as above, and let $D = 4N$ in Lemma 1.14. Then $\ker(\chi)$ consists *exactly* of Euler's $\pm\alpha$'s (when $N > 0$, the lemma also implies that $-1 \in \ker(\chi)$, which explains the $\pm$ signs). The second thing to note is that when N is odd and squarefree, $K = \ker(\chi)$ is uniquely characterized by the following four properties:

(i) K is a subgroup of index 2 in $(\mathbb{Z}/4N\mathbb{Z})^*$.

(ii) $-1 \in K$ when $N > 0$ and $-1 \notin K$ when $N < 0$.

(iii) K has period N if $N \equiv 1 \bmod 4$ and period $4N$ otherwise. (Having period $P > 0$ means that if $[a],[b] \in (\mathbb{Z}/4N\mathbb{Z})^*$, $[a] \in K$ and $a \equiv b \bmod P$, then $[b] \in K$.)

(iv) K does not have any smaller period.

For a proof of this characterization, see Weil [106, pp. 287–291]. In the Annotations to his 1744 paper, Euler gives very clear statements of (i)–(iii) (see Annotations 13–16 in [33, Vol. II, pp. 216–217]), and as for (iv), he notes that N is not a period when $N \not\equiv 1 \bmod 4$, but says nothing about the possibility of smaller periods (see Annotation 20 in [33, Vol. II, p. 219]). So Euler doesn't quite give a complete characterization of $\ker(\chi)$, but he comes incredibly close. It is a tribute to Euler's insight that he could deduce this underlying structure on the basis of examples like (1.8).

D. Beyond Quadratic Reciprocity

We will next discuss some of Euler's conjectures concerning primes of the form $x^2 + ny^2$ for $n > 3$. We start with the cases $n = 5$ and 14 (taken from his 1744 paper), for each will have something unexpected to offer us.

When $n = 5$, Euler conjectured that for odd primes $p \neq 5$,

$$\begin{aligned} p = x^2 + 5y^2 &\iff p \equiv 1, 9 \bmod 20 \\ 2p = x^2 + 5y^2 &\iff p \equiv 3, 7 \bmod 20. \end{aligned} \tag{1.20}$$

Recall from (1.8) that $p \mid x^2 + 5y^2$ is equivalent to $p \equiv 1, 3, 7, 9 \bmod 20$. Hence these four congruence classes break up into two groups $\{1, 9\}$ and $\{3, 7\}$ which have quite different representability properties. This is a new phenomenon, not encountered for $x^2 + ny^2$ when $n \leq 3$. Note also that the classes $3, 7$ modulo 20 are the ones that entered into Fermat's speculations on $x^2 + 5y^2$, so something interesting is going on here. In §2 we will see that this is one of the examples that led Lagrange to discover genus theory.

The case $n = 14$ is yet more complicated. Here, Euler makes the following conjecture for odd primes $\neq 7$:

$$\begin{aligned} p = \left\{ \begin{matrix} x^2 + 14y^2 \\ 2x^2 + 7y^2 \end{matrix} \right\} &\iff p \equiv 1, 9, 15, 23, 25, 39 \bmod 56 \\ 3p = x^2 + 14y^2 &\iff p \equiv 3, 5, 13, 19, 27, 45 \bmod 56. \end{aligned} \tag{1.21}$$

As with (1.20), the union of the two groups of congruence classes in (1.21) describe those primes for which $(-14/p) = 1$. The new puzzle here is that we don't seem to be able to separate $x^2 + 14y^2$ from $2x^2 + 7y^2$. In §2, we will see that this is not an oversight on Euler's part, for the two quadratic forms $x^2 + 14y^2$ and $2x^2 + 7y^2$ are in the same genus and hence can't be separated by congruence classes. Another puzzle is why (1.20) uses $2p$ while (1.21) uses $3p$. In §2 we will use composition to explain these facts. One could also ask what extra condition is needed to insure $p = x^2 + 14y^2$. This lies much deeper, for as we will see in §5, it involves the Hilbert class field of $\mathbb{Q}(\sqrt{-14})$.

The final examples we want to discuss come from quite a different source, the *Tractatus de numerorum doctrina capita sedecim quae supersunt*, which Euler wrote in the period 1748–1750 [33, Vol. V, pp. 182–283]. Euler intended this work to be a basic text for number theory, in the same way that his *Introductio in analysin infinitorum* [33, Vol. VIII–IX] was the first real textbook in analysis. Unfortunately, Euler never completed the *Tractatus*, and it was first published only in 1849. Weil [106, pp. 192–196] gives a description of what's in the *Tractatus* (see also [33, Vol. V, pp. XIX–XXVI]). For us, the most interesting chapters are the two that deal with cubic and biquadratic residues. Recall that a number a is a cubic (resp. biquadratic) residue modulo p if the congruence $x^3 \equiv a \bmod p$ (resp. $x^4 \equiv a \bmod p$) has an integer solution. Euler makes the following conjectures about when 2 is a cubic or biquadratic residue modulo an odd prime p:

$$p = x^2 + 27y^2 \Longleftrightarrow \begin{cases} p \equiv 1 \bmod 3 \text{ and } 2 \text{ is a} \\ \text{cubic residue modulo } p \end{cases} \tag{1.22}$$

$$p = x^2 + 64y^2 \Longleftrightarrow \begin{cases} p \equiv 1 \bmod 4 \text{ and } 2 \text{ is a} \\ \text{biquadratic residue modulo } p \end{cases} \tag{1.23}$$

(see [33, Vol. V, pp. 250 and 258]). In §4, we will see that both of these conjectures were proved by Gauss as consequences of his work on cubic and biquadratic reciprocity.

The importance of the examples (1.20)–(1.23) is hard to overestimate. Thanks to Euler's amazing ability to find patterns, we now see some of the serious problems to be tackled (in (1.20) and (1.21)), and we have our first hint of what the final solution will look like (in (1.22) and (1.23)). Much of the next three sections will be devoted to explaining and proving these conjectures. In particular, it should be clear that we need to learn a lot more about quadratic forms. Euler left us with a magnificent series of examples and conjectures, but it remained for Lagrange to develop the language which would bring the underlying structure to light.

E. Exercises

1.1. In this exercise, we prove some identities used by Euler.

(a) Prove (1.3) and its generalization (1.6).

(b) Generalize (1.6) to find an identity of the form

$$(ax^2 + cy^2)(az^2 + cw^2) = (?)^2 + ac(?)^2.$$

This is due to Euler [33, Vol. I, p. 424].

1.2. Let p be prime, and let $f(x)$ be a monic polynomial of degree $d < p$. This exercise will describe Euler's proof that the congruence $f(x) \not\equiv 0 \bmod p$ has a solution. Let $\Delta f(x) = f(x+1) - f(x)$ be the difference operator.

(a) For any $k \geq 1$, show that $\Delta^k f(x)$ is an integral linear combination of $f(x), f(x+1), \ldots, f(x+k)$.

(b) Show that $\Delta^d f(x) = d!$.

(c) Euler's argument is now easy to state: if $f(x) \not\equiv 0 \bmod p$ has no solutions, then $p \mid \Delta^d f(x)$ follows from (a). By (b), this is impossible.

1.3. Let n be a positive integer.

(a) Formulate and prove a version of Lemma 1.4 when a prime $q = x^2 + ny^2$ divides a number $N = a^2 + nb^2$.

(b) Show that your proof of (a) works when $n = 3$ and $q = 4$.

1.4. In this exercise, we will prove the Descent Steps for $x^2 + 2y^2$ and $x^2 + 3y^2$.

(a) If a prime p divides $x^2 + 2y^2$, $\gcd(x, y) = 1$, then adapt the argument of Theorem 1.2 to show that $p = x^2 + 2y^2$. Hint: use Exercise 1.3.

(b) Prove that if an odd prime p divides $x^2 + 3y^2$, $\gcd(x, y) = 1$, then $p = x^2 + 3y^2$. The argument is more complicated because the Descent Step fails for $p = 2$. Thus, if it fails for some odd prime p, you have to produce an *odd* prime $q < p$ where it also fails. Hint: part (b) of Exercise 1.3 will be useful.

1.5. If $p = 3k + 1$ is prime, prove that $(-3/p) = 1$. Hint:

$$\begin{aligned} 4(x^{3k} - 1) &= (x^k - 1) \cdot 4(x^{2k} + x^k + 1) \\ &= (x^k - 1)((2x^k + 1)^2 + 3). \end{aligned}$$

Note that Exercises 1.4(b) and 1.5 prove Fermat's theorem for $x^2 + 3y^2$.

1.6. Prove Lemma 1.7.

1.7. Use the properties (1.11) of the Legendre symbol to prove the quadratic reciprocity is equivalent to (1.12).

1.8. Prove (1.13).

1.9. In this exercise we will see how the Reciprocity Steps for $x^2 + y^2$, $x^2 + 2y^2$ and $x^2 + 3y^2$ relate to quadratic reciprocity.

(a) Use Lemma 1.7 to show that for a prime $p > 3$,

$$p \mid x^2 + 3y^2,\ \gcd(x, y) = 1 \iff p \equiv 1 \bmod 3$$

is equivalent to

$$\left(\frac{-3}{p}\right) = \left(\frac{p}{3}\right).$$

By (1.12), we recognize this as part of quadratic reciprocity.

(b) Use Lemma 1.7 and the bottom line of (1.11) to show that the statements

$$p \mid x^2 + y^2,\ \gcd(x,y) = 1 \Longleftrightarrow p \equiv 1 \bmod 4$$
$$p \mid x^2 + 2y^2,\ \gcd(x,y) = 1 \Longleftrightarrow p \equiv 1,3 \bmod 8$$

are equivalent to the statements

$$\left(\frac{-1}{p}\right) = (-1)^{(p-1)/2}$$
$$\left(\frac{2}{p}\right) = (-1)^{(p^2-1)/8}.$$

1.10. This exercise is concerned with the properties of the Jacobi symbol (M/m) defined in the proof of Lemma 1.14.

(a) Prove that $(M/m) = (N/m)$ when $M \equiv N \bmod m$.

(b) Prove (1.15).

(c) Prove (1.16) using quadratic reciprocity and the two supplementary laws $(-1/p) = (-1)^{(p-1)/2}$ and $(2/p) = (-1)^{(p^2-1)/8}$. Hint: if r and s are odd, show that

$$(rs-1)/2 \equiv (r-1)/2 + (s-1)/2 \bmod 2$$
$$(r^2s^2-1)/8 \equiv (r^2-1)/8 + (s^2-1)/8 \bmod 2.$$

(d) If M is a quadratic residue modulo m, show that $(M/m) = 1$. Give an example to show that the converse is not true.

1.11. Use (1.15) and (1.16) to complete the proof of (1.17) begun in the text.

1.12. This exercise is concerned with the map $\chi : (\mathbb{Z}/D\mathbb{Z})^* \to \{\pm 1\}$ of Lemma 1.14. When m is odd and positive, we define $\chi([m])$ to be the Jacobi symbol (D/m).

(a) Show that any class in $(\mathbb{Z}/D\mathbb{Z})^*$ may be written as $[m]$, where m is odd and positive, and then use (1.17) to show that χ is a well-defined homomorphism on $(\mathbb{Z}/D\mathbb{Z})^*$.

(b) Show that

$$\chi([-1]) = \begin{cases} 1 & \text{if } D > 0 \\ -1 & \text{if } D < 0. \end{cases}$$

(c) If $D \equiv 1 \bmod 4$, show that

$$\chi([2]) = \begin{cases} 1 & \text{if } D \equiv 1 \bmod 8 \\ -1 & \text{if } D \equiv 5 \bmod 8. \end{cases}$$

1.13. In this exercise, we will assume that Lemma 1.14 holds for all nonzero integers $D \equiv 0, 1 \bmod 4$, and we will prove quadratic reciprocity and the supplementary laws.

(a) Let p and q be distinct odd primes, and let $q^* = (-1)^{(p-1)/2}$. By applying the lemma with $D = q^*$, show that $(q^*/\cdot)$ induces a homomorphism from $(\mathbf{Z}/q\mathbf{Z})^*$ to $\{\pm 1\}$. Since $(\cdot/q)$ can be regarded as a homomorphism between the same two groups and $(\mathbf{Z}/q\mathbf{Z})^*$ is cyclic, conclude that the two are equal.

(b) Use similar arguments to prove the supplementary laws. Hint: apply the lemma with $D = -4$ and 8 respectively.

1.14. Use Lemma 1.14 to prove that when $n \equiv 3 \bmod 4$, there are integers $\alpha, \beta, \gamma, \ldots$ such that for an odd prime p not dividing n, $p \mid x^2 + ny^2$, $\gcd(x, y) = 1$ if and only if $p \equiv \alpha, \beta, \gamma, \ldots \bmod n$.

1.15. Use quadratic reciprocity to determine those classes in $(\mathbf{Z}/84\mathbf{Z})^*$ with $(-21/p) = 1$. This tells us when $p \mid x^2 + 21y^2$, and thus solves Reciprocity Step when $n = 21$.

1.16. In the discussion following the proof of Lemma 1.14, we stated that $K = \ker(\chi)$ is characterized by the four properties (i)–(iv). When $D = 4q$, where q is an odd prime, prove that (i) and (ii) suffice to determine K uniquely.

§2. LAGRANGE, LEGENDRE AND QUADRATIC FORMS

The study of integral quadratic forms in two variables

$$f(x, y) = ax^2 + bxy + cy^2, \qquad a, b, c \in \mathbf{Z}$$

began with Lagrange, who introduced the concepts of discriminant, equivalence and reduced form. When these are combined with Gauss' notion of proper equivalence, one has all of the ingredients necessary to develop the basic theory of quadratic forms. We will concentrate on the special case of positive definite forms. Here, Lagrange's theory of reduced forms is especially nice, and in particular we will get a complete solution of the Descent Step from §1. When this is combined with the solution of the Reciprocity Step given by quadratic reciprocity, we will get immediate proofs of Fermat's theorems (1.1) as well as several new results. We will then describe an elementary form of genus theory due to Lagrange, which will enable us to prove some of Euler's conjectures from §1, and we will also be able to solve our basic question of $p = x^2 + ny^2$ for quite a few n. The section will end with some historical remarks concerning Lagrange and Legendre.

A. Quadratic Forms

Our treatment of quadratic forms is taken primarily from Lagrange's "Recherches d'Arithmétique" of 1773–1775 [69, pp. 695–795] and Gauss' *Disquisitiones Arithmeticae* of 1801 [41, §§153–226]. Most of the terminology is due to Gauss, though many of the terms he introduced refer to concepts used implicitly by Lagrange (with some important exceptions).

A first definition is that a form $ax^2 + bxy + cy^2$ is *primitive* if its coefficients a, b and c are relatively prime. Note that any form is an integer multiple of a primitive form. We will deal exclusively with primitive forms.

An integer m is *represented* by a form $f(x,y)$ if the equation

$$m = f(x,y) \tag{2.1}$$

has an integer solution in x and y. If the x and y in (2.1) are relatively prime, we say that m is *properly represented* by $f(x,y)$. Note that the basic question of the book can be restated as: which primes are represented by the quadratic form $x^2 + ny^2$?

Next, we say that two forms $f(x,y)$ and $g(x,y)$ are *equivalent* if there are integers p, q, r and s such that

$$f(x,y) = g(px + qy, rx + sy) \quad \text{and} \quad ps - qr = \pm 1. \tag{2.2}$$

Since $\det\begin{pmatrix} p & q \\ r & s \end{pmatrix} = ps - qr = \pm 1$, this means that $\begin{pmatrix} p & q \\ r & s \end{pmatrix}$ is in the group of 2×2 invertible integer matrices $\mathrm{GL}(2,\mathbb{Z})$, and it follows easily that the equivalence of forms is an equivalence relation (see Exercise 2.2). An important observation is that equivalent forms represent the same numbers, and the same is true for proper representations (see Exercise 2.2). Note also that any form equivalent to a primitive form is itself primitive (see Exercise 2.2). Following Gauss, we say that an equivalence is a *proper equivalence* if $ps - qr = 1$, i.e., $\begin{pmatrix} p & q \\ r & s \end{pmatrix} \in \mathrm{SL}(2,\mathbb{Z})$, and it is an *improper equivalence* if $ps - qr = -1$ [41, §158]. Since $\mathrm{SL}(2,\mathbb{Z})$ is a subgroup of $\mathrm{GL}(2,\mathbb{Z})$, it follows that proper equivalence is also an equivalence relation (see Exercise 2.2).

The notion of equivalence is due to Lagrange, though he simply said that one form "can be transformed into another of the same kind" [69, p. 723]. Neither Lagrange nor Legendre made use of proper equivalence. The terms "equivalence" and "proper equivalence" are due to Gauss [41, §157], and after stating their definitions, Gauss promises that "the usefulness of these distinctions will soon be made clear" [41, §158]. In §3 we will see that he was true to his word.

As an example of these concepts, note that the forms $ax^2+bxy+cy^2$ and $ax^2-bxy+cy^2$ are always improperly equivalent via the substitution $(x,y)\mapsto(x,-y)$. But are they properly equivalent? This is not obvious. We will see below that the answer is sometimes yes (for $2x^2\pm 2xy+3y^2$) and sometimes no (for $3x^2\pm 2xy+5y^2$).

There is a very nice relation between proper representations and proper equivalence:

Lemma 2.3. *A form $f(x,y)$ properly represents an integer m if and only if $f(x,y)$ is properly equivalent to the form $mx^2+bxy+cy^2$ for some $b,c\in\mathbf{Z}$.*

Proof. First, suppose that $f(p,q)=m$, where p and q are relatively prime. We can find integers r and s so that $ps-qr=1$, and then

$$\begin{aligned} f(px+ry,qx+sy) &= f(p,q)x^2+(f(p,s)+f(r,q))xy+f(r,s)y^2 \\ &= mx^2+bxy+cy^2 \end{aligned}$$

is of the desired form. To prove the converse, note that $mx^2+bxy+cy^2$ represents m properly by taking $(x,y)=(1,0)$, and the lemma is proved. Q.E.D.

We define the *discriminant* of $ax^2+bxy+cy^2$ to be $D=b^2-4ac$. To see how this definition relates to equivalence, suppose the forms $f(x,y)$ and $g(x,y)$ have discriminants D and D' respectively, and that

$$f(x,y)=g(px+qy,rx+sy),\qquad p,q,r,s\in\mathbf{Z}.$$

Then a straightforward calculation shows that

$$D=(ps-qr)^2D',$$

(see Exercise 2.3), so that the two forms have the same discriminant whenever $ps-qr=\pm 1$. Thus equivalent forms have the same discriminant.

The sign of the discriminant D has a strong effect on the behavior of the form. If $f(x,y)=ax^2+bxy+cy^2$, then we have the identity

$$4af(x,y)=(2ax+by)^2-Dy^2. \tag{2.4}$$

If $D>0$, then $f(x,y)$ represents both positive and negative integers, and we call the form indefinite, while if $D<0$, then the form represents only positive integers or only negative ones, depending on the sign of a, and $f(x,y)$ is accordingly called positive definite or negative definite (see Exercise 2.4). Note that all of these notions are invariant under equivalence.

The discriminant D influences the form in one other way: since $D=b^2-4ac$, we have $D\equiv b^2 \bmod 4$, and it follows that the middle coefficient b is even (resp. odd) if and only if $D\equiv 0$ (resp. 1) mod 4.

We have the following necessary and sufficient condition for a number m to be represented by a form of discriminant D:

Lemma 2.5. *Let $D \equiv 0, 1 \bmod 4$ be an integer and m be an odd integer relatively prime to D. Then m is properly represented by a primitive form of discriminant D if and only if D is a quadratic residue modulo m.*

Proof. If $f(x,y)$ properly represents m, then by Lemma 2.3, we may assume $f(x,y) = mx^2 + 2bxy + cy^2$. Thus $D = b^2 - 4mc$, and $D \equiv b^2 \bmod m$ follows immediately.

Conversely, suppose that $D \equiv b^2 \bmod m$. Since m is odd, we can assume that D and b have the same parity (replace b by $b + m$ if necessary), and then $D \equiv 0, 1 \bmod 4$ implies that $D \equiv b^2 \bmod 4m$. This means that $D = b^2 - 4mc$ for some c. Then $mx^2 + bxy + cy^2$ represents m properly and has discriminant D, and the coefficients are relatively prime since m is relatively prime to D. Q.E.D.

For our purposes, the most useful version of Lemma 2.5 will be the following corollary:

Corollary 2.6. *Let n be an integer and let p be an odd prime not dividing n. Then $(-n/p) = 1$ if and only if p is represented by a primitive form of discriminant $-4n$.*

Proof. This follows immediately from Lemma 2.5 because $-4n$ is a quadratic residue modulo p if and only if $(-4n/p) = (-n/p) = 1$. Q.E.D.

This corollary is relevant to the question raised in §1 when we tried to generalize the Descent Step of Euler's strategy. Recall that we asked how to represent prime divisors of $x^2 + ny^2$, $\gcd(x,y) = 1$. Note that Corollary 2.6 gives a first answer to this question, for such primes satisfy $(-n/p) = 1$, and hence are represented by forms of discriminant $-4n$. The problem is that there are too many quadratic forms of a given discriminant. For example, if the proof of Lemma 2.5 is applied to $(-3/13) = 1$, then we see that 13 is represented by the form $13x^2 + 12xy + 3y^2$ of discriminant -12. This is not very enlightening. So to improve Corollary 2.6, we need to show that every form is equivalent to an especially simple one. Lagrange's theory of reduced forms does this and a lot more.

So far, we've dealt with arbitrary quadratic forms, but from this point on, we will specialize to the positive definite case. These forms include the ones we're most interested in (namely, $x^2 + ny^2$ for $n > 0$), and their theory has a classical simplicity and elegance. In particular, there is an especially nice notion of reduced form.

A primitive positive definite form $ax^2 + bxy + cy^2$ is said to be *reduced* if

$$|b| \leq a \leq c, \text{ and } b \geq 0 \text{ if either } |b| = a \text{ or } a = c. \tag{2.7}$$

(Note that a and c are positive since the form is positive definite.) The basic theorem is the following:

Theorem 2.8. *Every primitive positive definite form is properly equivalent to a unique reduced form.*

Proof. The first step is to show that a given form is properly equivalent to one satisfying $|b| \leq a \leq c$. Among all forms properly equivalent to the given one, pick $f(x,y) = ax^2 + bxy + cy^2$ so that $|b|$ is as small as possible. If $a < |b|$, then

$$g(x,y) = f(x+my, y) = ax^2 + (2am+b)xy + c'y^2$$

is properly equivalent to $f(x,y)$. Since $a < |b|$, we can choose $m \in \mathbf{Z}$ so that $|2am + b| < |b|$, which contradicts our choice of $f(x,y)$. Thus $a \geq |b|$, and $c \geq |b|$ follows similarly. If $a > c$, we need to interchange the outer coefficients, which is accomplished by the proper equivalence $(x,y) \mapsto (-y,x)$. The resulting form satisfies $|b| \leq a \leq c$.

The next step is to show that such a form is properly equivalent to a reduced one. By definition (2.7), the form is already reduced unless $b < 0$ and $a = -b$ or $a = c$. In these exceptional cases, $ax^2 - bxy + cy^2$ is reduced, so that we need only show that the two forms $ax^2 \pm bxy + cy^2$ are properly equivalent. This is done as follows:

$$a = -b : (x,y) \mapsto (x+y,y) \text{ takes } ax^2 - axy + cy^2 \text{ to } ax^2 + axy + cy^2.$$

$$a = c \ : (x,y) \mapsto (-y,x) \text{ takes } ax^2 + bxy + ay^2 \text{ to } ax^2 - bxy + ay^2.$$

The final step in the proof is to show that different reduced forms cannot be properly equivalent. This is the uniqueness part of the theorem. If $f(x,y) = ax^2 + bxy + cy^2$ satisfies $|b| \leq a \leq c$, then one easily shows that

$$f(x,y) \geq (a - |b| + c)\min(x^2, y^2) \tag{2.9}$$

(see Exercise 2.7). Thus $f(x,y) \geq a - |b| + c$ whenever $xy \neq 0$, and it follows that a is the smallest nonzero value of $f(x,y)$. Furthermore, if $c > a$, then c is the next smallest number represented properly by $f(x,y)$, so that in this case the outer coefficients of a reduced form give the minimum values properly represented by any equivalent form. These observations are due to Legendre [74, Vol. I, pp. 77–78].

We can now prove uniqueness. For simplicity, assume that $f(x,y) = ax^2 + bxy + cy^2$ is a reduced form that satisfies the strict inequalities $|b| < a < c$. The above considerations imply that

$$a < c < a - |b| + c \tag{2.10}$$

are the three smallest numbers properly represented by $f(x,y)$. Using these inequalities and (2.9), it follows that

$$\begin{aligned} f(x,y) = a,\ \gcd(x,y) = 1 &\Longleftrightarrow (x,y) = \pm(1,0) \\ f(x,y) = c,\ \gcd(x,y) = 1 &\Longleftrightarrow (x,y) = \pm(0,1) \end{aligned} \tag{2.11}$$

(see Exercise 2.8). Now let $g(x,y)$ be a reduced form equivalent to $f(x,y)$. Since these forms represent the same numbers and are reduced, they must have the same first coefficient a by Legendre's observation. Now consider the third coefficient c' of $g(x,y)$. We know that $a \leq c'$ since $g(x,y)$ is reduced. If equality occurred, then the equation $g(x,y) = a$ would have four proper solutions $\pm(1,0)$ and $\pm(0,1)$. Since $f(x,y)$ is equivalent to $g(x,y)$, this would contradict (2.11). Thus $a < c'$, and then Legendre's observation shows that $c = c'$. Hence the outer coefficients of $f(x,y)$ and $g(x,y)$ are the same, and since they have the same discriminant, it follows that $g(x,y) = ax^2 \pm bxy + cy^2$.

It remains to show that $f(x,y) = g(x,y)$ when we make the stronger assumption that the forms are properly equivalent. If we assume that

$$g(x,y) = f(px + qy, rx + sy), \qquad ps - qr = 1,$$

then $a = g(1,0) = f(p,q)$ and $c = g(0,1) = f(r,s)$ are proper representations. By (2.11), it follows that $(p,q) = \pm(1,0)$ and $(r,s) = \pm(0,1)$. Then $ps - qr = 1$ implies $\left(\begin{smallmatrix} p & q \\ r & s \end{smallmatrix}\right) = \pm\left(\begin{smallmatrix} 1 & 0 \\ 0 & 1 \end{smallmatrix}\right)$, and $f(x,y) = g(x,y)$ follows easily.

When $a = |b|$ or $a = c$, the above argument breaks down, because the values in (2.10) are no longer distinct. Nevertheless, one can still show that $f(x,y)$ and $g(x,y)$ reduce to $ax^2 \pm bxy + cy^2$, and then the restriction $b \geq 0$ in definition (2.7) implies equality. (See Exercise 2.8, or for the complete details, Scharlau and Opolka [86, pp. 36–38].) Q.E.D.

Note that we can now answer our earliers question about equivalence versus proper equivalence. Namely, the forms $3x^2 \pm 2xy + 5y^2$ are clearly equivalent, but since they are both reduced, Theorem 2.8 implies that they are not properly equivalent. On the other hand, of $2x^2 \pm 2xy + 3y^2$, only $2x^2 + 2xy + 3y^2$ is reduced (because $a = |b|$), and by the proof of Theorem 2.8, it is properly equivalent to $2x^2 - 2xy + 3y^2$.

In order to complete the elementary theory of reduced forms, we need one more observation. Suppose that $ax^2 + bxy + cy^2$ is a reduced form of discriminant $D < 0$. Then $b^2 \leq a^2$ and $a \leq c$, so that

$$-D = 4ac - b^2 \geq 4a^2 - a^2 = 3a^2$$

and thus

$$a \leq \sqrt{(-D)/3}. \tag{2.12}$$

If D is fixed, then $|b| \leq a$ and (2.12) imply that there are only finitely many choices for a and b. Since $b^2 - 4ac = D$, the same is true for c, so that there are only a finite number of reduced forms of discriminant D. Then Theorem 2.8 implies that the number of proper equivalence classes is also finite. Following Gauss [41, §223], we say that two forms are in the same *class* if they are properly equivalent. We will let $h(D)$ denote the number of classes of primitive positive definite forms of discriminant D, which by Theorem 2.8 is just the number of reduced forms. We have thus proved the following theorem:

Theorem 2.13. *Let $D < 0$ be fixed. Then the number $h(D)$ of classes of primitive positive definite forms of discriminant D is finite, and furthermore $h(D)$ is equal to the number of reduced forms of discriminant D.* Q.E.D.

The above discussion also shows that there is an algorithm for computing reduced forms and class numbers which, for small discriminants, is easily implemented on a computer (see Exercise 2.9). Here are some examples which will prove useful later on:

(2.14)

D	$h(D)$	Reduced Forms of Discriminant D
-4	1	$x^2 + y^2$
-8	1	$x^2 + 2y^2$
-12	1	$x^2 + 3y^2$
-20	2	$x^2 + 5y^2, 2x^2 + 2xy + 3y^2$
-28	1	$x^2 + 7y^2$
-56	4	$x^2 + 14y^2, 2x^2 + 7y^2, 3x^2 \pm 2xy + 5y^2$
-108	3	$x^2 + 27y^2, 4x^2 \pm 2xy + 7y^2$
-256	4	$x^2 + 64y^2, 4x^2 + 4xy + 17y^2, 5x^2 \pm 2xy + 13y^2$

Note, by the way, that $x^2 + ny^2$ is always a reduced form! For a further discussion of the computational aspects of class numbers, see Buell [12] and Shanks [89] (the algorithm described in [89] makes nice use of the theory to be described in §3).

This completes our discussion of positive definite forms. We should also mention that there is a corresponding theory for indefinite forms. Its roots reach back to Fermat and Euler (both considered special cases, such as $x^2 - 2y^2$), and Lagrange and Gauss each developed a general theory of such forms. There are notions of reduced form, class number, etc., but the uniqueness problem is much more complicated. As Gauss notes, "it can happen that many reduced forms are properly equivalent among themselves" [41, §184]. Determining exactly which reduced forms are properly equivalent is not easy (see Lagrange [69, pp. 728–740] and Gauss [41, §§183–193]). There are also connections with continued fractions and Pell's equation (see [41, §§183–205]), so that the indefinite case has a very different flavor. Two modern references are Flath [36, Chapter IV] and Zagier [111, §§8, 13 and 14].

B. $p = x^2 + ny^2$ and Quadratic Forms

We can now apply the theory of positive definite quadratic forms to solve some of the problems encountered in §1. We start by giving a complete solution of the Descent Step of Euler's strategy:

Proposition 2.15. *Let n be a positive integer and p be an odd prime not dividing n. Then $(-n/p) = 1$ if and only if p is represented by one of the $h(-4n)$ reduced forms of discriminant $-4n$.*

Proof. This follows immediately from Corollary 2.6 and Theorem 2.8. Q.E.D.

In §1 we showed how quadratic reciprocity gives a general solution of the Reciprocity Step of Euler's strategy. Having just solved the Descent Step, it makes sense to put the two together and see what we get. But rather than just treat the case of forms of discriminant $-4n$, we will state a result that applies to *all* negative discriminants $D < 0$. Recall from Lemma 1.14 that there is a homomorphism $\chi : (\mathbf{Z}/D\mathbf{Z})^* \to \{\pm 1\}$ such that $\chi([p]) = (D/p)$ for odd primes not dividing D. Note that $\ker(\chi) \subset (\mathbf{Z}/D\mathbf{Z})^*$ is a subgroup of index 2. We then have the following general theorem:

Theorem 2.16. *Let $D \equiv 0, 1 \bmod 4$ be negative, and let $\chi : (\mathbf{Z}/D\mathbf{Z})^* \to \{\pm 1\}$ be the homomorphism from Lemma 1.14. Then, for an odd prime p not dividing D, $[p] \in \ker(\chi)$ if and only if p is represented by one of the $h(D)$ reduced forms of discriminant D.*

Proof. The definition of χ tells us that $[p] \in \ker(\chi)$ if and only if $(D/p) = 1$. By Lemma 2.5, this last condition is equivalent to being represented by a primitive positive definite form of discriminant D, and then we are done by Theorem 2.8. Q.E.D.

The basic content of this theorem is that there is a congruence $p \equiv \alpha, \beta, \gamma, \ldots \bmod D$ which gives necessary and sufficient conditions for an odd prime p to be represented by a reduced form of discriminant D. This result is very computational, for we know how to find the reduced forms, and quadratic reciprocity makes it easy to find the congruence classes $\alpha, \beta, \gamma, \ldots$ $\bmod D$ such that $(D/p) = 1$.

For an example of how Theorem 2.16 works, note that $x^2 + y^2$, $x^2 + 2y^2$ and $x^2 + 3y^2$ are the only reduced forms of discriminants -4, -8 and -12 respectively (this is from (2.14)). Using quadratic reciprocity to find the congruence classes for which $(-1/p)$, $(-2/p)$ and $(-3/p)$ equal 1, we get immediate proofs of Fermat's three theorems (1.1) (see Exercise 2.11). This shows just how powerful a theory we have: Fermat's theorems are now reduced to the status of an exercise. We can also go beyond Fermat, for notice that by (2.14), $x^2 + 7y^2$ is the only reduced form of discriminant -28, and it follows easily that

$$(2.17) \qquad p = x^2 + 7y^2 \iff p \equiv 1, 9, 11, 15, 23, 25 \bmod 28$$

for primes $p \neq 7$ (see Exercise 2.11). Thus we have made significant progress in answering our basic question of when $p = x^2 + ny^2$.

Unfortunately, this method for characterizing $p = x^2 + ny^2$ works only when $h(-4n) = 1$. In 1903, Landau proved a conjecture of Gauss that there are very few n's with this property:

Theorem 2.18. *Let n be a positive integer. Then*

$$h(-4n) = 1 \iff n = 1, 2, 3, 4 \text{ or } 7.$$

Proof. We will follow Landau's proof [70]. The basic idea is very simple: $x^2 + ny^2$ is a reduced form, and for $n \notin \{1, 2, 3, 4, 7\}$, we will produce a second reduced form of the same discriminant, showing that $h(-4n) > 1$. We may assume $n > 1$.

First suppose that n is not a prime power. Then n can be written $n = ac$, where $1 < a < c$ and $\gcd(a, c) = 1$ (see Exercise 2.12), and the form

$$ax^2 + cy^2$$

is reduced of disciminant $-4ac = -4n$. Thus $h(-4n) > 1$ when n is not a prime power.

Next suppose that $n = 2^r$. If $r \geq 4$, then

$$4x^2 + 4xy + (2^{r-2} + 1)y^2$$

has relatively prime coefficients and is reduced since $4 \leq 2^{r-2} + 1$. Furthermore, it has discriminant $4^2 - 4 \cdot 4(2^{r-2} + 1) = -16 \cdot 2^{r-2} = -4n$. Thus $h(-4n) > 1$ when $n = 2^r$, $r \geq 4$. One computes directly that $h(-4 \cdot 8) = 2$ (see Exercise 2.12), which leaves us with the known cases $n = 2$ and 4.

Finally, assume that $n = p^r$, where p is an odd prime. If $n+1$ can be written $n+1 = ac$, where $2 \le a < c$ and $\gcd(a,c) = 1$, then

$$ax^2 + 2xy + cy^2$$

is reduced of discriminant $2^2 - 4ac = 4 - 4(n+1) = -4n$. Thus $h(-4n) > 1$ when $n+1$ is not a prime power. But $n = p^r$ is odd, so that $n+1$ is even, and hence it remains to consider the case $n+1 = 2^s$. If $r \ge 6$, then

$$8x^2 + 6xy + (2^{s-3} + 1)y^2$$

has relatively prime coefficients and is reduced since $8 \le 2^{s-3} + 1$. Furthermore, it has discriminant $6^2 - 4 \cdot 8(2^{s-3} + 1) = 4 - 4 \cdot 2^s = 4 - 4(n+1) = -4n$, and hence $h(-4n) > 1$ when $s \ge 6$. The cases $s = 1$, 2, 3, 4 and 5 correspond to $n = 1$, 3, 7, 15 and 31 respectively. Now $n = 15$ is not a prime power, and one computes that $h(-4 \cdot 31) = 3$ (see Exercise 2.12). This leaves us with the three known cases $n = 1$, 3 and 7, and completes the proof of the theorem. Q.E.D.

Note that we've already discussed the cases $n = 1$, 2, 3 and 7, and the case $n = 4$ was omitted since $p = x^2 + 4y^2$ is a trivial corollary of $p = x^2 + y^2$ (p is odd, so that one of x or y must be even). One could also ask if there is a similar finite list of *odd* discriminants $D < 0$ with $h(D) = 1$. The answer is yes, but the proof is *much* more difficult. We will discuss this problem in §7 and give a proof in §12.

C. Elementary Genus Theory

One consequence of Theorem 2.18 is that we need some new ideas to characterize $p = x^2 + ny^2$ when $h(-4n) > 1$. To get a sense of what's involved, consider the example $n = 5$. Here, Theorem 2.16, quadratic reciprocity and (2.14) tell us that

$$\text{(2.19)} \qquad \begin{aligned} p \equiv 1,3,7,9 \bmod 20 &\iff \left(\frac{-5}{p}\right) = 1 \\ &\iff p = x^2 + 5y^2 \text{ or } 2x^2 + 2xy + 3y^2. \end{aligned}$$

We need a method of separating reduced forms of the same discriminant, and this is where genus theory comes in. The basic idea is due to Lagrange, who, like us, used quadratic forms to prove conjectures of Fermat and Euler. But rather than working with reduced forms collectively, as we did in Theorem 2.16, Lagrange considers the congruence classes represented in $(\mathbb{Z}/D\mathbb{Z})^*$ by a single form, and he groups together forms that represent the same classes. This turns out to be the basic idea of genus theory!

Let's work out some examples to see how this grouping works. When $D = -20$, one easily computes that

$$(2.20)\quad \begin{array}{lll} x^2 + 5y^2 & \text{represents} & 1,9 \text{ in } (\mathbf{Z}/20\mathbf{Z})^* \\ 2x^2 + 2xy + 3y^2 & \text{represents} & 3,7 \text{ in } (\mathbf{Z}/20\mathbf{Z})^* \end{array}$$

while for $D = -56$ one has

$$(2.21)\quad \begin{array}{lll} x^2 + 14y^2,\ 2x^2 + 7y^2 & \text{represent} & 1,9,15,23,25,29 \text{ in } (\mathbf{Z}/56\mathbf{Z})^* \\ 3x^2 \pm 2xy + 5y^2 & \text{represent} & 3,5,13,19,27,45 \text{ in } (\mathbf{Z}/56\mathbf{Z})^* \end{array}$$

(see Exercise 2.14—the reduced forms are taken from (2.14)). In his memoir on quadratic forms, Lagrange gives a systematic procedure for determining the congruence classes in $(\mathbf{Z}/D\mathbf{Z})^*$ represented by a form of discriminant D [69, pp. 759–765], and he includes a table listing various reduced forms together with the corresponding congruence classes [69, pp. 766–767]. The examples in Lagrange's table show that this is a very natural way to group forms of the same discriminant.

In general, we say that two primitive positive definite forms of discriminant D are in the same *genus* if they represent the same values in $(\mathbf{Z}/D\mathbf{Z})^*$. Note that equivalent forms represent the same numbers and hence are in the same genus. In particular, each genus consists of a finite number of classes of forms. The above examples show that when $D = -20$, there are two genera, each consisting of a single class, and when $D = -56$, there are again two genera, but this time each genus consists of two classes.

The real impact of this theory becomes clear when we combine it with Theorem 2.16. The basic idea is that genus theory refines our earlier correspondence between congruence classes and representations by reduced forms. For example, when $D = -20$, (2.19) tells us that $p \equiv 1,3,7,9 \bmod 20 \iff x^2 + 5y^2$ or $2x^2 + 2xy + 3y^2$. If we combine this with (2.20), we obtain

$$(2.22)\quad \begin{array}{l} p = x^2 + 5y^2 \iff p \equiv 1,9 \bmod 20 \\ p = 2x^2 + 2xy + 3y^2 \iff p \equiv 3,7 \bmod 20. \end{array}$$

Notice that the top line of (2.22) solves Euler's conjecture (1.20) for when $p = x^2 + 5y^2$! The thing that makes this work is that the two genera represent disjoint values in $(\mathbf{Z}/20\mathbf{Z})^*$. Looking at (2.21), we see that the same thing happens when $D = -56$, and then using Theorem 2.16 it is straightforward to prove that

$$(2.23)\quad \begin{array}{l} p = x^2 + 14y^2 \text{ or } 2x^2 + 7y^2 \iff p \equiv 1,9,15,23,25,29 \bmod 56 \\ p = 3x^2 \pm 2xy + 5y^2 \iff p \equiv 3,5,13,19,27,45 \bmod 56 \end{array}$$

(see Exercise 2.15). Note that the top line proves part of Euler's conjecture (1.21) concerning $x^2 + 14y^2$.

In order to combine Theorem 2.16 and genus theory into a general theorem, we must show that the above examples reflect the general case. We first introduce some terminology. Given a negative integer $D \equiv 0, 1 \bmod 4$, the *principal form* is defined by

$$x^2 - \frac{D}{4}y^2, \qquad D \equiv 0 \bmod 4$$

$$x^2 + xy + \frac{1-D}{4}y^2, \qquad D \equiv 1 \bmod 4.$$

It is easy to check that the principal form has discriminant D and is reduced (see Exercise 2.16). Note that when $D = -4n$, we get our friend $x^2 + ny^2$. Using the principal form, we can characterize the congruence classes in $(\mathbb{Z}/D\mathbb{Z})^*$ represented by a form of discriminant D:

Lemma 2.24. *Given a negative integer $D \equiv 0, 1 \bmod 4$, let $\ker(\chi) \subset (\mathbb{Z}/D\mathbb{Z})^*$ be as in Theorem 2.16, and let $f(x, y)$ be a form of discriminant D.*

(i) *The values in $(\mathbb{Z}/D\mathbb{Z})^*$ represented by the principal form of discriminant D form a subgroup $H \subset \ker(\chi)$.*

(ii) *The values in $(\mathbb{Z}/D\mathbb{Z})^*$ represented by $f(x, y)$ form a coset of H in $\ker(\chi)$.*

Proof. We first show that if a number m is prime to D and is represented by a form of discriminant D, then $[m] \in \ker(\chi)$. By Exercise 2.1, we can write $m = d^2 m'$, where m' is properly represented by $f(x, y)$. Then $\chi([m]) = \chi([d^2 m']) = \chi([d])^2 \chi([m']) = \chi([m'])$. Thus we may assume that m is properly represented by $f(x, y)$, and then Lemma 2.5 implies that D is a quadratic residue modulo m, i.e., $D = b^2 - km$ for some b and k. When m is odd, the properties of the Jacobi symbol (see Lemma 1.14) imply that

$$\chi([m]) = \left(\frac{D}{m}\right) = \left(\frac{b^2 - km}{m}\right) = \left(\frac{b^2}{m}\right) = \left(\frac{b}{m}\right)^2 = 1$$

and our claim is proved. The case when m is even is covered in Exercise 2.17.

We now turn to statements (i) and (ii) of the lemma. Concerning (i), the above paragraph shows that $H \subset \ker(\chi)$. When $D = -4n$, the identity (1.6) shows that H is closed under multiplication, and hence H is a subgroup. When $D \equiv 1 \bmod 4$, the argument is slightly different: here, notice that

$$4\left(x^2 + xy + \frac{1-D}{4}y^2\right) \equiv (2x + y)^2 \bmod D,$$

which makes it easy to show that H is in fact the subgroup of squares in $(\mathbb{Z}/D\mathbb{Z})^*$ (see Exercise 2.17).

To prove (ii), we need the following observation of Gauss [41, §228]:

Lemma 2.25. *Given a form $f(x,y)$ and an integer M, then $f(x,y)$ properly represents numbers relatively prime to M.*

Proof. See Exercise 2.18. Q.E.D.

Now suppose that $D = -4n$. If we apply Lemma 2.25 with $M = 4n$ and then use Lemma 2.3, we may assume that $f(x,y) = ax^2 + bxy + cy^2$, where a is prime to $4n$. Since $f(x,y)$ has discriminant $-4n$, b is even and can be written as $2b'$, and then (2.4) implies that

$$af(x,y) = (ax + b'y)^2 + ny^2.$$

Since a is relatively prime to $4n$, it follows that the values of $f(x,y)$ in $(\mathbb{Z}/4n\mathbb{Z})^*$ lie in the coset $[a]^{-1}H$. Conversely, if $[c] \in [a]^{-1}H$, then $ac \equiv z^2 + nw^2 \bmod 4n$ for some z and w. Using the above identity, it is easy to solve the congruence $f(x,y) \equiv c \bmod 4n$, and thus the coset $[a]^{-1}H$ consists exactly of the values represented in $(\mathbb{Z}/D\mathbb{Z})^*$ by $f(x,y)$. The case $D \equiv 1 \bmod 4$ is similar (see Exercise 2.17), and Lemma 2.24 is proved. Q.E.D.

Since distinct cosets of H are disjoint, Lemma 2.24 implies that different genera represent disjoint values in $(\mathbb{Z}/D\mathbb{Z})^*$. This allows us to describe genera by cosets H' of H in $\ker(\chi)$. We define the *genus of H'* to consist of all forms of discriminant D which represent the values of H' modulo D. Then Lemma 2.24 immediately implies the following refinement of Theorem 2.16:

Theorem 2.26. *Let $D \equiv 0,1 \bmod 4$ be negative, and let $H \subset \ker(\chi)$ be as in Lemma 2.24. If H' is a coset of H in $\ker(\chi)$ and p is an odd prime not dividing D, then $[p] \in H'$ if and only if p is represented by a reduced form of discriminant D in the genus of H'.* Q.E.D.

This theorem is the main result of our elementary genus theory. It generalizes examples (2.22) and (2.23), and it shows that there are always congruence conditions which characterize when a prime is represented by some form in a given genus.

For us, the most interesting genus is the one containing the principal form, which following Gauss, we call the *principal genus*. When $D = -4n$, the principal form is $x^2 + ny^2$, and since $x^2 + ny^2$ is congruent modulo $4n$ to x^2 or $x^2 + n$, depending on whether y is odd or even, we get the following explicit congruence conditions for this case:

Corollary 2.27. *Let n be a positive integer and p an odd prime not dividing n. Then p is represented by a form of discriminant $-4n$ in the principal genus if and only if for some integer β,*

$$p \equiv \beta^2 \text{ or } \beta^2 + n \bmod 4n. \qquad \text{Q.E.D.}$$

There is also a version of this for discriminants $D \equiv 1 \bmod 4$—see Exercise 2.20.

The nicest case of Corollary 2.27 is when the principal genus consists of a single class, for then we get congruence conditions that characterize $p = x^2 + ny^2$. This is what happened when $n = 5$ (see (2.22)), and this isn't the only case. For example, the table of reduced forms in Lagrange's memoir [69, pp. 766–767] shows that the same thing happens for n=6, 10, 13, 15, 21, 22 and 30—for each of these n's, the principal genus consists of only one class (see Exercise 2.21). Corollary 2.27 then gives us the following theorems for primes p:

$$\begin{aligned}
&p = x^2 + 6y^2 \iff p \equiv 1,7 \bmod 12\\
&p = x^2 + 10y^2 \iff p \equiv 1,9,11,19 \bmod 40\\
&p = x^2 + 13y^2 \iff p \equiv 1,9,17,25,29,49 \bmod 52\\
&p = x^2 + 15y^2 \iff p \equiv 1,19,31,49 \bmod 60\\
&p = x^2 + 21y^2 \iff p \equiv 1,25,37 \bmod 84\\
&p = x^2 + 22y^2 \iff p \equiv 1,9,15,23,25,31,47,49,71,81 \bmod 88\\
&p = x^2 + 30y^2 \iff p \equiv 1,31,49,79 \bmod 120.
\end{aligned} \tag{2.28}$$

It should be clear that this is a powerful theory! A natural question to ask is how often does the principal genus consist of only one class, i.e., how many theorems like (2.28) do we get? We will explore this question in more detail in §3.

The genus theory just discussed has been very successful, but it hasn't solved all of the problems posed in §1. In particular, we have yet to prove Fermat's conjecture concerning $pq = x^2 + 5y^2$, and we've only done parts of Euler's conjectures (1.20) and (1.21) concerning $x^2 + 5y^2$ and $x^2 + 14y^2$. To complete the proofs, we again turn to Lagrange for help.

Let's begin with $x^2 + 5y^2$. We've already proved the part concerning when a prime p can equal $x^2 + 5y^2$ (see (2.22)), but it remains to show that for primes p and q,

$$\begin{aligned}
p, q \equiv 3,7 \bmod 20 &\Rightarrow pq = x^2 + 5y^2 \qquad \text{(Fermat)}\\
p \equiv 3,7 \bmod 20 &\Rightarrow 2p = x^2 + 5y^2 \qquad \text{(Euler).}
\end{aligned} \tag{2.29}$$

Lagrange's argument [69, pp. 788–789] is as follows. He first notes that primes congruent to 3 or 7 modulo 20 can be written as $2x^2 + 2xy + 3y^2$ (this is (2.22)), so that both parts of (2.29) can be proved by showing that the product of two numbers represented by $2x^2 + 2xy + 3y^2$ is of the form $x^2 + 5y^2$. He then states the identity

$$(2.30) \qquad (2x^2 + 2xy + 3y^2)(2z^2 + 2zw + 3w^2) = (2xz + xw + yz + 3yw)^2 + 5(xw - yz)^2$$

(see Exercise 2.22), and everything is proved!

Turning to Euler's conjecture (1.21) for $x^2 + 14y^2$, we proved part of it in (2.23), but we still need to show that

$$p \equiv 3, 5, 13, 19, 27, 45 \bmod 56 \iff 3p = x^2 + 14y^2.$$

Using (2.23), it suffices to show that 3 times a number represented by $3x^2 \pm 2xy + 5y^2$, or more generally the product of any two such numbers, is of the form $x^2 + 14y^2$. So what we need is another identity of the form (2.30), and in fact there is a version of (2.30) that holds for any form of discriminant $-4n$:

$$(2.31) \qquad (ax^2 + 2bxy + cy^2)(az^2 + 2bzw + cw^2) = (axz + bxw + byz + cyw)^2 + n(xw - yz)^2$$

(see Exercise 2.21). Applying this to $3x^2 + 2xy + 5y^2$ and $n = 14$, we are done.

We can also explain one other aspect of Euler's conjectures (1.20) and (1.21), for recall that we wondered why (1.20) used $2p$ while (1.21) used $3p$. The answer again involves the identities (2.30) and (2.31): they show that 2 (resp. 3) can be replaced by any value represented by $2x^2 + 2xy + 3y^2$ (resp. $3x^2 \pm 2xy + 5y^2$). But Legendre's observation from the proof of Theorem 2.8 shows that 2 (resp. 3) is the best choice because it's the smallest nonzero value represented by the form in question. We will see below and in §3 that identities like (2.30) and (2.31) are special cases of the composition of quadratic forms.

We now have complete proofs of Euler's conjectures (1.20) and (1.21) for $x^2 + 5y^2$ and $x^2 + 14y^2$. Notice that we've used a lot of mathematics: quadratic reciprocity, reduced quadratic forms, genus theory and the composition of quadratic forms. This amply justifies the high estimate of Euler's insight that was made in §1, and Lagrange is equally impressive for providing the proper tools to understand what lay behind Euler's conjectures.

D. Lagrange and Legendre

We've already described parts of Lagrange's memoir "Recherches d'Arithmétique", but there are some further comments we'd like to add. First, although we credit Lagrange with the discovery of genus theory, it appears only implicitly in his work. The groupings that appear in his tables of reduced forms are striking, but Lagrange's comments on genus theory are a different matter. On the page before the tables begin, Lagrange explains his grouping of forms as follows: "when two different [forms] give the same values of b [in $(\mathbf{Z}/4n\mathbf{Z})^*$], one combines these [forms] into the same case" [69, p. 765]. This is the sum total of what Lagrange says about genus theory!

After completing the basic theory of quadratic forms (both definite and indefinite), Lagrange gives some applications to number theory. To motivate his results, he turns to Fermat and Euler, and he quotes from two of our main sources of inspiration: Fermat's 1658 letter to Digby and Euler's 1744 paper on prime divisors of $paa \pm qyy$. Lagrange explicitly states Fermat's results (1.1) on primes of the form $x^2 + ny^2$, $n = 1,2$ or 3, and he notes Fermat's speculation that $pq = x^2 + 5y^2$ whenever p and q are primes congruent to 3 or 7 modulo 20. Lagrange also mentions several of Euler's conjectures, including (1.20), and he adds "one finds a very large number of similar theorems in Volume XIV of the old *Commentaires de Pétersbourg* [where Euler's 1744 paper appeared], but none of them have been demonstrated until now" [69, pp. 775–776].

The last section of Lagrange's memoir is titled "Prime numbers of the form $4nm + b$ which are at the same time of the form $x^2 \pm ny^2$" [69, p. 775]. It's clear that Lagrange wanted to prove Theorem 2.26, so that he could read off corollaries like (2.17), (2.22), (2.23) and (2.28). The problem is that these proofs depend on quadratic reciprocity, which Lagrange didn't know in general—he could only prove some special cases. For example, he was able to determine $(\pm 2/p)$, $(\pm 3/p)$ and $(\pm 5/p)$, but he had only partial results for $(\pm 7/p)$. Thus, he could prove all of (2.22) but only parts of the others (see [69, pp. 784–793] for the full list of his results). To get the flavor of Lagrange's arguments, the reader should see Exercise 2.23 or Scharlau and Opolka [86, pp. 41–43]. At the end of the memoir, Lagrange summarizes what he could prove about quadratic reciprocity, stating his results in terms of Euler's criterion

$$a^{(p-1)/2} \equiv \left(\frac{a}{p}\right) \bmod p.$$

For example, for $(2/p)$, Lagrange states [69, p. 794]:

> Thus, if p is a prime number of one of the forms $8n \pm 1$, $2^{(p-1)/2} - 1$ will be divisible by p, and if p is of the form $8n \pm 3$, $2^{(p-1)/2} + 1$ will thus be divisible by p.

We next turn to Legendre. In his 1785 memoir "Recherches d'Analyse Indeterminée" [75], the two major results are first, a necessary and sufficient criterion for the equation

$$ax^2 + by^2 + cz^2 = 0, \qquad a, b, c \in \mathbf{Z}$$

to have a nontrivial integral solution, and second, a proof of quadratic reciprocity. Legendre was clearly influenced by Lagrange, but he replaces Lagrange's "$2^{(p-1)/2} - 1$ will be divisible by p" by the simpler phrase "$2^{(p-1)/2} = 1$", where, as he warns the reader, "one has thrown out the multiples of p in the first member" [75, p. 516]. He then goes on to state quadratic reciprocity in the following form [75, p. 517]:

> c and d being two [odd] prime numbers, the expressions $c^{(d-1)/2}$, $d^{(c-1)/2}$ do not have different signs except when c & d are both of the form $4n - 1$; in all other cases, these expressions will always have the same sign.

Except for the notation, this is a thoroughly modern statement of quadratic reciprocity. Legendre's proof is a different matter, for it is quite incomplete. We won't examine the proof in detail—this is done in Weil [106, pp. 328–330 and 344–345]. Suffice it to say that some of the cases are proved rigorously (see Exercise 2.24), some depend on Dirichlet's theorem on primes in arithmetic progressions, and some are a tangle of circular reasoning.

In 1798 Legendre published a more ambitious work, the *Essai sur la Théorie des Nombres*. (The third edition [74], published 1830, was titled *Théorie des Nombres*, and all of our references will be to this edition.) Lagrange must have been dissatisfied with the notation of the "Recherches", for in the *Essai* he introduces the Legendre symbol (a/p). Then, in a section titled "Theorem containing a law of reciprocity which exists between two arbitrary prime numbers," Legendre states that if n and m are distinct odd primes, then

$$\left(\frac{n}{m}\right) = (-1)^{(n-1)/2 \cdot (m-1)/2} \left(\frac{m}{n}\right)$$

(see [74, Vol. I, p. 230]). This is where our notation and terminology for quadratic reciprocity come from. Unfortunately, the *Essai* repeats Legendre's incomplete proof from 1785, although by the 1830 edition there had been enough criticism of this proof that Legendre added Gauss' third proof of reciprocity as well as one communicated to him by Jacobi (still maintaining that his original proof was valid).

The *Essai* also contains a treatment of quadratic forms. Like Lagrange, one of Legendre's goals was to prove theorems in number theory using quadratic forms. The difference is that Legendre knows quadratic reciprocity (or at least he thinks he does), and this allows him to state a version of our main result, Theorem 2.26. Legendre calls it his "Theoremé Generalé"

[74, Vol. I, p. 299], and it goes as follows: if $[a]$ is a congruence class lying in $\ker(\chi)$, then

> every prime number comprised of the form $4nx + a$ ...will consequently be given by one of the quadratic forms $py^2 + 2qyz \pm rz^2$ which correspond to the linear form $4nx + a$.

The terminology here is interesting. Euler and Lagrange would speak of numbers "of the form" $4nx + a$ or "of the form" $ax^2 + bxy + cy^2$. As the above quote indicates, Legendre distinguished these two by calling them linear forms and quadratic forms respectively. This is where we get the term "quadratic form".

While Legendre's "Theoremé" makes no explicit reference to genus theory, the context shows that it's there implicitly. Namely, Legendre's book has tables similar to Lagrange's, with the forms grouped according to the values they represent in $(\mathbf{Z}/D\mathbf{Z})^*$. Since the explanation of the tables immediately precedes the statement of the "Theoremé" [74, Vol. I, pp. 286–298], it's clear that Legendre's correspondence between linear forms and quadratic forms is exactly that given by Theorem 2.26.

To Legendre, this theorem "is, without contradiction, one of the most general and most important in the theory of numbers" [74, Vol. I, p. 302]. It's main consequence is that every entry in his tables becomes a theorem, and Legendre gives several pages of explicit examples [74, Vol. I, pp. 305–307]. This is a big advance over what Lagrange could do, and Legendre notes that quadratic reciprocity was the key to his success [74, Vol. I, p. 307]:

> Lagrange is the first who opened the way for the study of these sorts of theorems. ... But the methods which served the great geometer are not applicable ... except in very few cases; and the difficulty in this regard could not be completely resolved without the aid of the law of reciprocity.

Besides completing Lagrange's program, Legendre also tried to understand some of the other ideas implicit in Lagrange's memoir. We will discuss one Legendre's attempts that is particularly relevant to our purposes: his theory of composition. Legendre's basic idea was to generalize the identity (2.30)

$$\begin{aligned}(2x^2 + 2xy + 3y^2)&(2z^2 + 2zw + 3w^2)\\ &= (2xz + xw + yz + 3yw)^2 + 5(xw - yz)^2\end{aligned}$$

used by Lagrange in proving the conjectures of Fermat and Euler concerning $x^2 + 5y^2$. We gave one generalization in (2.31), but Legendre saw that something more general was going on. More precisely, let $f(x,y)$ and $g(x,y)$ be forms of discriminant D. Then a form $F(x,y)$ of the same discriminant is their *composition* provided that

$$f(x,y)g(z,w) = F(B_1(x,y;z,w), B_2(x,y;z,w))$$

where

$$B_i(x,y;z,w) = a_i xz + b_i xw + c_i yz + d_i yw, \qquad i = 1,2$$

are bilinear forms in x,y and z,w. Thus Lagrange's identity shows that $x^2 + 5y^2$ is the composition of $2x^2 + 2xy + 3y^2$ with itself. And this is not the only example we've seen—the reader can check that (1.3), (1.6) and (2.31) are also examples of the composition of forms.

A useful consequence of composition is that whenever $F(x,y)$ is composed of $f(x,y)$ and $g(x,y)$, then the product of numbers represented by $f(x,y)$ and $g(x,y)$ will be represented by $F(x,y)$. This was the idea that enabled us to complete the conjectures of Fermat and Euler for $x^2 + 5y^2$ and $x^2 + 14y^2$.

The basic question is whether any two forms of the same discriminant can be composed, and Legendre showed that the answer is yes [74, Vol. II, pp. 27–30]. For simplicitly, let's discuss the case where the forms $f(x,y) = ax^2 + 2bxy + cy^2$ and $g(x,y) = a'x^2 + 2b'xy + c'y^2$ have discriminant $-4n$, and a and a' are relatively prime (we can always arrange the last condition by changing the forms by a proper equivalence). Then the Chinese remainder theorem shows that there is a number B such that

$$\begin{aligned} B &\equiv \pm b \bmod a \\ B &\equiv \pm b' \bmod a'. \end{aligned} \tag{2.32}$$

It follows that $B^2 + n \equiv b^2 + (ac - b^2) \equiv 0 \bmod a$, so that $a \mid B^2 + n$. The same holds for a', and thus $aa' \mid B^2 + n$. Then Legendre shows that the form

$$F(x,y) = aa'x^2 + 2Bxy + \frac{B^2 + n}{aa'}y^2$$

is the composition of $f(x,y)$ and $g(x,y)$. A modern account of Legendre's argument may be found in Weil [106, pp. 332–335]), and we will consider this problem (from a slightly different point of view) in §3 when we discuss composition in more detail.

Because of the $\pm$ signs in (2.32), two forms in general may be composed in four different ways. For example, the forms $14x^2 + 10xy + 21y^2$ and $9x^2 + 2xy + 30y^2$ compose to the four forms

$$126x^2 \pm 38xy + 5y^2, \qquad 126x^2 \pm 74xy + 13y^2,$$

and it is easy to show that these forms all lie in different classes (see Exercise 2.26). Since Legendre used equivalence rather than proper equivalence, he sees two rather than four forms here—for him, this operation "leads in general to two solutions" [74, Vol. II, p. 28].

One of Legendre's important ideas is that since every form is equivalent to a reduced one, it suffices to work out the compositions of reduced forms. The resulting table would then give the compositions of all possible forms of that discriminant. Let's look at the case $n = 41$, which Legendre does in detail in [74, Vol. II, pp. 39–40]. He labels the reduced forms as follows:

$$\begin{aligned} A &= x^2 + 41y^2 \\ B &= 2x^2 + 2xy + 21y^2 \\ C &= 5x^2 + 6xy + 10y^2 \\ D &= 3x^2 + 2xy + 14y^2 \\ E &= 6x^2 + 2xy + 7y^2. \end{aligned} \tag{2.33}$$

(Legendre writes the forms slightly differently, but it's more convenient to work with reduced forms.) He then gives the following table of compositions:

(2.34)

$AA = A$	$BB = A$	$CC = A$ or B	$DD = A$ or C	$EE = A$ or C
$AB = B$	$BC = C$	$CD = D$ or E	$DE = B$ or C	
$AC = C$	$BD = E$	$CE = D$ or E		
$AD = D$	$BE = D$			
$AE = E$				

This almost looks like the multiplication table for a group, but the binary operation isn't single-valued. To the modern reader, it's clear that Legendre must be doing something slightly wrong.

One problem is that (2.33) lists 5 forms, while the class number is 8. (C, D and E each give two reduced forms, while A and B each give only one.) This is closely related to the ambiguity in Legendre's operation: as long as we work with equivalence rather than proper equivalence, we can't fix the sign of the middle coefficient $2b$ of a reduced form, so that the $\pm$ signs in (2.32) are forced upon us.

This suggests that composition might give a group operation on the *classes* of forms of discriminant D. However, there remain serious problems to be solved. Composition, as defined above, is still a multiple-valued operation. Thus one has to show that the signs in (2.32) can be chosen *uniformly* so that as we vary $f(x, y)$ and $g(x, y)$ within their proper equivalance classes, the resulting compositions are all properly equivalent. Then one has to worry about associativity, inverses, etc. There's a lot of work to be done!

This concludes our discussion of Lagrange and Legendre. While the last few pages have raised more questions than answers, the reader should still be convinced of the richness of the theory of quadratic forms. The surprising fact is that we have barely reached the really interesting part of the theory, for we have yet to consider the work of Gauss.

E. Exercises

2.1. If a form $f(x,y)$ represents an integer m, show that m can be written $m = d^2m'$, where $f(x,y)$ properly represents m'.

2.2. In this exercise we study equivalence and proper equivalence.

(a) Show that equivalence and proper equivalence are equivalence relations.

(b) Show that improper equivalence is not an equivalence relation.

(c) Show that equivalent forms represent the same numbers, and show that the same holds for proper representations.

(d) Show that any form equivalent to a primitive form is itself primitive. Hint: use (c).

2.3. Let $f(x,y)$ and $g(x,y)$ be forms of discriminants D and D' respectively, and assume that there are integers p, q, r and s such that

$$f(x,y) = g(px+qy, rx+sy).$$

Prove that $D = (ps-qr)^2D'$.

2.4. Let $f(x,y)$ be a form of discriminant $D \neq 0$.

(a) If $D > 0$, then use (2.4) to prove that $f(x,y)$ represents both positive and negative numbers.

(b) If $D < 0$, then show that $f(x,y)$ represents only positive or only negative numbers, depending on the sign of the coefficient of x^2.

2.5. Formulate and prove a version of Corollary 2.6 which holds for arbitrary discriminants.

2.6. Find a reduced form that is properly equivalent to the form $126x^2 + 74xy + 13y^2$. Hint: make the middle coefficient small—see the proof of Theorem 2.8.

2.7. Prove (2.9) for forms that satisfy $|b| \leq a \leq c$.

2.8. This exercise is concerned with the uniqueness part of Theorem 2.8.

(a) Prove (2.11).

(b) Prove a version of (2.11) that holds in the exceptional cases $|b| = a$ or $a = c$, and use this to complete the uniqueness part of the proof of Theorem 2.8.

2.9. Write a computer program that computes all reduced forms for a given discriminant in the range $-32768 < D < 0$. This range is easily implemented using the integer arithmetic of standard languages such as BASIC or Pascal. For example, one finds that $h(-32767) = 52$. If you don't write a computer program, you should check the following examples by hand.

(a) Verify the entries in table (2.14).

(b) Compute all reduced forms of discriminants -3, -15, -24, -31, and -52.

2.10. This exercise is concerned with indefinite forms of discriminant $D > 0$, D not a perfect square. The last condition implies that the outer coefficients of a form with discriminant D are nonzero.

(a) Adapt the proof of Theorem 2.8 to show that any form of discriminant D is properly equivalent to $ax^2 + bxy + cy^2$, where

$$|b| \leq |a| \leq |c|.$$

(b) If $ax^2 + bxy + cy^2$ satifies the above inequalities, prove that

$$|a| \leq \frac{\sqrt{D}}{2}.$$

(c) Conclude that there are only finitely many proper equivalence classes of forms of discriminant D. This proves that the class number $h(D)$ is finite.

2.11. Use Theorem 2.16, quadratic reciprocity and table (2.14) to prove Fermat's three theorems (1.1) and the new result (2.17) for $x^2 + 7y^2$.

2.12. This exercise is concerned with the proof of Theorem 2.18.

(a) If $m > 1$ is an integer which is not a prime power, prove that m can be written $m = ac$ where $1 < a < c$ and $\gcd(a,c) = 1$.

(b) Show that $h(-32) = 2$ and $h(-124) = 3$.

2.13. Use Theorem 2.16, quadratic reciprocity and table (2.14) to prove (2.19), and work out similar results for discriminants -3, -15, -24, -31 and -52.

2.14. Prove (2.20) and (2.21). Hint: use Lemma 2.24.

2.15. Prove (2.23).

2.16. Let D be a number congruent to 1 modulo 4. Show that the form $x^2 + xy + (1 - D)/4y^2$ has discriminant D, and show that it is reduced when $D < 0$.

2.17. In this exercise, we will complete the proof of Lemma 2.24 for discriminants $D \equiv 1 \bmod 4$. Let $\chi : (\mathbf{Z}/D\mathbf{Z})^* \to \{\pm 1\}$ be as in Lemma 1.14.

(a) If an even number is properly represented by a form of discriminant D, then show that $D \equiv 1 \bmod 8$. Hint: use Lemma 2.3.

(b) If m is relatively prime to D and is represented by a form of discriminant D, then show that $[m] \in \ker(\chi)$. Hint: use Lemma 2.5 and, when m is even, (a) and Exercise 1.12(c).

(c) Let $H \subset (\mathbf{Z}/D\mathbf{Z})^*$ be the subgroup of squares. Show that H consists of the values represented by $x^2 + xy + (1-D)/4y^2$. Hint: use

$$4\left(x^2 + xy + \frac{1-D}{4}y^2\right) \equiv (2x+y)^2 \bmod D.$$

(d) If $f(x,y)$ is a form of discriminant D, then show that the values in $(\mathbf{Z}/D\mathbf{Z})^*$ represented by $f(x,y)$ form a coset of H in $\ker(\chi)$. Hint: use (2.4).

2.18. Let $f(x,y) = ax^2 + bxy + cy^2$, where as usual we assume $\gcd(a,b,c) = 1$.

(a) Given a prime p, prove that at least one of $f(1,0)$, $f(0,1)$ and $f(1,1)$ is relatively prime to p.

(b) Prove Lemma 2.25. Hint: use (a) and the Chinese Remainder Theorem.

2.19. Work out the genus theory of Theorem 2.26 for discriminants -15, -24, -31 and -52. Your answers should be similar to (2.22) and (2.23).

2.20. Formulate and prove a version of Corollary 2.27 for negative discriminants $D \equiv 1 \bmod 4$. Hint: by Exercise 2.17(c), H is the subgroup of squares.

2.21. Prove (2.28). Hint: for each n, find the reduced forms and use Lemma 2.24.

2.22. Prove (2.30) and its generalization (2.31).

2.23. The goal of this exercise is to prove that $(-2/p) = 1$ when $p \equiv 1, 3 \bmod 8$. The argument below is due to Lagrange, and is similar to the one used by Euler in his proof of the Reciprocity Step for $x^2 + 2y^2$ [33, Vol. II, pp. 240–281].

(a) When $p \equiv 1 \bmod 8$, write $p = 8k+1$, and then use the identity

$$x^{8k} - 1 = (x^{4k} - 1)^2 + 2x^{4k}$$

to show that $(-2/p) = 1$.

(b) When $p \equiv 3 \bmod 8$, assume that $(-2/p) = -1$. Show that $(2/p) = 1$, and thus by Corollary 2.6, p is represented by a form of discriminant 8.

(c) Use Exercise 2.10(a) to show that any form of discriminant 8 is properly equivalent to $\pm(x^2 - 2y^2)$.

(d) Show that an odd prime $p = \pm(x^2 - 2y^2)$ must be congruent to ± 1 modulo 8.

From (a)–(d), it follows easily that $(-2/p) = 1$ when $p \equiv 1, 3 \bmod 8$.

2.24. One of the main theorems is Legendre's 1785 memoir [74, pp. 509–513] states that the equation

$$ax^2 + by^2 + cz^2 = 0,$$

where abc is squarefree, has a nontrivial integral solution if and only if

(i) a, b and c are not all of the same sign, and

(ii) $-bc$, $-ac$ and $-ab$ are quadratic residues modulo $|a|$, $|b|$ and $|c|$ respectively.

As we've already noted, Legendre tried to use this result to prove quadratic reciprocity. In this problem, we will treat one of the cases where he succeeded. Let p and q be primes which satisfy $p \equiv 1 \bmod 4$ and $q \equiv 3 \bmod 4$, and assume that $(p/q) = -1$ and $(q/p) = 1$. We will derive a contradiction as follows:

(a) Use Legendre's theorem to show that $x^2 + py^2 - qz^2 = 0$ has a nontrivial integral solution.

(b) Working modulo 4, show that $x^2 + py^2 - qz^2 = 0$ has no nontrivial integral solutions.

In [106, pp. 339–345], Weil explains why this argument works.

2.25. Recall that the opposite of the form $ax^2 + bxy + cy^2$ is the form $ax^2 - bxy + cy^2$. Prove that two forms are properly equivalent if and only if their opposites are.

2.26. Verify that $14x^2 + 10xy + 21y^2$ and $9x^2 + 2xy + 30y^2$ compose to the four forms $126x^2 \pm 74xy + 13y^2$ and $126x^2 \pm 38xy + 5y^2$, and show that they all lie in different classes. Hint: use Exercises 2.6 and 2.25.

2.27. Let p be a prime number which is represented by forms $f(x,y)$ and $g(x,y)$ of the same discriminant.

(a) Show that $f(x,y)$ and $g(x,y)$ are equivalent. Hint: use Lemma 2.3, and examine the middle coefficient modulo p.

(b) If $f(x,y) = x^2 + ny^2$, and $g(x,y)$ is reduced, then show that $f(x,y)$ and $g(x,y)$ are equal.

§3. GAUSS, COMPOSITION AND GENERA

While genus theory and composition were implicit in Lagrange's work, these concepts are still primarily linked to Gauss, and for good reason: he may not have been the first to use them, but he was the first to understand their astonishing depth and interconnection. In this section we will prove Gauss' major results on composition and genus theory for the special case of positive definite forms. We will then apply this theory to our question concerning primes of the form $x^2 + ny^2$, and we will also discuss Euler's convenient numbers. These turn out to be those n's for which each genus consists of a single class, and it is still not known exactly how many there are. The section will end with a discussion of Gauss' *Disquisitiones Arithmeticae*.

A. Composition and the Class Group

The basic definition of composition was given in §2: if $f(x,y)$ and $g(x,y)$ are primitive positive definite forms of discriminant D, then a form $F(x,y)$ of the same type is their *composition* provided that

$$f(x,y)g(z,w) = F(B_1(x,y;z,w), B_2(x,y;z,w)),$$

where

$$B_i(x,y;z,w) = a_i xz + b_i xw + c_i yz + d_i yw, \qquad i = 1,2$$

are integral bilinear forms. Two forms can be composed in many different ways, and the resulting forms need not be properly equivalent. In §2 we gave an example of two forms whose compositions lay in four distinct classes. So if we want a well-defined operation on classes of forms, we must somehow *restrict* the notion of composition. Gauss does this as follows: given the above composition data, he proves that

$$(3.1) \qquad a_1 b_2 - a_2 b_1 = \pm f(1,0), \qquad a_1 c_2 - a_2 c_1 = \pm g(1,0)$$

(see [41, §235] or Exercise 3.1), and then he defines the composition to be a *direct composition* provided that both of the signs in (3.1) are $+$.

The main result of Gauss' theory of composition is that for a fixed discriminant, direct composition makes the set of classes of forms into a finite Abelian group [41, §§236–40, 245 and 249]. Unfortunately, direct composition is an awkward concept to work with, and Gauss' proof of the group structure is long and complicated. So rather than follow Gauss, we will take a different approach to the study of composition. The basic idea is due to Dirichlet [28, Supplement X], though his treatment was clearly influenced by Legendre. Before giving Dirichlet's definition, we will need the following lemma:

Lemma 3.2. *Assume that* $f(x,y) = ax^2 + bxy + cy^2$ *and* $g(x,y) = a'x^2 + b'xy + c'y^2$ *have discriminant* D *and satisfy* $\gcd(a,a',(b+b')/2) = 1$ *(since* b *and* b' *have the same parity,* $(b+b')/2$ *is an integer). Then there is a unique integer* B *modulo* $2aa'$ *such that*

$$\begin{aligned} B &\equiv b \bmod 2a \\ B &\equiv b' \bmod 2a' \\ B^2 &\equiv D \bmod 4aa'. \end{aligned}$$

Proof. The first step is to put these congruences into a standard form. If a number B satisfies the first two, then

$$B^2 - (b+b')B + bb' \equiv (B-b)(B-b') \equiv 0 \bmod 4aa',$$

so that the third congruence can be written as

$$(b+b')B \equiv bb' + D \bmod 4aa'.$$

Dividing by 2, this becomes

$$(b+b')/2 \cdot B \equiv (bb'+D)/2 \bmod 2aa'. \tag{3.3}$$

If we multiply the first two congruences by a' and a respectively and combine them with (3.3), we see that the three congruences in the statement of the lemma are equivalent to

$$\begin{aligned} a' \cdot B &\equiv a'b \bmod 2aa' \\ a \cdot B &\equiv ab' \bmod 2aa' \\ (b+b')/2 \cdot B &\equiv (bb'+D)/2 \bmod 2aa'. \end{aligned} \tag{3.4}$$

The following lemma tells us about the solvability of these congruences:

Lemma 3.5. *Let* $p_1, q_1, \ldots, p_r, q_r, m$ *be numbers with* $\gcd(p_1, \ldots, p_r, m) = 1$. *Then the congruences*

$$p_i B \equiv q_i \bmod m, \qquad i = 1, \ldots, r$$

have a unique solution modulo m *if and only if for all* $i, j = 1, \ldots, r$ *we have*

$$p_i q_j \equiv p_j q_i \bmod m. \tag{3.6}$$

Proof. See Exercise 3.3. Q.E.D.

Since we are assuming $\gcd(a, a', (b+b')/2) = 1$, the congruences (3.4) satisfy the gcd condition of the above lemma, and the compatibility conditions (3.6) are easy to verify (see Exercise 3.4). The existence and uniqueness of the desired B follow immediately. Q.E.D.

We can now give Dirichlet's definition of composition. Let $f(x,y) = ax^2 + bxy + cy^2$ and $g(x,y) = a'x^2 + b'xy + c'y^2$ be primitive positive definite forms of discriminant $D < 0$ which satisfy $\gcd(a, a', (b+b')/2) = 1$. Then the *Dirichlet composition* of $f(x,y)$ and $g(x,y)$ is the form

$$F(x,y) = aa'x^2 + 2Bxy + \frac{B^2 - D}{4aa'}y^2, \tag{3.7}$$

where B is the integer determined by Lemma 3.2. The basic properties of $F(x,y)$ are:

Proposition 3.8. *Given $f(x,y)$ and $g(x,y)$ as above, the Dirichlet composition $F(x,y)$ defined in (3.7) is a primitive positive definite form of discriminant D, and $F(x,y)$ is the direct composition of $f(x,y)$ and $g(x,y)$ in the sense of (3.1).*

Proof. An easy calculation shows that $F(x,y)$ has discriminant D, and the form is consequently positive definite.

The next step is to prove that $F(x,y)$ is the composition of $f(x,y)$ and $g(x,y)$. We will sketch the argument and leave the details to the reader. Let $C = (B^2 - D)/4aa'$, so that $F(x,y) = aa'x^2 + Bxy + Cy^2$. Then, using the first two congruences of Lemma 3.2, it is easy to show that $f(x,y)$ and $g(x,y)$ are properly equivalent to the forms $ax^2 + Bxy + a'Cy^2$ and $a'x^2 + Bxy + aCy^2$ respectively. However, for these last two forms one has the composition identity

$$(ax^2 + Bxy + a'Cy^2)(a'z^2 + Bzw + aCw^2) = aa'X^2 + BXY + CY^2,$$

where $X = xz - Czw$ and $Y = axw + a'yz + Byw$. It follows easily that $F(x,y)$ is the composition of $f(x,y)$ and $g(x,y)$. With a little more effort, it can be checked that this is a direct composition in Gauss' sense (3.1). The details of these arguments are covered in Exercise 3.5.

It remains to show that $F(x,y)$ is primitive, i.e., that its coefficients are relatively prime. Suppose that some prime p divided all of the coefficients. This would imply that p divided all numbers represented by $F(x,y)$. Since $F(x,y)$ is the composition of $f(x,y)$ and $g(x,y)$, this implies that p divides all numbers of the form $f(x,y)g(z,w)$. But we know that $f(x,y)$ and $g(x,y)$ are primitive, and from here it is easy to derive a contradiction (see Exercise 3.5 for the details). This completes the proof of the proposition. Q.E.D.

While Dirichlet composition is not as general as direct composition (not all direct compositions satisfy $\gcd(a,a',(b+b')/2)=1$), it is easier to use in practice since there is an explicit formula (3.7) for the composition. Notice also that the congruence conditions in Lemma 3.2 are similar to the ones (2.32) used by Legendre. This is no accident, for when $D=-4n$ and $\gcd(a,a')=1$, Dirichlet's formula reduces exactly to the one given by Legendre (see Exercise 3.6).

We can now state our main result on composition:

Theorem 3.9. *Let $D \equiv 0,1 \bmod 4$ be negative, and let $C(D)$ be the set of classes of primitive positive definite forms of discriminant D. Then Dirichlet composition induces a well-defined binary operation on $C(D)$ which makes $C(D)$ into a finite Abelian group whose order is the class number $h(D)$.*

Furthermore, the identity element of $C(D)$ is the class containing the principal form

$$x^2 - \frac{D}{4}y^2 \qquad \text{if } D \equiv 0 \bmod 4$$

$$x^2 + xy + \frac{1-D}{4}y^2 \quad \text{if } D \equiv 1 \bmod 4,$$

and the inverse of the class containing the form $ax^2+bxy+cy^2$ is the class containing $ax^2-bxy+cy^2$.

Remarks. Some terminology is in order here.

(i) The group $C(D)$ is called the *class group*, though we will sometimes refer to $C(D)$ as the *form class group* to distinguish it from the ideal class group to be defined later.

(ii) The principal form of discriminant D was introduced in §2. The class it lies in is called the *principal class*. When $D=-4n$, the principal form is x^2+ny^2.

(iii) The form $ax^2-bxy+cy^2$ is called the *opposite* of $ax^2+bxy+cy^2$, so that the opposite form gives the inverse under Dirichlet composition.

Proof. Let $f(x,y)=ax^2+bxy+cy^2$ and $g(x,y)$ be forms of the given type. Using Lemmas 2.3 and 2.25, we can replace $g(x,y)$ by a properly equivalent form $a'x^2+b'xy+c'y^2$ where $\gcd(a,a')=1$. Then the Dirichlet composition of these forms is defined, which proves that Dirichlet composition is defined for any pair of classes in $C(D)$. To get a group structure out of this, we must then prove that:

(i) This operation is well-defined on the level of classes, and
(ii) The induced binary operation makes $C(D)$ into an Abelian group.

The proofs of (i) and (ii) can be done directly using the definition of Dirichlet composition (see Dirichlet [28, Supplement X] or Flath [36, §V.2]), but the argument is much easier using ideal class groups (to be studied in §7). We will therefore postpone this part of the proof until then. For now, we will assume that (i) and (ii) are true.

Let's next show that the principal class is the identity element of $C(D)$. To compose the principal form with $f(x,y) = ax^2 + bxy + cy^2$, first note that the gcd condition is clearly met, and thus the Dirichlet composition is defined. Then observe that $B = b$ satisfies the conditions of Lemma 3.2, so that by formula (3.7), the Dirichlet composition $F(x,y)$ reduces to the given form $f(x,y)$. This proves that the principal class is the identity.

Finally, given $f(x,y) = ax^2 + bxy + cy^2$, its opposite is $f'(x,y) = ax^2 - bxy + cy^2$. Since $\gcd(a,a,(b+(-b))/2) = a$ may be > 1, we can't apply Dirichlet composition directly. But if we use the proper equivalence $(x,y) \mapsto (-y,x)$, then we can replace $f'(x,y)$ by $g(x,y) = cx^2 + bxy + ay^2$. Since $\gcd(a,c,(b+b)/2) = \gcd(a,c,b) = 1$, we can apply Dirichlet's formulas to $f(x,y)$ and $g(x,y)$. One checks easily that $B = b$ satisfies the conditions of Lemma 3.2, so that the Dirichlet composition is $acx^2 + bxy + y^2$. We leave it to the reader to show that this form is properly equivalent to the principal form (see Exercise 3.7). This completes the proof of the theorem. Q.E.D.

We can now complete the discussion (begun in §2) of Legendre's theory of composition. To prevent confusion, we will distinguish between a *class* (all forms properly equivalent to a given form) and a *Lagrangian class* (all forms equivalent to a given one). In Theorem 3.9, we studied the composition of classes, while Legendre was concerned with the composition of Lagrangian classes. It is an easy exercise to show that the Lagrangian class of a form is the union of its class and the class of its opposite (see Exercise 3.8). By Theorem 3.9, this means that a Lagrangian class is the union of a class and its inverse in the class group $C(D)$. Thus Legendre's "operation" is the multiple-valued operation that multiplication induces on the set $C(D)/\sim$, where $\sim$ is the equivalence relation that identifies $x \in C(D)$ with x^{-1} (see Exercise 3.9). In Legendre's example (2.33), which dealt with forms of discriminant -164, we will see shortly that $C(-164) \simeq \mathbf{Z}/8\mathbf{Z}$, and it is then an easy exercise to show that $C(-164)/\sim$ is isomorphic to the structure given in (2.34) (see Exercise 3.9).

The elements of order ≤ 2 in the class group $C(D)$ play a special role in composition and genus theory. The reduced forms that lie in such classes are easy to find:

Lemma 3.10. *A reduced form $f(x,y) = ax^2 + bxy + cy^2$ of discriminant D has order ≤ 2 in the class group $C(D)$ if and only if $b = 0$, $a = b$ or $a = c$.*

Proof. Let $f'(x,y)$ be the opposite of $f(x,y)$. By Theorem 3.9, the class of $f(x,y)$ has order ≤ 2 if and only if the forms $f(x,y)$ and $f'(x,y)$ are properly equivalent. There are two cases to consider:

$|b| < a < c$: Here, $f'(x,y)$ is also reduced, so that by Theorem 2.8,

the two forms are properly equivalent $\Longleftrightarrow b = 0$.

$a = b$ or $a = c$: In these cases, the proof of Theorem 2.8 shows that

the two forms are always properly equivalent.

The lemma now follows immediately. Q.E.D.

For an example of how this works, consider Legendre's example from §2 of forms of discriminant -164. The reduced forms are listed in (2.33), and Lemma 3.10 shows that only $2x^2 + 2xy + 21y^2$ has order 2. Since the class number is 8, the structure theorem for finite Abelian groups shows that the class group $C(-164)$ must be $\mathbf{Z}/8\mathbf{Z}$.

A surprising fact is that one doesn't need to list the reduced forms in order to determine the number of elements of order 2 in the class group:

Proposition 3.11. *Let $D \equiv 0, 1 \bmod 4$ be negative, and let r be the number of odd primes dividing D. Define the number μ as follows: if $D \equiv 1 \bmod 4$, then $\mu = r$, and if $D \equiv 0 \bmod 4$, then $D = -4n$, where $n > 0$, and μ is determined by the following table:*

n	μ
$n \equiv 3 \bmod 4$	r
$n \equiv 1, 2 \bmod 4$	$r + 1$
$n \equiv 4 \bmod 8$	$r + 1$
$n \equiv 0 \bmod 8$	$r + 2$

Then the class group $C(D)$ has exactly $2^{\mu-1}$ elements of order ≤ 2.

Proof. For simplicity, we will treat only the special case $D = -4n$, where $n \equiv 1 \bmod 4$. Recall that a form of discriminant $-4n$ may be written as $ax^2 + 2bxy + cy^2$. The basic idea of the proof is to count the number of reduced forms that satisfy $2b = 0$, $a = 2b$ or $a = c$, for by Lemma 3.10, this gives the number of classes of order ≤ 2 in $C(-4n)$. Since n is odd, note that r is the number of prime divisors of n.

First, consider forms with $2b = 0$, i.e., the forms $ax^2 + cy^2$, where $ac = n$. Since a and c must be relatively prime and positive, there are 2^r choices for a. To be reduced, we must also have $a < c$, so that we get 2^{r-1} reduced forms of this type.

Next consider forms with $a = 2b$ or $a = c$. Write $n = bk$, where b and k are relatively prime and $0 < b < k$. As above, there are 2^{r-1} such b's. Set $c = (b+k)/2$, and consider the form $2bx^2 + 2bxy + cy^2$. One computes that it has discriminant $-4n$, and since $n \equiv 1 \bmod 4$, its coefficients are relatively prime. We then get 2^{r-1} reduced forms as follows:

$2b < c$: Here, $2bx^2 + 2bxy + cy^2$ is a reduced form.

$2b > c$: Here, $2bx^2 + 2bxy + cy^2$ is properly equivalent to

$cx^2 + 2(c-b)xy + cy^2$ via $(x,y) \mapsto (-y, x+y)$.

Since $2b > c \Rightarrow 2(c-b) < c$, the latter is reduced.

The next step is to check that this process gives *all* reduced forms with $a = 2b$ or $a = c$. We leave this to the reader (see Exercise 3.10).

We thus have $2^{r-1} + 2^{r-1} = 2^r$ elements of order ≤ 2, which shows that $\mu = r+1$ in this case. The remaining cases are similar and are left to the reader (see Exercise 3.10, Flath [36, §V.5], Gauss [41, §257–258] or Mathews [78, pp. 171–173]). Q.E.D.

This is not the last we will see of the number μ, for it also plays an important role in genus theory.

B. Genus Theory

As in §2, we define two forms of discriminant D to be in the same genus if they represent the same values in $(\mathbb{Z}/D\mathbb{Z})^*$. Let's recall the classification of genera given in §2. Consider the subgroups $H \subset \ker(\chi) \subset (\mathbb{Z}/D\mathbb{Z})^*$, where H consists of the values represented by the principal form, and $\chi : (\mathbb{Z}/D\mathbb{Z})^* \longrightarrow \{\pm 1\}$ is defined by $\chi([p]) = (D/p)$ for $p \nmid D$ prime. Then the key result was Lemma 2.24, where we proved that the values represented in $(\mathbb{Z}/D\mathbb{Z})^*$ by a given form $f(x,y)$ are a coset of H in $\ker(\chi)$. This coset determines which genus $f(x,y)$ is in.

Our first step is to relate this theory to the class group $C(D)$. Since all forms in a given class represent the same numbers, sending the class to the coset of $H \subset \ker(\chi)$ it represents defines a map

$$\Phi : C(D) \longrightarrow \ker(\chi)/H. \tag{3.12}$$

Note that a given fiber $\Phi^{-1}(H')$, $H' \in \ker(\chi)/H$, consists of all classes in a given genus (this is what we called the *genus of* H' in Theorem 2.26), and the image of Φ may thus be identified with the set of genera. A crucial observation is that Φ is a group homomorphism:

Lemma 3.13. *The map Φ which maps a class in $C(D)$ to the coset of values represented in* $\ker(\chi)/H$ *is a group homomorphism.*

Proof. Let $f(x,y)$ and $g(x,y)$ be two forms of discriminant D taking values in the cosets H' and H'' respectively. We can assume that their Dirichlet composition $F(x,y)$ is defined, so that a product of values represented by $f(x,y)$ and $g(x,y)$ is represented by $F(x,y)$. Then $F(x,y)$ represents values in $H'H''$, which proves that $H'H''$ is the coset associated to the composition of $f(x,y)$ and $g(x,y)$. Thus Φ is a homomorphism. Q.E.D.

This lemma has the following consequences:

Corollary 3.14. *Let* $D \equiv 0,1 \bmod 4$ *be negative. Then:*

(i) *All genera of forms of discriminant D consist of the same number of classes.*

(ii) *The number of genera of forms of discriminant D is a power of two.*

Proof. The first statement follows since all fibers of a homomorphism have the same number of elements. To prove the second, first note that the subgroup H contains all squares in $(\mathbb{Z}/D\mathbb{Z})^*$. This is obvious because if $f(x,y)$ is the principal form, then $f(x,0) = x^2$. Thus every element in $\ker(\chi)/H$ has order ≤ 2, and it follows from the structure theorem for finite Abelian groups that $\ker(\chi)/H \simeq \{\pm 1\}^m$ for some m. Thus the image of Φ, being a subgroup of $\ker(\chi)/H$, has order 2^k for some k. Since $\Phi(C(D))$ tells us the number of genera, we are done. Q.E.D.

Note also that $\Phi(C(D))$ gives a natural group structure on the set of genera, or as Gauss would say, one can define the composition of genera [41, §§246–247].

These elementary facts are nice, but they aren't the whole story. The real depth of the relation between composition and genera is indicated by the following theorem:

Theorem 3.15. *Let* $D \equiv 0,1 \bmod 4$ *be negative. Then:*

(i) *There are 2^{μ} genera of forms of discriminant D, where μ is the number defined in Proposition 3.11.*

(ii) *The principal genus (the genus containing the principal form) consists of the classes in $C(D)^2$, the subgroup of squares in the class group $C(D)$. Thus every form in the principal genus arises by duplication.*

Proof. We first need to give a more efficient method for determining when two forms are in the same genus. The basic idea is to use certain *assigned characters*, which are defined as follows. Let $p_1, \ldots, p_r$ be the distinct odd primes dividing D. Then consider the functions:

$$\chi_i(a) = \left(\frac{a}{p_i}\right) \qquad \text{defined for } a \text{ prime to } p_i,\ i = 1, \ldots, r$$

$$\delta(a) = (-1)^{(a-1)/2} \qquad \text{defined for } a \text{ odd}$$

$$\epsilon(a) = (-1)^{(a^2-1)/8} \qquad \text{defined for } a \text{ odd.}$$

Rather than using all of these functions, we assign only certain ones, depending on the discriminant D. When $D \equiv 1 \bmod 4$, we define $\chi_1, \ldots, \chi_r$ to be the *assigned characters*, and when $D \equiv 0 \bmod 4$, we write $D = -4n$, and then the *assigned characters* are defined by the following table:

n	assigned characters
$n \equiv 3 \bmod 4$	$\chi_1, \ldots, \chi_r$
$n \equiv 1 \bmod 4$	$\chi_1, \ldots, \chi_r, \delta$
$n \equiv 2 \bmod 8$	$\chi_1, \ldots, \chi_r, \delta\epsilon$
$n \equiv 6 \bmod 8$	$\chi_1, \ldots, \chi_r, \epsilon$
$n \equiv 4 \bmod 8$	$\chi_1, \ldots, \chi_r, \delta$
$n \equiv 0 \bmod 8$	$\chi_1, \ldots, \chi_r, \delta, \epsilon$

Note that the number of assigned characters is exactly the number μ given in Proposition 3.11. It is easy to see that the assigned characters give a homomorphism

$$\Psi : (\mathbb{Z}/D\mathbb{Z})^* \longrightarrow \{\pm 1\}^{\mu}. \tag{3.16}$$

The crucial property of Ψ is the following:

Lemma 3.17. *The homomorphism $\Psi : (\mathbb{Z}/D\mathbb{Z})^* \longrightarrow \{\pm 1\}^{\mu}$ of (3.16) is surjective and its kernel is the subgroup H of values represented by the principal form. Thus Ψ induces an isomorphism*

$$(\mathbb{Z}/D\mathbb{Z})^*/H \xrightarrow{\sim} \{\pm 1\}^{\mu}.$$

Proof. When $D \equiv 1 \bmod 4$, the proof is quite easy. First note that if p is an odd prime, then for any $m \in 1$, the Legendre symbol (a/p) induces a surjective homomorphism

$$(\cdot/p) : (\mathbb{Z}/p^m\mathbb{Z})^* \longrightarrow \{\pm 1\} \tag{3.18}$$

whose kernel is exactly the subgroup of squares of $(\mathbb{Z}/p^m\mathbb{Z})^*$ (see Exercise 3.11). Now let $D = -\prod_{i=1}^{\mu} p_i^{m_i}$ be the prime factorization of D. The

Chinese Remainder Theorem tells us that

$$(\mathbf{Z}/D\mathbf{Z})^* \xrightarrow{\sim} \prod_{1=i}^{\mu} (\mathbf{Z}/p_i^{m_i}\mathbf{Z})^*,$$

so that the map Ψ can be interpreted as the map

$$\prod_{1=i}^{\mu} (\mathbf{Z}/p_i^{m_i}\mathbf{Z})^* \longrightarrow \{\pm 1\}^{\mu}$$

given by $([a_1],\ldots,[a_\mu]) \mapsto ((a_1/p_1),\ldots,(a_\mu/p_\mu))$. By the analysis of (3.18), it follows that Ψ is surjective and its kernel is exactly the subgroup of squares of $(\mathbf{Z}/D\mathbf{Z})^*$. By part (c) of Exercise 2.17, this equals the subgroup H of values represented by the principal form $x^2 + xy + ((1-D)/4)y^2$, and we are done.

The proof is more complicated when $D = -4n$, mainly because the subgroup H represented by $x^2 + ny^2$ may be slightly larger than the subgroup of squares. However, the above argument using the Chinese Remainder Theorem can be adapted to this case. The odd primes dividing n are no problem, but 2 causes considerable difficulty (see Exercise 3.11 for the details). Q.E.D.

We can now prove Theorem 3.15. To prove (i), note that $\ker(\chi)$ has index 2 in $(\mathbf{Z}/D\mathbf{Z})^*$. By Lemma 3.17, it follows that $\ker(\chi)/H$ has order $2^{\mu-1}$. We know that the number of genera is the order of $\Phi(C(D)) \subset \ker(\chi)/H$, so that it suffices to show $\Phi(C(D)) = \ker(\chi)/H$. Since Φ maps a class to the coset of values it represents, we need to show that every congruence class in $\ker(\chi)$ contains a number represented by a form of discriminant D. This is easy: Dirichlet's theorem on primes in arithmetic progressions tells us that any class in $\ker(\chi)$ contains an odd prime p. But $[p] \in \ker(\chi)$ means that $\chi([p]) = (D/p) = 1$, so that by Lemma 2.5, p is represented by a form of discriminant D, and (i) is proved.

To prove (ii), let C denote the class group $C(D)$. Since $\Phi : C \to \ker(\chi)/H \simeq \{\pm 1\}^{\mu-1}$ is a homomorphism, it follows that $C^2 \subset \ker(\Phi)$, and we get an induced map

$$C/C^2 \longrightarrow \{\pm 1\}^{\mu-1}. \tag{3.19}$$

We compute the order of C/C^2 as follows. The squaring map from C to itself gives a short exact sequence

$$0 \to C_0 \to C \to C^2 \to 0$$

where C_0 is the subgroup of elements of order ≤ 2. It follows that the index $[C : C^2]$ equals the order of C_0, which is $2^{\mu-1}$ by Proposition 3.11.

Thus, in map given in (3.19), both the domain and the range have the same order. But from (i) we know that the map is surjective, so that it must be an isomorphism. Hence C^2 is exactly the kernel of the map Φ. Since $\ker(\Phi)$ consists of the classes in the principal genus, the theorem is proved. Q.E.D.

We have now proved the main theorems of genus theory for primitive positive definite forms. These results are due to Gauss and appear in the fifth section of *Disquisitiones Arithmeticae* [41, §§229–287]. Gauss' treatment is more general than ours, for he considers both the definite and indefinite forms, and in particular, he shows that Proposition 3.11 and Theorem 3.15 are true for *any* nonsquare discriminant, positive or negative. His proofs are quite difficult, and at the end of this long series of arguments, Gauss makes the following comment about genus theory [41, §287]:

> these theorems are among the most beautiful in the theory of binary forms, especially because, despite their extreme simplicity, they are so profound that a rigorous demonstration requires the help of many other investigations.

Besides these theorems, there is another component to Gauss' genus theory not mentioned so far: Gauss' second proof of quadratic reciprocity [41, §262], which uses the genus theory developed above. We will not discuss Gauss' proof since it uses forms of positive discriminant, though the main ideas of the proof are outlined in Exercises 3.12 and 3.13. Many people regard this as the deepest of Gauss' many proofs of quadratic reciprocity.

Gauss' approach to genus theory is somewhat different from ours. In *Disquisitiones*, genera are defined in terms of the *assigned characters* introduced in the proof of Theorem 3.15. Given a form $f(x,y)$ of discriminant D, let $f(x,y)$ represent a number a relatively prime to D. If the μ assigned characters are evaluated at a, then Gauss calls the resulting μ-tuple the *complete character* of $f(x,y)$, and he defines two forms of discriminant D to be in the same genus if they have the same complete character [41, §231]. The following lemma shows that this is equivalent to our previous definition of genus:

Lemma 3.20. *The complete character depends only on the form $f(x,y)$, and two forms of discriminant D lie in the same genus (as defined in §2) if and only if they have the same complete character.*

Proof. Suppose that $f(x,y)$ represents a, where a is relatively prime to D. Then Gauss' complete character is nothing other than $\Psi([a])$, where Ψ is the map defined in (3.16). By Lemma 2.24, the possible a's lie in a coset H' of H in $(\mathbf{Z}/D\mathbf{Z})^*$, and this coset determines the genus of $f(x,y)$. Using Lemma 3.17, it follows that the complete character is uniquely determined by H', and Lemma 3.20 is proved. Q.E.D.

We should mention that Gauss' use of the word "character" is where the modern term "group character" comes from. Also, it is interesting to note that Gauss never mentions the connection between his characters and Lagrange's implicit genus theory. While Gauss' characters make it easy to decide when two forms belong to the same genus (see Exercise 3.14 for an example), they are not very intuitive. Unfortunately, most of Gauss' successors followed his presentation of genus theory, so that readers were presented with long lists of characters and no motivation whatsoever. The simple idea of grouping forms according to the congruence classes they represent was usually not mentioned. This happens in Dirichlet [28, pp. 313–316] and in Mathews [78, pp. 132–136], although Smith [95, pp. 202–207] does discuss congruence classes.

So far we have discussed two ways to formulate genera, Lagrange's and Gauss'. There are many other ways to state the definition, but before we can discuss them, we need some terminology. We say that two forms $f(x,y)$ and $g(x,y)$ are *equivalent over a ring* R if there is a matrix $\left(\begin{smallmatrix} p & q \\ r & s \end{smallmatrix}\right) \in \mathrm{GL}(2,R)$ such that $f(x,y) = g(px+qy, rx+sy)$. If $R = \mathbf{Z}/m\mathbf{Z}$, we say that $f(x,y)$ and $g(x,y)$ are *equivalent modulo* m. We then have the following theorem:

Theorem 3.21. *Let $f(x,y)$ and $g(x,y)$ be primitive forms of discriminant $D \neq 0$, positive definite if $D < 0$. Then the following statements are equivalent:*

(i) *$f(x,y)$ and $g(x,y)$ are in the same genus, i.e., they represent the same values in $(\mathbf{Z}/D\mathbf{Z})^*$.*

(ii) *$f(x,y)$ and $g(x,y)$ represent the same values in $(\mathbf{Z}/m\mathbf{Z})^*$ for all nonzero integers m.*

(iii) *$f(x,y)$ and $g(x,y)$ are equivalent modulo m for all nonzero integers m.*

(iv) *$f(x,y)$ and $g(x,y)$ are equivalent over the p-adic integers $\mathbf{Z}_p$ for all primes p.*

(v) *$f(x,y)$ and $g(x,y)$ are equivalent over $\mathbf{Q}$ via a matrix in $\mathrm{GL}(2,\mathbf{Q})$ whose entries have denominators prime to $2D$.*

(vi) *$f(x,y)$ and $g(x,y)$ are equivalent over $\mathbf{Q}$ without essential denominator, i.e., given any nonzero m, a matrix in $\mathrm{GL}(2,\mathbf{Q})$ can be found which takes one form to the other and whose entries have denominators prime to m.*

Proof. It is easy to prove (vi) $\Rightarrow$ (iii) $\Rightarrow$ (ii) $\Rightarrow$ (i) and (vi) $\Rightarrow$ (v) $\Rightarrow$ (i) (see Exercise 3.15), and (iii) $\Leftrightarrow$ (iv) is a standard argument using the compactness of $\mathbf{Z}_p$ (see Borevich and Shafarevich [8, p. 41] for an analagous case). A proof of (i) $\Rightarrow$ (iii) appears in Hua [57, §12.5, Exercise 4], and (i) $\Rightarrow$ (iv) is in Jones [63, pp. 103–104]. Finally, the implication (iv) $\Rightarrow$ (vi) uses the

Hasse principle for the equivalence of forms over $\mathbb{Q}$ and may be found in Jones [63, Theorem 40] or Siegel [91]. Q.E.D.

Some modern texts give yet a different definition, saying that two forms are in the same genus if and only if they are equivalent over $\mathbb{Q}$ (see, for example, Borevich and Shafarevich [8, p. 241]). This characterization doesn't hold in general ($x^2 + 18y^2$ and $2x^2 + 9y^2$ are rationally equivalent but belong to different genera—see Exercise 3.16), but it does work for *field discriminants*, which means that $D \equiv 1 \bmod 4$, D squarefree, or $D = 4k$, $k \not\equiv 1 \bmod 4$, k squarefree (see Exercise 3.17—we will study such discriminants in more detail in §5). According to Dickson [26, Vol. III, pp. 216 and 236], Eisenstein suggested in 1852 that genera could be defined using rational equivalence, and only later, in 1867, did Smith point out that extra assumptions are needed on the denominators.

C. $p = x^2 + ny^2$ and Euler's Convenient Numbers

Our discussion of genus theory has distracted us from our problem of determining when a prime p can be written as $x^2 + ny^2$. Recall from Corollary 2.27 that genus theory gives us congruence conditions for p to be represented by a reduced form in the principal genus. The nicest case is when every genus of discriminant $-4n$ consists of a single class, for then we get congruence conditions that characterize $p = x^2 + ny^2$ (this is what made the examples in (2.28) work). Let's see if the genus theory developed in this section can shed any light on this special case. We have the following result:

Theorem 3.22. *Let n be a positive integer. Then the following statements are equivalent:*

(i) *Every genus of forms of discriminant $-4n$ consists of a single class.*
(ii) *If $ax^2 + bxy + cy^2$ is a reduced form of discriminant $-4n$, then either $b = 0$, $a = b$ or $a = c$.*
(iii) *Two forms of discriminant $-4n$ are equivalent if and only if they are properly equivalent.*
(iv) *The class group $C(-4n)$ is isomorphic to $(\mathbb{Z}/2\mathbb{Z})^m$ for some integer m.*
(v) *The class number $h(-4n)$ equals $2^{\mu-1}$, where μ is as in Proposition 3.11.*

Proof. We will prove (i) $\Rightarrow$ (ii) $\Rightarrow$ (iii) $\Rightarrow$ (iv) $\Rightarrow$ (v) $\Rightarrow$ (i). Let C denote the class group $C(-4n)$.

Since the principal genus is C^2 by Theorem 3.15, (i) implies that $C^2 = \{1\}$, so that every element of C has order ≤ 2. Then Lemma 3.10 shows that (i) $\Rightarrow$ (ii).

Next assume (ii), and suppose that two forms of discriminant $-4n$ are equivalent. By Exercise 3.8, we know that one is properly equivalent to the other or its opposite. We may assume that the forms are reduced, so that by assumption $b = 0$, $a = b$ or $a = c$. The proof of Theorem 2.8 shows that forms of this type are always properly equivalent to their opposites, so that the forms are properly equivalent. This proves (ii) $\Rightarrow$ (iii).

Recall that any form is equivalent to its opposite via $(x,y) \mapsto (x,-y)$. Thus (iii) implies that any form and its opposite lie in the same class in C. Since the opposite gives the inverse in C by Theorem 3.9, we see that every class is its own inverse. The structure theorem for finite Abelian groups shows that the only groups with this property are $(\mathbf{Z}/2\mathbf{Z})^m$, and (iii) $\Rightarrow$ (iv) is proved.

Next, Theorem 3.15 implies that the number of genera is $[C : C^2] = 2^{\mu-1}$, so that

$$h(-4n) = |C| = [C : C^2]|C^2| = 2^{\mu-1}|C^2|. \tag{3.23}$$

If (iv) holds, then $C^2 = \{1\}$, and then (v) follows immediately from (3.23). Finally, given (v), (3.23) implies that $C^2 = \{1\}$, so that by Theorem 3.15, the principal genus consists of a single class. Since every genus consists of the same number of classes, (i) follows, and the theorem is proved. Q.E.D.

Notice how this theorem runs the full gamut of what we've done so far: the conditions of Theorem 3.22 involve genera, reduced forms, the class number, the structure of the class group and the relation between equivalence and proper equivalence. For computational purposes, the last condition (v) is especially useful, for it only requires knowing the class number. This makes it much easier to verify that the examples in (2.28) have only one class per genus.

Near the end of the fifth section of *Disquisitiones*, Gauss lists 65 discriminants that satisfy this theorem [41, §303]. Grouped according to class number, they are:

$h(-4n)$	n's with one class per genus
1	1,2,3,4,7
2	5,6,8,9,10,12,13,15,16,18,22,25,28,37,38
4	21,24,30,33,40,42,45,48,57,60,70,72,78,85,88,93,102,112 130,133,177,190,232,253
8	105,120,165,168,210,240,273,280,312,330,345,357,385 408,462,520,760
16	840,1320,1365,1848

Gauss was interested in these 65 n's not for their relation to the question of when $p = x^2 + ny^2$, but rather because they had been discovered earlier by Euler in a different context. Euler called a number n a *convenient number* (numerus idoneus) if it satisfies the following criterion:

> Let m be an odd number relatively prime to n which is properly represented by $x^2 + ny^2$. If the equation $m = x^2 + ny^2$ has only one solution with $x, y \geq 0$, then m is a prime number.

Euler was interested in convenient numbers because they helped him find large primes. For example, working with $n = 1848$, he was able to show that

$$18{,}518{,}809 = 197^2 + 1848 \cdot 100^2$$

is prime, a large one for Euler's time. Convenient numbers are a fascinating topic, and the reader should consult Frei [38] or Weil [106, pp. 219–226] for a fuller discussion. We will confine ourselves to the following remarkable observation of Gauss:

Proposition 3.24. *A positive integer n is a convenient number if and only if for forms of discriminant $-4n$, every genus consists of a single class.*

Proof. We begin with a lemma:

Lemma 3.25. *Let m be a positive odd number relatively prime to $n > 1$. Then the number of ways that m is properly represented by a reduced form of discriminant $-4n$ is*

$$2\prod_{p|m}\left(1 + \left(\frac{-n}{p}\right)\right).$$

Proof. See Exercise 3.20 or Landau [71, Vol. 1, p. 144]. Q.E.D.

This classical lemma belongs to an area of quadratic forms that we have ignored, namely the study of the *number* of representations of a number by a form. To see what this has to do with genus theory, note that two forms representing m must lie in the same genus, for the values they represent in $(\mathbf{Z}/4n\mathbf{Z})^*$ are not disjoint. We thus get the following corollary of Lemma 3.25:

Corollary 3.26. *Let m be properly represented by a primitive positive definite form $f(x,y)$ of discriminant $-4n$, $n > 1$, and assume that m is odd and relatively prime to n. If r denotes the number of prime divisors of m, then m is properly represented in exactly 2^{r+1} ways by a reduced form in the genus of $f(x,y)$.* Q.E.D.

Now we can prove the proposition. First, assume that there is only one class per genus. If m is properly represented by $x^2 + ny^2$ and $m = x^2 + ny^2$ has a unique solution when $x, y \geq 0$, then we need to prove that m is prime. The above corollary shows that m is properly represented by $x^2 + ny^2$ in 2^{r+1} ways since $x^2 + ny^2$ is the only reduced form in its genus. At least 2^{r-1} of these representations satisfy $x, y \geq 0$, and then our assumption on m implies that $r = 1$, i.e., m is a prime power p^a. If $a \geq 2$, then Lemma 3.25 shows that p^{a-2} also has a proper representation, and it follows easily that m has at least two representations in nonnegative integers. This contradiction proves that m is prime, and hence n is a convenient number.

Conversely, assume that n is convenient. Let $f(x, y)$ be a form of discriminant $-4n$, and let $g(x, y)$ be the composition of $f(x, y)$ with itself. We can assume that $g(x, y)$ is reduced, and it suffices to show that $g(x, y) = x^2 + ny^2$ (for then every element in the class group has order ≤ 2, which implies one class per genus by Theorem 3.22).

Assume that $g(x, y) \neq x^2 + ny^2$, and let p and q be distinct odd primes not dividing n which are represented by $f(x, y)$. (In §9 we will prove that $f(x, y)$ represents infinitely many primes.) Then $g(x, y)$ represents pq, and formula (2.31) shows that $x^2 + ny^2$ does too. By Corollary 3.26, pq has only 8 proper representations by reduced forms of discriminant $-4n$. At least one comes from $g(x, y)$, leaving at most 7 for $x^2 + ny^2$. It follows that pq is uniquely represented by $x^2 + ny^2$ when we restrict to nonnegative integers. This contradicts our assumption that n is convenient. Q.E.D.

Gauss never states Proposition 3.24 formally, but it is implicit in the methods he discusses for factoring large numbers [41, §§329–334].

In §2 we asked how many such n's there were. Gauss suggests [41, §303] that the 65 given by Euler are the only ones. In 1934 Chowla [17] proved that the number of such n's is finite, and by 1973 it was known that Euler's list is complete except for possibly one more n (see Weinberger [108]). Whether or not this last n actually exists is still an open question.

From our point of view, the upshot is that there are only finitely many theorems like (2.28) where $p = x^2 + ny^2$ is characterized by simple congruences modulo $4n$. Thus genus theory cannot solve our basic question for all n. In some cases, such as $D = -108$, it's completely useless (all three reduced forms $x^2 + 27y^2$ and $4x^2 \pm 2xy + 7y^2$ lie in the same genus), and even when it's a partial help, such as $D = -56$, we're still stuck (we can separate $x^2 + 14y^2$ and $2x^2 + 7y^2$ from $3x^2 \pm 2xy + 5y^2$, but we can't distinguish between the first two). And notice that by part (iii) of Theorem 3.21, forms in the same genus are equivalent modulo m for *all* $m \neq 0$, so that no matter how m is chosen, there are no congruences $p \equiv a, b, c, \ldots \bmod m$ which can separate forms in the same genus. Something new is needed. In 1833, Dirichlet described the situation as follows [27, Vol. I, p. 201]:

> there lies in the mentioned [genus] theory an incompleteness, in that it certainly shows that a prime number, as soon as it is contained in a linear form [congruence class], necessarily must assume one of the corresponding quadratic forms, only without giving any a priori method for deciding which quadratic form it will be. ... It becomes clear that the characteristic property of a single quadratic form belonging to a group [genus] cannot be expressed through the prime numbers in the corresponding linear forms, but necessarily must be expressed by another theory not depending on the elements at hand.

As we already know from Euler's conjectures concerning $x^2 + 27y^2$ and $x^2 + 64y^2$ (see (1.22) and (1.23)), the new theory we're seeking involves residues of higher powers. Gauss rediscovered Euler's conjectures in 1805, and he proved them in the course of his work on cubic and biquadratic reciprocity. In §4 we will give careful statements of these reciprocity theorems and show how they can be used to prove Euler's conjectures.

D. Disquisitiones Arithmeticae

Gauss' *Disquisitiones Arithmeticae* covers a wide range of topics in number theory, including congruences, quadratic reciprocity, quadratic forms (in two and three variables), and the cyclotomic fields $\mathbf{Q}(\zeta_n)$, $\zeta_n = e^{2\pi i/n}$. There are several excellent accounts of what's in *Disquisitiones*, notably Bühler [13, Chapter 3], Bachmann [42, Vol. X.2.1, pp. 8–40] and Reiger [84], and translations into English and German are available (see item [41] in the references). Rather than try to survey the whole book, we will instead make some comments on Gauss' treatment of quadratic reciprocity and quadratic forms, for in each case he does things slightly different from the theory presented in §§2 and 3.

Disquisitiones contains the first published (valid) proof of the law of quadratic reciprocity. One surprise is that Gauss never uses the term "quadratic reciprocity". Instead, Gauss uses the phrase "fundamental theorem", which he explains as follows [41, §131]:

> Since almost everything that can be said about quadratic residues depends on this theorem, the term *fundamental theorem* which we will use from now on should be acceptable.

In the more informal setting of his mathematical diary, Gauss uses the term "golden theorem" to describe his high regard for quadratic reciprocity [42, Vol. X.1, entries 16, 23 and 30 on pp. 496–501] (see Gray [44] for an English translation). Likewise absent from *Disquisitiones* is the Legendre symbol, for Gauss uses the notation aRb or aNb to indicate whether or not a was a quadratic residue modulo b [41, §131]. (The Legendre symbol does appear in some of his handwritten notes—see [42, Vol. X.1, p. 53]—but this doesn't happen very often.)

One reason why Gauss ignored Legendre's terminology is that Gauss discovered quadratic reciprocity independent of his predecessors. In a marginal note in his copy of *Disquisitiones*, Gauss states that "we discovered the fundamental theorem by induction in March 1795. We found our first proof, the one contained in this section, April 1796" [41, p. 468, English editions] or [42, Vol. I, p. 476]. In 1795 Gauss was still a student at the Collegium Carolinum in Brunswick, and only later, while at Göttingen, did he discover the earlier work of Euler and Legendre on reciprocity.

Gauss' proof from April 1796 appears in §§135–144 of *Disquisitiones*. The theorem is stated in two forms: the usual version of quadratic reciprocity appears in [41, §131], and the more general version that holds for the Jacobi symbol (which we used in the proof of Lemma 1.14) is given in [41, §133]. The proof uses complete induction on the prime p, and there are many cases to consider, some of which use reciprocity for the Jacobi symbol (which would hold for numbers smaller than p). As Gauss wrote in 1808, the proof "proceeds by laborious steps and is burdened by detailed calculations" [42, Vol. II, p. 4]. In 1857, Dirichlet used the Jacobi symbol to simplify the proof and reduce the number of cases to just two [27, Vol. II, pp. 121–138]. It is interesting to note that what Gauss proves in *Disquisitiones* is actually a bit more general than the usual statment of quadratic reciprocity for the Jacobi symbol (see Exercise 3.24). Thus, when Jacobi introduced the Jacobi symbol in 1837 [61, Vol. VI, p. 262], he was simply giving a nicer but less general formulation of what was already in *Disquisitiones*.

As we mentioned in our discussion of genus theory, *Disquisitiones* also contains a second proof of reciprocity that is quite different in nature. The first proof is awkward but elementary, while the second uses Gauss' genus theory and is much more sophisticated.

Gauss' treatment of quadratic forms occupies the fifth (and longest) section of *Disquisitiones*. It is not easy reading, for many of the arguments are very complicated. Fortunately, there are more modern texts that cover pretty much the same material (in particular, see either Flath [36] or Mathews [78]). Gauss starts with the case of positive definite forms, and the theory he develops is similar to the first part of §2. Then, in [41, §182], he gives some applications to number theory, which are introduced as follows:

> Let us now consider certain particular cases both because of their remarkable elegance and because of the painstaking work done on them by Euler, who endowed them with an almost classical distinction.

As might be expected, Gauss first proves Fermat's three theorems (1.1), and then he proves Euler's conjecture for $p = x^2 + 5y^2$ using Lagrange's implicit genus theory (his proof is similar to what we did in (2.19), (2.20) and (2.22)). Interestingly enough, Gauss never mentions the relation between this example and genus theory. In contrast to Lagrange and Legendre, Gauss works out few examples. His one comment is that "the reader can derive this proposition [concerning $x^2 + 5y^2$] and an infinite number of other particular ones from the preceding and the following discussions" [41, §182].

Gauss always assumed that the middle coefficient was even, so that his forms were written $f(x,y) = ax^2 + 2bxy + cy^2$. He used the ordered triple (a,b,c) to denote $f(x,y)$ [41, §153], and he defined its *determinant* to be $b^2 - ac$ [41, §154]. Note that the discriminant of $ax^2 + 2bxy + cy^2$ is just 4 times Gauss' determinant.

Gauss did not assume that the coefficients of his forms were relatively prime, and he organized forms into *orders* according to the common divisors of the coefficients. More precisely, the forms $ax^2 + 2bxy + cy^2$ and $a'x^2 + 2b'xy + c'y^2$ are in the same *order* provided that $\gcd(a,b,c) = \gcd(a',b',c')$ and $\gcd(a,2b,c) = \gcd(a',2b',c')$ [41, §226]. To get a better idea of how this works, consider a primitive quadratic form $ax^2 + bxy + cy^2$. Here, a, b and c are relatively prime integers, and b may be even or odd. We can fit this form into Gauss' scheme as follows:

b is even: Then $b = 2b'$, and $ax^2 + 2b'xy + cy^2$ satisfies $\gcd(a,b',c)$ $= \gcd(a,2b',c) = 1$. Gauss called forms in this order *properly primitive.*

b is odd: Then $2ax^2 + 2bxy + 2cy^2$ satisfies $\gcd(2a,b,2c) = 1$, $\gcd(2a,2b,2c) = 2$. Gauss called forms in this order *improperly primitive.*

So all primitive forms are present, though the ones with b odd appear in disguised form. This doesn't affect the class number, but it does cause problems with composition.

Gauss' classification of forms thus consists of orders, which are made up of genera, which are in turn made up of classes. This is reminiscent of the Linnean classification in biology, where the categories are class, order, family, genus and species. Gauss' terms all appear on Linneaus' list, and it is thus likely that this is where Gauss got his terminology. Since our current term "equivalence class" comes from Gauss' example of *classes* of properly *equivalent* forms, we see that there is an unexpected link between modern set theory and eighteenth-century biology.

Finally, let's make one comment about composition. Gauss' theory of composition has always been one of the more difficult parts of *Disquisitiones* to read, and part of the reason is the complexity of Gauss' presentation. For example, the proof that composition is associative involves checking that 28 equations are satisfied [41, §240]. But a multiplicity of equations is not the only difficulty here—there is also an interesting conceptual issue. Namely, in order to define the class group, notice that Gauss has to put the structure of an abstract Abelian group on a set of equivalence classes. Considering that we're talking about the year 1801, this is an amazing level of abstraction. But then, *Disquisitiones* is an amazing book.

E. Exercises

3.1. Assume that $F(x,y) = Ax^2 + Bxy + Cy^2$ is the composition of the forms $f(x,y) = ax^2 + bxy + cy^2$ and $g(x,y) = a'x^2 + b'xy + c'y^2$ via

$$f(x,y)g(z,w) = F(a_1xz + b_1xw + c_1yz + d_1yw, a_2xz + b_2xw + c_2yz + d_2yw),$$

and suppose that all three forms have discriminant $D \neq 0$. The goal of this exercise is to prove Gauss' formulas (3.1).

(a) By specializing the variables x, y, z and w, prove that

$$\begin{aligned} aa' &= Aa_1^2 + Ba_1a_2 + Ca_2^2 \\ ac' &= Ab_1^2 + Bb_1b_2 + Cb_2^2 \\ ab' &= 2Aa_1b_1 + B(a_1b_2 + a_2b_1) + 2Ca_2b_2. \end{aligned}$$

Hint: for the first one, try $x = z = 1$ and $y = w = 0$.

(b) Prove that $a = \pm(a_1b_2 - a_2b_1)$. Hint: prove that

$$a^2(b'^2 - 4a'c') = (a_1b_2 - a_2b_1)^2(B^2 - 4AC).$$

(c) Prove that $a' = \pm(a_1c_2 - a_2c_1)$.

3.2. Show that the compositions given in (2.30) and (2.31) are not direct compositions.

3.3. Prove Lemma 3.5. Hint: there are $a, a_1, \ldots, a_r$ such that $am + \sum_{i=1}^{r} a_ip_i = 1$.

3.4. Verify that the congruences (3.4) satisfy the compatibility conditions of Lemma 3.5.

3.5. Let $f(x,y) = ax^2 + bxy + cy^2$, $g(x,y) = a'x^2 + b'xy + c'y^2$ and B be as in Lemma 3.2. We want to show that $aa'x^2 + Bxy + Cy^2$, $C = (B^2 - D)/4aa'$, is the direct composition of $f(x,y)$ and $g(x,y)$.

(a) Show that $f(x,y)$ (resp. $g(x,y)$) is properly equivalent to $ax^2 + Bxy + a'Cy^2$ (resp. $a'x^2 + Bxy + aCy^2$). Hint: for $f(x,y)$, use $B \equiv b \bmod 2a$.

(b) Let $X = xz - Cyw$ and $Y = axw + a'yz + Byw$. Then show that

$$(ax^2 + Bxy + a'Cy^2)(a'z^2 + Bzw + aCw^2)$$
$$= aa'X^2 + BXY + CY^2.$$

Furthermore, show that this is a direct composition in the sense of (3.1). Hint: first show that

$$(ax + (B + \sqrt{D})y/2)(a'z + (B + \sqrt{D})w/2)$$
$$= aa'X + (B + \sqrt{D})Y/2.$$

(c) Suppose that a form $G(x,y)$ is the direct composition of forms $h(x,y)$ and $k(x,y)$. If $\tilde{h}(x,y)$ is properly equivalent to $h(x,y)$, then show that $G(x,y)$ is also the direct composition of $\tilde{h}(x,y)$ and $k(x,y)$.

(d) Use (a)–(c) to show that the Dirichlet composition is a direct composition.

(e) Prove that the Dirichlet composition of primitive forms is primitive. Hint: since $F(x,y)$ represents any product $f(x,y)g(z,w)$, show that the gcd of all numbers represented by $F(x,y)$ is 1, and conclude that $F(x,y)$ is primitive.

3.6. This problem studies the relation between Legendre's and Dirichlet's formulas for composition.

(a) Suppose that $f(x,y) = ax^2 + 2bxy + cy^2$ and $g(x,y) = a'x^2 + 2b'xy + c'y^2$ have the same discriminant and satisfy $\gcd(a,a') = 1$. Show that the Dirichlet composition of these forms is the one given by Legendre's formula with both signs + in (2.32).

(b) In Exercise 2.26, we saw that the forms $14x^2 + 10xy + 21y^2$ and $9x^2 + 2xy + 30y^2$ compose to $126x^2 \pm 74xy + 13y^2$ and $126x^2 \pm 38xy + 5y^2$. Which one of these four is the direct composition of the original two forms?

3.7. Show that $acx^2 + bxy + y^2$ is properly equivalent to the principal form.

3.8. For us, a *class* consists of all forms properly equivalent to a given form. Let a *Lagrangian class* (this terminology is due to Weil [106, p. 319]) consist of all forms equivalent (properly or improperly) to a given form.

(a) Prove that the Lagrangian class of a form is the union of the class of the form and the class of its opposite.

(b) Show that the following statements are equivalent:

(i) The Lagrangian class of $f(x,y)$ equals the class of $f(x,y)$.

(ii) $f(x,y)$ is properly equivalent to its opposite.

(iii) $f(x,y)$ is properly and improperly equivalent to itself.

(iv) The class of $f(x,y)$ has order ≤ 2 in the class group.

3.9. In this problem we will describe the "almost" group structure given by Legendre's theory of composition. Let G be an Abelian group and let $\sim$ be the equivalence relation which identifies a^{-1} and a for all $a \in G$.

(a) Show that multiplication on G induces an operation on $G/\sim$ which takes either one or two values. Furthermore, if a, $b \in G$ and $[a]$, $[b]$ are their classes in $G/\sim$, then show that $[a] \cdot [b]$ takes on only one value if and only if a, b or ab has order ≤ 2 in G.

(b) If G is cyclic of order 8, show that $G/\sim$ is isomorphic (in the obvious sense) to the structure given by (2.33) and (2.34).

(c) If $C(D)$ is the class group of forms of discriminant D, show that $C(D)/\sim$ can be naturally identified with the set of Lagrangian classes of forms of discriminant D (see Exercise 3.8).

3.10. Complete the proof of Proposition 3.11 for the case $D = -4n$, $n \equiv 1 \bmod 4$, and prove all of the remaining cases.

3.11. This exercise is concerned with the proof of Lemma 3.17.

(a) Prove that the map (3.18) is surjective and its kernel is the subgroup of squares.

(b) We next want to prove the lemma when $D = -4n$, $n > 0$. Write $n = 2^a m$ where m is odd, so that we have an isomorphism

$$(\mathbf{Z}/D\mathbf{Z})^* \simeq (\mathbf{Z}/2^{a+2}\mathbf{Z})^* \times (\mathbf{Z}/m\mathbf{Z})^*.$$

Let H denote the subgroup of values represented by $x^2 + ny^2$.

(i) Show that $H = H_1 \times (\mathbf{Z}/m\mathbf{Z})^{*2}$ where $H_1 = H \cap (\{1\} \times (\mathbf{Z}/m\mathbf{Z})^*)$.

(ii) When $a \geq 4$, show that $H_1 = (\mathbf{Z}/2^{a+2}\mathbf{Z})^{*2}$, where H_1 is as in (i). Hint: the description of $(\mathbf{Z}/2^{a+2}\mathbf{Z})^*$ given in Ireland and Rosen [59, §4.1] will be useful.

(iii) Prove Lemma 3.17 when $D \equiv 0 \bmod 4$. Hint: treat the cases $a = 0$, 1, 2, 3 and ≥ 4 separately. See also Ireland and Rosen [59, §4.1].

3.12. In Exercises 3.12 and 3.13 we will sketch Gauss' second proof of quadratic reciprocity. There are two parts to the proof: first, one shows, without using quadratic reciprocity, that for any nonsquare discriminant D,

$(*)$ the number of genera of forms of discriminant D is $\leq 2^{\mu-1}$,

where μ is defined in Proposition 3.11, and second, one shows that $(*)$ implies quadratic reciprocity. This exercise will do the first step, and Exercise 3.13 will take care of the second.

We proved in Exercise 2.10 that when $D > 0$ is not a perfect square, there are only finitely many proper equivalence classes of primitive forms of discriminant D. The set of equivalence classes will be denoted $C(D)$, and as in the positive definite case, $C(D)$ becomes a finite Abelian group under Dirichlet composition (we will prove this in the exercises to §7). We will assume that Proposition 3.11 and Theorem 3.15 hold for all nonsquare discriminants D. This is where we pay the price for restricting ourselves to positive definite forms—the proofs in the text only work for $D < 0$. For proofs of these theorems when $D > 0$, see Flath [36, Chapter V], Gauss [41, §§257–258] or Mathews [78, pp. 171–173].

To prove $(*)$, let D be any nonsquare discriminant, and let C denote the class group $C(D)$. Let $H \subset (\mathbb{Z}/D\mathbb{Z})^*$ be the subgroup of values represented by the principal form.

(a) Show that genera can classified by cosets of H in $(\mathbb{Z}/D\mathbb{Z})^*$. Thus, instead of the map Φ of (3.12), we can use the map

$$\Phi' : C \longrightarrow (\mathbb{Z}/D\mathbb{Z})^*/H,$$

so that $\ker(\Phi')$ is the principal genus and $\Phi'(C)$ is the set of genera. Note that this argument does not use quadratic reciprocity.

(b) Since H contains all squares in $(\mathbb{Z}/D\mathbb{Z})^*$, it follows that $C^2 \subset \ker(\Phi')$. Now adapt the proof of Theorem 3.15 to show that

$$\text{the number of genera is } \leq [C : C^2] = 2^{\mu-1},$$

where the last equality follows from Proposition 3.11. This proves $(*)$.

3.13. In this exercise we will show that quadratic reciprocity follows from statement $(*)$ of Exercise 3.12. As we saw in §1, it suffices to show

$$\left(\frac{p^*}{q}\right) = 1 \iff \left(\frac{q}{p}\right) = 1,$$

where p and q are distinct odd primes and $p^* = (-1)^{(p-1)/2}p$.

(a) Show that Lemma 3.17 holds for all nonzero discriminants D, so that we can use the assigned characters to distinguish genera.

(b) Assume that $(p^*/q) = 1$. Applying Lemma 2.5 with $D = p^*$, q is represented by a form $f(x,y)$ of discriminant p^*. The number μ from Proposition 3.11 is 1, so that by $(*)$, there is only one genus. Hence the assigned character (there is only one in this case) must equal 1 on any number represented by $f(x,y)$, in particular q. Use this to prove that $(q/p) = 1$. This proves that $(p^*/q) = 1 \Rightarrow (q/p) = 1$.

(c) Next, assume that $(q/p) = 1$ and that either $p \equiv 1 \bmod 4$ or $q \equiv 1 \bmod 4$. Use part (b) to show that $(p^*/q) = 1$.

(d) Finally, assume that $(q/p) = 1$ and that $p \equiv q \equiv 3 \bmod 4$. This time we will consider forms of discriminant pq. Proposition 3.11 shows that $\mu = 2$, so that by $(*)$, there are at most two genera. Furthermore, the assigned characters are $\chi_1(a) = (a/p)$ and $\chi_2(a) = (a/q)$. Now consider the form $f(x,y) = px^2 + pxy + ((p-q)/4)y^2$, which is easily seen to have discriminant pq. Letting $(x,y) = (0,4)$, it represents $p - q$. Use this to compute the complete character of the forms $f(x,y)$ and $-f(x,y)$, and show that one of these must lie in the principal genus since there are at most two genera. Then show that $(-p/q) = 1$. Note that parts (c) and (d) imply that $(q/p) = 1 \Rightarrow (p^*/q) = 1$, which completes the proof of quadratic reciprocity.

(e) Gauss also used $(*)$ to show that $(2/p) = (-1)^{(p^2-1)/8}$. Adapt the argument given above to prove this. Hint: when $p \equiv 3, 5 \bmod 8$, show that p is properly represented by a form of discriminant 8. When $p \equiv 1 \bmod 8$, note that the form $2x^2 + xy + ((1-p)/8)y^2$ has discriminant p and represents 2, and the argument is similar when $p \equiv 7 \bmod 8$.

3.14. Use Gauss' definition of genus to divide the forms of discriminant -164 into genera. Hint: the forms are given in (2.33). Notice that this is much easier than working with our original definition!

3.15. Prove the implications (vi) $\Rightarrow$ (iii) $\Rightarrow$ (ii) $\Rightarrow$ (i) and (vi) $\Rightarrow$ (v) $\Rightarrow$ (i) of Theorem 3.21.

3.16. Prove that the forms $x^2 + 18y^2$ and $2x^2 + 9y^2$ are rationally equivalent but belong to different genera. Hint: if they represent the same values in $(\mathbb{Z}/72\mathbb{Z})^*$, then the same is true for any divisor of 72.

3.17. Let D be a field discriminant, i.e., $D \equiv 1 \bmod 4$, D squarefree, or $D = 4k$, $k \not\equiv 1 \bmod 4$, k squarefree. Let $f(x,y)$ and $g(x,y)$ be two forms of discriminant D which are rationally equivalent. We want to prove that they lie in the same genus.

(a) Let m be prime to D and represented by $g(x,y)$. Show that $f(x,y)$ represents d^2m for some nonzero integer d.

(b) Show that $f(x,y)$ and $g(x,y)$ lie in the same genus. Hint: by Exercise 2.1, $f(x,y)$ properly represents m' where $d'^2m' = d^2m$ for some integer d'. Show that m' is relatively prime to D. To do this, use Lemma 2.5 to write $f(x,y) = m'x^2 + bxy + cy^2$.

3.18. When $D = -4n$ is a field discriminant, we can use Theorem 3.21 to give a different proof that every form in the principal genus is a square (this is part (ii) of Theorem 3.15). Let $f(x,y)$ be a form of discriminant $-4n$ which lies in the principal genus.

(a) Show that $f(x,y)$ properly represents a number of the form a^2, where a is odd and relatively prime to n. Hint: use part (v) of Theorem 3.21.

(b) By (a), we may assume that $f(x,y) = a^2x^2 + 2bxy + cy^2$. Show that $\gcd(a,2b) = 1$, and conclude that $g(x,y) = ax^2 + 2bxy + acy^2$ has relatively prime coefficients and discriminant $-4n$.

(c) Show that $f(x,y)$ is the Dirichlet composition of $g(x,y)$ with itself.

This argument is due to Arndt (see Smith [95, pp. 254–256]), though Arndt proved (a) using the theorem of Legendre discussed in Exercise 2.24. Note that (a) can be restated in terms of ternary forms: if $f(x,y)$ is in the principal genus, then (a) proves that the ternary form $f(x,y) - z^2$ has a nontrivial zero. This result shows that there is a connection between ternary forms and genus theory. It is therefore not surprising that Gauss used ternary forms in his proof of Theorem 3.15.

3.19. Let $C(D)$ be the class group of forms of discriminant $D < 0$. Prove that the following statements are equivalent:

(i) Every genus of discriminant D consists of a single class.

(ii) $C(D) \simeq \{\pm 1\}^{\mu-1}$, where μ is as in Proposition 3.11.

(iii) Every genus of discriminant D consists of equivalent forms.

3.20. In this exercise we will prove Lemma 3.25. Let $m > 0$ be odd and prime to $n > 1$.

(a) Show that the number of solutions modulo m of the congruence

$$x^2 \equiv -n \bmod m$$

is given by the formula

$$\prod_{p \mid m} \left(1 + \left(\frac{-n}{p}\right)\right).$$

(b) Consider forms $g(x,y)$ of discriminant $-4n$ of the form

$$g(x,y) = mx^2 + 2bxy + cy^2, \qquad 0 \le b < m.$$

Show that the map sending $g(x,y)$ to $[b] \in (\mathbf{Z}/m\mathbf{Z})^*$ induces a bijection between the $g(x,y)$'s and the solutions modulo m of $x^2 \equiv -n \bmod m$.

(c) Let $f(x,y)$ have discriminant $-4n$ and let $f(u,v) = m$ be a proper representation. Pick r_0, s_0 so that $us_0 - vr_0 = 1$, and set $r = r_0 + uk$, $s = s_0 + vk$. Note that as $k \in \mathbf{Z}$ varies, we get all solutions of $us - vr = 1$. Then set

$$g(x,y) = f(ux + ry, vx + sy)$$

and show that there is a unique $k \in \mathbf{Z}$ such that $g(x,y)$ satisfies the condition of (b). This form is denoted $g_{u,v}(x,y)$.

(d) Show that the map sending a proper representation $f(u,v) = m$ to the form $g_{u,v}(x,y)$ is onto.

(e) If $g_{u',v'}(x,y) = g_{u,v}(x,y)$, let

$$\begin{pmatrix} \alpha & \beta \\ \gamma & \delta \end{pmatrix} = \begin{pmatrix} u' & v' \\ r' & s' \end{pmatrix}^{-1} \begin{pmatrix} u & v \\ r & s \end{pmatrix}.$$

Show that $f(\alpha x + \beta y, \gamma x + \delta y) = f(x,y)$ and, since $n > 1$, show that $\left(\begin{smallmatrix} \alpha & \beta \\ \gamma & \delta \end{smallmatrix}\right) = \pm\left(\begin{smallmatrix} 1 & 0 \\ 0 & 1 \end{smallmatrix}\right)$. Hint: assume that $f(x,y)$ is reduced, and use the arguments from the uniqueness part of the proof of Theorem 2.8.

(f) Conclude that $g_{u',v'}(x,y) = g_{u,v}(x,y)$ if and only if $(u',v') = \pm(u,v)$, so that the map of (d) is exactly two-to-one. Combining this with (a) and (b), we get a proof of Lemma 3.25.

3.21. This exercise will use Lemma 3.25 to study the equation $m^3 = a^2 + 2b^2$.

(a) If m is odd, use Lemma 3.25 to show that the equations $m = x^2 + 2y^2$ and $m^3 = x^2 + 2y^2$ have the same number of proper solutions.

(b) If $m = a^2 + 2b^2$ is a proper representation, then show that

$$m^3 = (a^3 - 6ab^2)^2 + 2(3a^2b - 2b^3)^2$$

is a proper representation.

(c) Show that the map sending (a,b) to $(a^3 - 6ab^2, 3a^2b - 2b^3)$ is injective. Hint: note that

$$(a + b\sqrt{-2})^3 = (a^3 - 6ab^2) + (3a^2b - 2b^3)\sqrt{-2}.$$

(d) Combine (a) and (c) to show that all proper representations of $m^3 = x^2 + 2y^2$, m odd, arise from (b).

3.22. Use Exercise 3.21 to prove Fermat's famous result that $(x, y) = (3, \pm 5)$ are the only integral solutions of the equation $x^3 = y^2 + 2$. Hint: first show that x must be odd, and then apply Exercise 3.21 to the proper representation $x^3 = y^2 + 2 \cdot 1^2$. It's likely that Fermat's original proof of this result was similar to the argument presented here, though he would have used a version of Lemma 1.4 to prove part (c) of Exercise. See Weil [106, pp. 68–69 and 71–73] for more details.

3.23. Let p be an odd prime of the form $x^2 + ny^2$, $n > 1$. Use Lemma 3.25 to show that the equation

$$p = x^2 + ny^2$$

has a unique solution once we require x and y to be nonnegative. Note also that Lemma 3.25 gives a very quick proof of Exercise 2.27.

3.24. This exercise will examine a generalization of the Jacobi symbol. Let P and Q be relatively prime nonzero integers, where Q is odd but possibly negative. Then define the extended Jacobi symbol (P/Q) via

$$\left(\frac{P}{Q}\right) = \begin{cases} (P/|Q|) & \text{when } |Q| > 1 \\ 1 & \text{when } |Q| = 1. \end{cases}$$

(a) Prove that when P and Q are odd and relatively prime, then

$$\left(\frac{P}{Q}\right)\left(\frac{Q}{P}\right) = (-1)^{(P-1)(Q-1)/4 + (\operatorname{sgn}(P)-1)(\operatorname{sgn}(Q)-1)/4}$$

where $\operatorname{sgn}(P) = P/|P|$.

(b) Gauss' version of (a) is more complicated to state. First, given P and Q as above, he lets p denote the number of prime factors of Q (counted with multiplicity) for which P is a quadratic residue. This relates to (P/Q) by the formula

$$\left(\frac{P}{Q}\right) = (-1)^p.$$

Interchanging P and Q, we get a similarly defined number q. To relate the parity of p and q, Gauss states a rule in [41, §133] which breaks up into 10 separate cases. Verify that the rule proved in (a) covers all 10 of Gauss' cases.

(c) Prove the supplementary laws:

$$\left(\frac{-1}{P}\right) = \operatorname{sgn}(P)(-1)^{(P-1)/2}$$

$$\left(\frac{2}{P}\right) = (-1)^{(P^2-1)/8}.$$

3.25. Let $p \equiv 1 \bmod 8$ be prime.

(a) If $C(-4p)$ is the class group of forms of discriminant $-4p$, then use genus theory to prove that

$$C(-4p) \simeq (\mathbf{Z}/2^a\mathbf{Z}) \times G$$

where $a \geq 1$ and G has odd order. Thus $2 \mid h(-4p)$.

(b) Let $f(x,y) = 2x^2 + 2xy + ((p+1)/2)y^2$. Use Gauss's definition of genus to show that $f(x,y)$ is in the principal genus.

(c) Use Theorem 3.15 to show that $C(-4p)$ has an element of order 4. Thus $4 \mid h(-4p)$.

§4. CUBIC AND BIQUADRATIC RECIPROCITY

In this section we will study cubic and biquadratic reciprocity and use them to prove Euler's conjectures for $p = x^2 + 27y^2$ and $p = x^2 + 64y^2$ (see (1.22) and (1.23)). An interesting feature of these reciprocity theorems is that each one requires that we extend the notion of integer: for cubic reciprocity we will use the ring

$$\text{(4.1)} \qquad \mathbf{Z}[\omega] = \{a + b\omega : a,b \in \mathbf{Z}\}, \qquad \omega = e^{2\pi i/3} = (-1+\sqrt{-3})/2,$$

and for biquadratic reciprocity we will use the Gaussian integers

$$\text{(4.2)} \qquad \mathbf{Z}[i] = \{a + bi : a,b \in \mathbf{Z}\}, \qquad i = \sqrt{-1}.$$

Both $\mathbf{Z}[\omega]$ and $\mathbf{Z}[i]$ are subrings of the complex numbers (see Exercise 4.1). Our first task will be to describe the arithmetic properties of these rings and determine their units and primes. We will then define the generalized Legendre symbols $(\alpha/\pi)_3$ and $(\alpha/\pi)_4$ and state the laws of cubic and biquadratic reciprocity. The proofs will be omitted since excellent proofs are already available in print (see especially Ireland and Rosen [59, Chapter 9]). At the end of the section we will discuss Gauss' work on reciprocity and say a few words about the origins of class field theory.

A. $\mathbf{Z}[\omega]$ and Cubic Reciprocity

The law of cubic reciprocity is intimately bound up with the ring $\mathbf{Z}[\omega]$ of (4.1). The main tool used to study the arithmetic of $\mathbf{Z}[\omega]$ is the norm function: if $\alpha = a + b\omega$ is in $\mathbf{Z}[\omega]$, then its *norm* $N(\alpha)$ is the positive integer

$$N(\alpha) = \alpha\overline{\alpha} = a^2 - ab + b^2,$$

where $\overline{\alpha}$ is the complex conjugate of α (in Exercise 4.1 we will see that $\overline{\alpha} \in \mathbf{Z}[\omega]$). Note that the norm is multiplicative, i.e., for $\alpha, \beta \in \mathbf{Z}[\omega]$, we have

$$N(\alpha\beta) = N(\alpha)N(\beta)$$

(see Exercise 4.2). Using the norm, one can prove that $\mathbf{Z}[\omega]$ is a Euclidean ring:

Proposition 4.3. *Given $\alpha, \beta \in \mathbf{Z}[\omega]$, $\beta \neq 0$, there are $\gamma, \delta \in \mathbf{Z}[\omega]$ such that*

$$\alpha = \gamma\beta + \delta \qquad \textit{and} \qquad N(\delta) < N(\beta).$$

Thus $\mathbf{Z}[\omega]$ is a Euclidean ring.

Proof. Note that the norm function $N(\alpha) = \alpha\overline{\alpha}$ is defined on $\mathbf{Q}(\omega) = \{r + s\omega : r, s \in \mathbf{Q}\}$ and satisfies $N(uv) = N(u)N(v)$ for $u, v \in \mathbf{Q}(\omega)$ (see Exercise 4.2). Then

$$\frac{\alpha}{\beta} = \frac{\alpha\overline{\beta}}{\beta\overline{\beta}} = \frac{\alpha\overline{\beta}}{N(\beta)} \in \mathbf{Q}(\omega),$$

so that $\alpha/\beta = r + s\omega$ for some $r, s \in \mathbf{Q}$. Let r_1, s_1 be integers such that $|r - r_1| \le 1/2$ and $|s - s_1| \le 1/2$, and then set $\gamma = r_1 + s_1\omega$ and $\delta = \alpha - \gamma\beta$. Note that $\gamma, \delta \in \mathbf{Z}[\omega]$ and $\alpha = \gamma\beta + \delta$. It remains to show that $N(\delta) < N(\beta)$. To see this, let $\epsilon = \alpha/\beta - \gamma = (r - r_1) + (s - s_1)\omega$, and note that

$$\delta = \alpha - \gamma\beta = \beta(\alpha/\beta - \gamma) = \beta\epsilon.$$

Since the norm is multiplicative, it suffices to prove that $N(\epsilon) < 1$. But

$$N(\epsilon) = N((r - r_1) + (s - s_1)\omega) = (r - r_1)^2 - (r - r_1)(s - s_1) + (s - s_1)^2,$$

and the desired inequality follows from $|r - r_1|, |s - s_1| \le 1/2$. By the standard definition of a Euclidean ring (see, for example, Herstein [54, §3.7]), we are done. Q.E.D.

Corollary 4.4. *$\mathbf{Z}[\omega]$ is a* PID *(principal ideal domain) and a* UFD *(unique factorization domain).*

Proof. It is well known that any Euclidean ring is a PID and a UFD—see, for example, Herstein [54, Theorems 3.7.1 and 3.7.2]. Q.E.D.

For completeness, let's recall the definitions of PID and UFD. Let R be an integral domain. An ideal of R is *principal* if it can be written in the form $\alpha R = \{\alpha\beta : \beta \in R\}$ for some $\alpha \in R$, and R is a PID if every ideal of R is principal. To explain what a UFD is, we first need to define units, associates and irreducibles:

(i) $\alpha \in R$ is a *unit* if $\alpha\beta = 1$ for some $\beta \in R$.

(ii) $\alpha, \beta \in R$ are *associates* if α is a unit times β. This is equivalent to $\alpha R = \beta R$.

(iii) A nonunit $\alpha \in R$ is *irreducible* if $\alpha = \beta\gamma$ in R implies that β or γ is a unit.

Then R is a UFD if every nonunit $\alpha \neq 0$ can be written as a product of irreducibles, and given two such factorizations of α, each irreducible in the first factorization can be matched up in an one-to-one manner with an associate irreducible in the second. Thus factorization is unique up to order and associates.

It turns out that being a PID is the stronger property: every PID is a UFD (see Ireland and Rosen [59, §1.3]), but the converse is not true (see Exercise 4.3). Given an element $\alpha \neq 0$ in a PID R, the following statements are equivalent:

(i) α is irreducible.

(ii) α is prime (an element α of R is *prime* if $\alpha \mid \beta\gamma$ implies $\alpha \mid \beta$ or $\alpha \mid \gamma$).

(iii) αR is a prime ideal (an ideal $\mathfrak{p}$ of R is *prime* if $\beta\gamma \in \mathfrak{p}$ implies $\beta \in \mathfrak{p}$ or $\gamma \in \mathfrak{p}$).

(iv) αR is a maximal ideal.

(See Exercise 4.4 for the proof.)

Since $\mathbf{Z}[\omega]$ is a PID and a UFD, the next step is to determine the units and primes of $\mathbf{Z}[\omega]$. Let's start with the units:

Lemma 4.5.

(i) *An element $\alpha \in \mathbf{Z}[\omega]$ is a unit if and only if $N(\alpha) = 1$.*

(ii) *The units of $\mathbf{Z}[\omega]$ are $\mathbf{Z}[\omega]^* = \{\pm 1, \pm\omega, \pm\omega^2\}$.*

Proof. See Exercise 4.5. Q.E.D.

The next step is to describe the primes of $\mathbf{Z}[\omega]$. The following lemma will be useful:

Lemma 4.6. *If $\alpha \in \mathbf{Z}[\omega]$ and $N(\alpha)$ is a prime in $\mathbf{Z}$, then α is prime in $\mathbf{Z}[\omega]$.*

Proof. Since $\mathbb{Z}[\omega]$ is a PID, it suffices to prove that α is irreducible. So suppose that $\alpha = \beta\gamma$ in $\mathbb{Z}[\omega]$. Taking norms, we obtain the integer equation

$$N(\alpha) = N(\beta\gamma) = N(\beta)N(\gamma)$$

(recall that the norm is multiplicative). Since $N(\alpha)$ is prime by assumption, this implies that $N(\beta)$ or $N(\gamma)$ is 1, so that β or γ is a unit by Lemma 4.5. Q.E.D.

We can now determine all primes in $\mathbb{Z}[\omega]$:

Proposition 4.7. *Let p be a prime in $\mathbb{Z}$. Then:*

(i) *If $p = 3$, then $1-\omega$ is prime in $\mathbb{Z}[\omega]$ and $3 = -\omega^2(1-\omega)^2$.*

(ii) *If $p \equiv 1 \bmod 3$, then there is a prime $\pi \in \mathbb{Z}[\omega]$ such that $p = \pi\overline{\pi}$, and the primes π and $\overline{\pi}$ are nonassociate in $\mathbb{Z}[\omega]$.*

(iii) *If $p \equiv 2 \bmod 3$, then p remains prime in $\mathbb{Z}[\omega]$.*

Furthermore, every prime in $\mathbb{Z}[\omega]$ is associate to one of the primes listed in (i)–(iii) above.

Proof. Since $N(1-\omega) = 3$, Lemma 4.6 implies that $1-\omega$ is prime in $\mathbb{Z}[\omega]$, and (i) follows. To prove (ii), suppose that $p \equiv 1 \bmod 3$. Then $(-3/p) = 1$, so that p is represented by a reduced form of discriminant -3 (this is Theorem 2.16). The only such form is $x^2 + xy + y^2$, so that p can be written as $a^2 - ab + b^2$. Then $\pi = a + b\omega$ and $\overline{\pi} = a + b\omega^2$ have norms $N(\pi) = N(\overline{\pi}) = p$ and hence are prime in $\mathbb{Z}[\omega]$ by Lemma 4.6. In Exercise 4.7 we will prove that π and $\overline{\pi}$ are nonassociate. The proof of (iii) is left to the reader (see Exercise 4.7).

It remains to show that all primes in $\mathbb{Z}[\omega]$ are associate to one of the above. Let's temporarily call the primes given in (i)–(iii) the *known primes* of $\mathbb{Z}[\omega]$, and let α be any prime of $\mathbb{Z}[\omega]$. Then $N(\alpha) = \alpha\overline{\alpha}$ is an ordinary integer and may be factored into integer primes. But (i)–(iii) imply that any integer prime is a product of known primes in $\mathbb{Z}[\omega]$, and consequently $\alpha\overline{\alpha} = N(\alpha)$ is also a product of known primes. The proposition then follows since $\mathbb{Z}[\omega]$ is a UFD. Q.E.D.

Given a prime π of $\mathbb{Z}[\omega]$, we get the maximal ideal $\pi\mathbb{Z}[\omega]$ of $\mathbb{Z}[\omega]$. The quotient ring $\mathbb{Z}[\omega]/\pi\mathbb{Z}[\omega]$ is a thus a field. We can describe this field more carefully as follows:

Lemma 4.8. *If π is a prime of $\mathbb{Z}[\omega]$, then the quotient field $\mathbb{Z}[\omega]/\pi\mathbb{Z}[\omega]$ is a finite field with $N(\pi)$ elements. Furthermore, $N(\pi) = p$ or p^2 for some integer prime p, and:*

(i) *If* $p = 3$ *or* $p \equiv 1 \bmod 3$, *then* $N(\pi) = p$ *and* $\mathbf{Z}/p\mathbf{Z} \simeq \mathbf{Z}[\omega]/\pi\mathbf{Z}[\omega]$.
(ii) *If* $p \equiv 2 \bmod 3$, *then* $N(\pi) = p^2$ *and* $\mathbf{Z}/p\mathbf{Z}$ *is the unique subfield of order* p *of the field* $\mathbf{Z}[\omega]/\pi\mathbf{Z}[\omega]$ *of* p^2 *elements.*

Proof. In §7 we will prove the that if π is a nonzero element of $\mathbf{Z}[\omega]$, then $\mathbf{Z}[\omega]/\pi\mathbf{Z}[\omega]$ is a finite ring with $N(\pi)$ elements (see Lemma 7.14 or Ireland and Rosen [59, §§9.2 and 14.1]). Then (i) and (ii) follow easily (see Exercise 4.8). Q.E.D.

Given α, β and π in $\mathbf{Z}[\omega]$, we will write $\alpha \equiv \beta \bmod \pi$ to indicate that α and β differ by a multiple of π, i.e., that they give the same element in $\mathbf{Z}[\omega]/\pi\mathbf{Z}[\omega]$. Using this notation, Lemma 4.8 gives us the following analog of Fermat's Little Theorem:

Corollary 4.9. *If* π *is prime in* $\mathbf{Z}[\omega]$ *and doesn't divide* $\alpha \in \mathbf{Z}[\omega]$, *then*

$$\alpha^{N(\pi)-1} \equiv 1 \bmod \pi.$$

Proof. This follows because $(\mathbf{Z}[\omega]/\pi\mathbf{Z}[\omega])^*$ is a finite group with $N(\pi)-1$ elements. Q.E.D.

Given these properties of $\mathbf{Z}[\omega]$, we can now define the generalized Legendre symbol $(\alpha/\pi)_3$. Let π be a prime of $\mathbf{Z}[\omega]$ not dividing 3 (i.e., not associate to $1-\omega$). It is straightforward to check that $3 \mid N(\pi)-1$ (see Exercise 4.9). Now suppose that $\alpha \in \mathbf{Z}[\omega]$ is not divisible by π. It follows from Corollary 4.9 that $x = \alpha^{(N(\pi)-1)/3}$ is a root of $x^3 \equiv 1 \bmod \pi$. Since

$$x^3 - 1 \equiv (x-1)(x-\omega)(x-\omega^2) \bmod \pi$$

and π is prime, it follows that

$$\alpha^{(N(\pi)-1)/3} \equiv 1, \omega, \omega^2 \bmod \pi.$$

However, the cube roots of unity $1, \omega, \omega^2$ are incongruent modulo π. To see this, note that if any two were congruent, then we would have $1 \equiv \omega \bmod \pi$, which would contradict π not associate to $1-\omega$ (see Exercise 4.9 for the details). Then we define the *Legendre symbol* $(\alpha/\pi)_3$ to be the unique cube root of unity such that

$$\alpha^{(N(\pi)-1)/3} \equiv \left(\frac{\alpha}{\pi}\right)_3 \bmod \pi. \tag{4.10}$$

The basic properties of the Legendre symbol are easy to work out. First, from (4.10), one can show

$$\left(\frac{\alpha\beta}{\pi}\right)_3 = \left(\frac{\alpha}{\pi}\right)_3 \left(\frac{\beta}{\pi}\right)_3,$$

and second, $\alpha \equiv \beta \bmod \pi$ implies that

$$\left(\frac{\alpha}{\pi}\right)_3 = \left(\frac{\beta}{\pi}\right)_3$$

(see Exercise 4.10). The Legendre symbol may thus be regarded as a group homomorphism from $(\mathbb{Z}[\omega]/\pi\mathbb{Z}[\omega])^*$ to $\mathbb{C}^*$.

An important fact is that the multiplicative group of any finite field is cyclic (see Ireland and Rosen [59, §7.1]). In particular, $(\mathbb{Z}[\omega]/\pi\mathbb{Z}[\omega])^*$ is cyclic, which implies that

$$\begin{aligned}\left(\frac{\alpha}{\pi}\right)_3 = 1 &\iff \alpha^{(N(\pi)-1)/3} \equiv 1 \bmod \pi \\ &\iff x^3 \equiv \alpha \bmod \pi \text{ has a solution in } \mathbb{Z}[\omega]\end{aligned} \tag{4.11}$$

(see Exercise 4.11). This establishes the link between the Legendre symbol and cubic residues. Note that one-third of $(\mathbb{Z}[\omega]/\pi\mathbb{Z}[\omega])^*$ consists of cubic residues (where the Legendre symbol equals 1), and the remaining two-thirds consist of nonresidues (where the symbol equals ω or ω^2). Later on we will explain how this relates to the more elementary notion of cubic residues of integers.

To state the law of cubic reciprocity, we need one final definition: a prime π is called *primary* if $\pi \equiv \pm 1 \bmod 3$. Given any prime π not dividing 3, one can show that exactly two of the six associates $\pm\pi$, $\pm\omega\pi$ and $\pm\omega^2\pi$ are primary (see Exercise 4.12). Then the law of cubic reciprocity states the following:

Theorem 4.12. *If π and θ are primary primes in $\mathbb{Z}[\omega]$ of unequal norm, then*

$$\left(\frac{\theta}{\pi}\right)_3 = \left(\frac{\pi}{\theta}\right)_3.$$

Proof. See Ireland and Rosen [59, §§9.4–9.5] or Smith [95, pp. 89–91]. Q.E.D.

Notice how simple the statement of the theorem is—it's among the most elegant of all reciprocity theorems (biquadratic reciprocity, to be stated below, is a bit more complicated). The restriction to primary primes is a normalization analogous to the normalization $p > 0$ that we make for ordinary primes. Some books (such as Ireland and Rosen [59]) define primary to mean $\pi \equiv -1 \bmod 3$. Since $(-1/\pi)_3 = 1$, this doesn't affect the statement of cubic reciprocity.

There are also supplementary formulas for $(\omega/\pi)_3$ and $(1-\omega/\pi)_3$. Let π be prime and not associate to $1-\omega$. Then we may assume that $\pi \equiv -1 \bmod$

3 (if π is primary, one of $\pm\pi$ satisfies this condition). Writing $\pi = -1 + 3m + 3n\omega$, it can be shown that

$$(4.13)\qquad \begin{aligned} \left(\frac{\omega}{\pi}\right)_3 &= \omega^{m+n} \\ \left(\frac{1-\omega}{\pi}\right)_3 &= \omega^{2m}. \end{aligned}$$

The first line of (4.13) is easy to prove (see Exercise 4.13), while the second is more difficult (see Ireland and Rosen [59, p. 114] or Exercise 9.13).

Let's next discuss cubic residues of integers. If p is a prime, the basic question is: when does $x^3 \equiv a \bmod p$ have an integer solution? If $p = 3$, then Fermat's Little Theorem tells us that $a^3 \equiv a \bmod 3$ for all a, so that we always have a solution. If $p \equiv 2 \bmod 3$, then the map $a \mapsto a^3$ induces an automorphism of $(\mathbf{Z}/p\mathbf{Z})^*$ since $3 \nmid p-1$ (see Exercise 4.14), and consequently $x^3 \equiv a \bmod p$ is again always solvable. If $p \equiv 1 \bmod 3$, things are more interesting. In this case, $p = \pi\bar{\pi}$ in $\mathbf{Z}[\omega]$, and there is a natural isomorphism $\mathbf{Z}/p\mathbf{Z} \simeq \mathbf{Z}[\omega]/\pi\mathbf{Z}[\omega]$ by Lemma 4.8. Thus, for $p \nmid a$, (4.11) implies that

$$(4.14)\qquad x^3 \equiv a \bmod p \ \text{ is solvable in } \mathbf{Z} \iff \left(\frac{a}{\pi}\right)_3 = 1.$$

Furthermore, $(\mathbf{Z}/p\mathbf{Z})^*$ breaks up into three pieces of equal size, one of cubic residues and two of nonresidues.

We can now use cubic reciprocity to prove Euler's conjecture for primes of the form $x^2 + 27y^2$:

Theorem 4.15. *Let p be a prime. Then $p = x^2 + 27y^2$ if and only if $p \equiv 1 \bmod 3$ and 2 is a cubic residue modulo p.*

Proof. First, suppose that $p = x^2 + 27y^2$. This clearly implies that $p \equiv 1 \bmod 3$, so that we need only show that 2 is a cubic residue modulo p. Let $\pi = x + 3\sqrt{-3}y$, so that $p = \pi\bar{\pi}$ in $\mathbf{Z}[\omega]$. It follows that π is prime, and then by (4.14), 2 is a cubic residue modulo p if and only if $(2/\pi)_3 = 1$. However, both 2 and $\pi = x + 3\sqrt{-3}y$ are primary primes, so that cubic reciprocity implies

$$(4.16)\qquad \left(\frac{2}{\pi}\right)_3 = \left(\frac{\pi}{2}\right)_3.$$

It thus suffices to prove that $(\pi/2)_3 = 1$. However, from (4.10), we know that

$$(4.17)\qquad \left(\frac{\pi}{2}\right)_3 \equiv \pi \bmod 2$$

since $(N(2)-1)/3 = 1$. So we need only show that $\pi \equiv 1 \bmod 2$. Since $\sqrt{-3} = 1 + 2\omega$, $\pi = x + 3\sqrt{-3}y = x + 3y + 6y\omega$, so that $\pi \equiv x + 3y \equiv x + y \bmod 2$. But x and y must have opposite parity since $p = x^2 + 27y^2$, and we are done.

Conversely, suppose that $p \equiv 1 \bmod 3$ is prime and 2 is a cubic residue modulo p. We can write p as $p = \pi\overline{\pi}$, and we can assume that π is a primary prime in $\mathbf{Z}[\omega]$. This means that $\pi = a + 3b\omega$ for some integers a and b. Thus

$$4p = 4\pi\overline{\pi} = 4(a^2 - 3ab + 9b^2) = (2a - 3b)^2 + 27b^2.$$

Once we show b is even, it will follow immediately that p is of the form $x^2 + 27y^2$.

We now can use our assumption that 2 is a cubic residue modulo p. From (4.14) we know that $(2/\pi)_3 = 1$, and then cubic reciprocity (4.16) tells us that $(\pi/2)_3 = 1$. But by (4.17), this implies $\pi \equiv 1 \bmod 2$, which we can write as $a + 3b\omega \equiv 1 \bmod 2$. This easily implies that a is odd and b is even, and $p = x^2 + 27y^2$ follows. The theorem is proved. Q.E.D.

B. $\mathbf{Z}[i]$ and Biquadratic Reciprocity

Our treatment of biquadratic reciprocity will be brief since the basic ideas are similar to what we did for cubic residues (for a complete discussion, see Ireland and Rosen [59, §§9.7–9.9]). Here, the appropriate ring is the ring of Gaussian integers $\mathbf{Z}[i]$ as defined in (4.2). The norm function $N(a+bi) = a^2 + b^2$ makes $\mathbf{Z}[i]$ into a Euclidean ring, and hence $\mathbf{Z}[i]$ is also a PID and a UFD. The analogs of Lemma 4.5 and 4.6 hold for $\mathbf{Z}[i]$, and it is easy to check that its units are ± 1 and $\pm i$ (see Exercise 4.16). The primes of $\mathbf{Z}[i]$ are described as follows:

Proposition 4.18. *Let p be a prime in $\mathbf{Z}$. Then:*

(i) *If $p = 2$, then $1 + i$ is prime in $\mathbf{Z}[i]$ and $2 = i^3(1+i)^2$.*

(ii) *If $p \equiv 1 \bmod 4$, then there is a prime $\pi \in \mathbf{Z}[i]$ such that $p = \pi\overline{\pi}$, and the primes π and $\overline{\pi}$ are nonassociate in $\mathbf{Z}[i]$.*

(iii) *If $p \equiv 3 \bmod 4$, then p remains prime in $\mathbf{Z}[i]$.*

Furthermore, every prime in $\mathbf{Z}[i]$ is associate to one of the primes listed in (i)–(iii) above.

Proof. See Exercise 4.16. Q.E.D.

We also have the following version of Fermat's Little Theorem: if π is prime in $\mathbf{Z}[i]$ and doesn't divide $\alpha \in \mathbf{Z}[i]$, then

$$\alpha^{N(\pi)-1} \equiv 1 \bmod \pi \tag{4.19}$$

(see Exercise 4.16).

These basic facts about the Gaussian integers appear in many elementary texts (e.g., Herstein [54, §3.8]), but such books rarely mention that the whole reason Gauss introduced the Gaussian integers was so that he could state biquadratic reciprocity. We will have more to say about this later.

We can now define the Legendre symbol $(\alpha/\pi)_4$. Given a prime π of $\mathbf{Z}[i]$ not associate to $1+i$, it can be proved that $\pm 1, \pm i$ are distinct modulo π and that $4 \mid N(\pi)-1$ (see Exercise 4.17). Then, for α not divisible by π, the *Legendre symbol* $(\alpha/\pi)_4$ is defined to be the unique fourth root of unity such that

$$\alpha^{(N(\pi)-1)/4} \equiv \left(\frac{\alpha}{\pi}\right)_4 \bmod \pi. \tag{4.20}$$

As in the cubic case, we see that

$$\left(\frac{\alpha}{\pi}\right)_4 = 1 \iff x^4 \equiv \alpha \bmod \pi \text{ is solvable in } \mathbf{Z}[i],$$

and furthermore, the Legendre symbol gives a character from $(\mathbf{Z}[i]/\pi\mathbf{Z}[i])^*$ to $\mathbf{C}^*$, so that $(\mathbf{Z}[i]/\pi\mathbf{Z}[i])^*$ is divided into four equal parts (see Exercise 4.18). When $p \equiv 1 \bmod 4$, we have $(\mathbf{Z}[i]/\pi\mathbf{Z}[i])^* \simeq (\mathbf{Z}/p\mathbf{Z})^*$, and the partition can be described as follows: one part consists of biquadratic residues (where the symbol equals 1), another consists of quadratic residues which aren't biquadratic residues (where the symbol equals -1), and the final two parts consist of quadratic nonresidues (where the symbol equals $\pm i$)—see Exercise 4.19.

A prime π of $\mathbf{Z}[i]$ is *primary* if $\pi \equiv 1 \bmod 2+2i$. Any prime not associate to $1+i$ has a unique associate which is primary (see Exercise 4.21). With this normalization, the law of biquadratic reciprocity can be stated as follows:

Theorem 4.21. *If π and θ are distinct primary primes in $\mathbf{Z}[i]$, then*

$$\left(\frac{\theta}{\pi}\right)_4 = \left(\frac{\pi}{\theta}\right)_4 (-1)^{(N(\theta)-1)(N(\pi)-1)/16}.$$

Proof. See Ireland and Rosen [59, §9.9] or Smith [95, pp. 76–87]. Q.E.D.

There are also supplementary laws which state that

$$\begin{aligned} \left(\frac{i}{\pi}\right)_4 &= i^{-(a-1)/2} \\ \left(\frac{1+i}{\pi}\right)_4 &= i^{(a-b-1-b^2)/4} \end{aligned} \tag{4.22}$$

where $\pi = a + bi$ is a primary prime. As in the cubic case, the first line of (4.22) is easy to prove (see Exercise 4.22), while the second is more difficult (see Ireland and Rosen [59, Exercises 32–37, p. 136]).

We can now prove Euler's conjecture about $p = x^2 + 64y^2$:

Theorem 4.23.

(i) *If* $\pi = a + bi$ *is a primary prime in* $\mathbf{Z}[i]$, *then*

$$\left(\frac{2}{\pi}\right)_4 = i^{ab/2}.$$

(ii) *If* p *is prime, then* $p = x^2 + 64y^2$ *if and only if* $p \equiv 1 \bmod 4$ *and* 2 *is a biquadratic residue modulo* p.

Proof. First note that (i) implies (ii). To see this, let $p \equiv 1 \bmod 4$ be prime. We can write $p = a^2 + b^2 = \pi\overline{\pi}$, where $\pi = a + bi$ is primary. Note that a is odd and b is even. Since $\mathbf{Z}/p\mathbf{Z} \simeq \mathbf{Z}[i]/\pi\mathbf{Z}[i]$, (i) shows that 2 is a biquadratic residue modulo p if and only if b is divisible by 8, and (ii) follows easily.

One way to prove (i) is via the supplementary laws (4.22) since $2 = i^3(1+i)^2$ (see Exercise 4.23). However, in 1857, Dirichlet found a proof of (i) that uses only quadratic reciprocity [27, Vol. II, pp. 261–262]. A version of this proof is given in Exercise 4.24 (see also Ireland and Rosen [59, Exercises 26–28, p. 64]). Q.E.D.

C. Gauss and Higher Reciprocity

Most of the above theorems were discovered by Gauss in the period 1805–1814, though the bulk of what he knew was never published. Only in 1828 and 1832, long after the research was completed, did Gauss publish his two memoirs on biquadratic residues [42, Vol. II, pp. 65–148] (see also [41, pp. 511–586, German editions] for a German translation). The first memoir treats the elementary theory of biquadratic residues of integers, and it includes a proof Euler's conjecture for $x^2 + 64y^2$. In the second memoir, Gauss begins with a careful discussion of the Gaussian integers, and he explains their relevance to biquadratic reciprocity as follows [42, Vol. II, §30, p. 102]:

> the theorems on biquadratic residues gleam with the greatest simplicity and genuine beauty only when the field of arithmetic is extended to **imaginary** numbers, so that without restriction, the numbers of the form $a + bi$ constitute the object [of study], where as usual

> i denotes $\sqrt{-1}$ and the indeterminates a, b denote integral real numbers between $-\infty$ and $+\infty$. We will call such numbers **integral complex numbers** (numeros integros complexos) ...

Gauss' treatment of $\mathbf{Z}[i]$ includes most of what we did above, and in particular the terms norm, associate and primary are due to Gauss.

Gauss' statment of biquadratic reciprocity differs slightly from Theorem 4.21. In terms of the Legendre symbol, his version goes as follows: given distinct primary primes π and θ of $\mathbf{Z}[i]$,

If either π or θ is congruent to 1 modulo 4, then $(\pi/\theta)_4 = (\theta/\pi)_4$.

If both π and θ are congruent to $3+2i$ modulo 4, then $(\pi/\theta)_4 = -(\theta/\pi)_4$.

In Exercise 4.25 we will see that this is equivalent to Theorem 4.21. As might be expected, Gauss doesn't use the Legendre symbol in his memoir. Rather, he defines the *biquadratic character* of α with respect to π to be the number $\lambda \in \{0,1,2,3\}$ satisfying $\alpha^{(N(\pi)-1)/4} \equiv i^\lambda \bmod \pi$ (so that $(\alpha/\pi)_4 = i^\lambda$), and he states biquadratic reciprocity using the biquadratic character. For Gauss, this theorem is "the Fundamental Theorem of biquadratic residues" [42, Vol. II, §67, p. 138], but instead of giving a proof, Gauss comments that

> In spite of the great simplicity of this theorem, the proof belongs to the most hidden mysteries of higher arithmetic, and at least as things now stand, [the proof] can be explained only by the most subtle investigations, which would greatly exceed the limits of the present memoir.

Later on, we will have more to say about Gauss' proof.

In the second memoir, Gauss also makes his only published reference to cubic reciprocity [42, Vol. II, §30, p. 102]:

> The theory of cubic residues must be based in a similar way on a consideration of numbers of the form $a+bh$, where h is an imaginary root of the equation $h^3-1=0$, say $h=(-1+\sqrt{-3})/2$, and similarly the theory of residues of higher powers leads to the introduction of other imaginary quantities.

So Gauss was clearly aware of the properties of $\mathbf{Z}[\omega]$, even if he never made them public.

Turning to Gauss' unpublished material, we find that one of the earliest fragments on higher reciprocity, dated around 1805, is the following "Beautiful Observation Made By Induction" [42, Vol. VIII, pp. 5 and 11]:

> 2 is a cubic residue or nonresidue of a prime number p of the form $3n+1$, according to whether p is representable by the form $xx+27yy$ or $4xx+2xy+7yy$.

This shows that Euler's conjecture for x^2+27y^2 was one of Gauss' starting points. And notice that Gauss was aware that he was separating forms in the same genus—the very problem we discussed in §3.

Around the same time, Gauss also rediscovered Euler's conjecture for x^2+64y^2 [42, Vol. X.1, p. 37]. But how did he come to make these conjectures? There are two aspects of Gauss' work that bear on this question. The first has to do with quadratic forms. Let's follow the treatment in Gauss' first memoir on biquadratic residues [42, Vol. II, §§12–14, pp. 75–78]. Let $p \equiv 1 \bmod 4$ be prime. If 2 is to be a biquadratic residue modulo p, it follows by quadratic reciprocity that $p \equiv 1 \bmod 8$ (see Exercise 4.26). By Fermat's theorem for x^2+2y^2, p can be written as $p=a^2+2b^2$, and Gauss proves the lovely result that 2 is a biquadratic residue modulo p if and only if $a \equiv \pm 1 \bmod 8$ (see Exercise 4.27). This is nice, but Gauss isn't satisfied:

> Since the decomposition of the number p into a single and double square is bound up so prominently with the classification of the number 2, it would be worth the effort to understand whether the decomposition into two squares, to which the number p is equally liable, perhaps promises a similar success.

Gauss then computes some numerical examples, and they show that when p is written as a^2+b^2, 2 is a biquadratic residue exactly when b is divisible by 8. This could be how Gauss was led to the conjecture in the first place, and the same thing could have happened in the cubic case, where primes $p \equiv 1 \bmod 3$ can be written as a^2+3b^2.

The cubic case most likely came first, for it turns out that Gauss describes a relation between x^2+27y^2 and cubic residues in the last section of *Disquisitiones*. This is where Gauss discusses the cyclotomic equation $x^p-1=0$ and proves his celebrated theorem on the constructibility of regular polygons. To see what this has to do with cubic residues, let's describe a little of what he does. Given an odd prime p, let $\zeta_p = e^{2\pi i/p}$ be a primitive pth root of unity, and let g be a primitive root modulo p, i.e., g is an integer such that $[g]$ generates the cyclic group $(\mathbb{Z}/p\mathbb{Z})^*$. Now suppose that $p-1=ef$, and let λ be an integer. Gauss then defines [41, §343] the *period* (f,λ) to be the sum

$$(f,\lambda) = \sum_{j=0}^{f-1} \zeta_p^{\lambda g^{ej}}.$$

These periods are the key to Gauss' study of the cyclotomic field $\mathbf{Q}(\zeta_p)$. In fact, if we fix f, then the periods $(f,1)$, (f,g), $(f,g^2),\ldots,(f,g^{e-1})$ are the roots of an irreducible integer polynomial of degree e, and consequently these periods are primitive elements of the unique subfield $\mathbf{Q} \subset K \subset \mathbf{Q}(\zeta_p)$ of degree e over $\mathbf{Q}$.

When $p \equiv 1 \bmod 3$, we can write $p-1=3f$, and then the three above periods are $(f,1)$, (f,g) and (f,g^2). Gauss studies this case in [41, §358], and by analyzing the products of the periods, he deduces the amazing result that

$$\text{(4.24)}\qquad \begin{array}{l}\text{If } 4p = a^2+27b^2 \text{ and } a \equiv 1 \bmod 3, \text{ then } N = p+a-2, \text{ where}\\ N \text{ is the number of solutions modulo } p \text{ of } x^3 - y^3 \equiv 1 \bmod p.\end{array}$$

To see how cubic residues enter into (4.24), note that $N = 9M+6$, where M is the number of nonzero cubic residues which, when increased by one, remain a nonzero cubic residue (see Exercise 4.29). Gauss conjectured this result in October 1796 and proved it in July 1797 [42, Vol. X.1, entries 39 and 67, pp. 505–506 and 519]. So Gauss was aware of cubic residues and quadratic forms in 1796. Gauss' proof of (4.24) is sketched in Exercise 4.29.

Statement (4.24) is similar to the famous last entry in Gauss' mathematical diary. In this entry, Gauss gives the following analog of (4.24) for the decomposition $p = a^2+b^2$ of a prime $p \equiv 1 \bmod 4$:

$$\begin{array}{l}\text{If } p = a^2+b^2 \text{ and } a+bi \text{ is primary, then } N = p-2a-3, \text{ where}\\ N \text{ is the number of solutions modulo } p \text{ of } x^2+y^2+x^2y^2 \equiv 1 \bmod p\end{array}$$

(see [42, Vol. X.1, entry 146, pp. 571–572]). In general, the study of the solutions of equations modulo p leads to the zeta function of a variety over a finite field. For an introduction to this extremely rich topic, see Ireland and Rosen [59, Chapter 11]. In §14 we will see how Gauss' results relate to elliptic curves with complex multiplication.

Going back to the cubic case, there is a footnote in [41, §358] which gives another interesting property of the periods $(f,1)$, (f,g) and (f,g^2):

$$\text{(4.25)}\qquad \begin{array}{l}\left((f,1)+\omega(f,g)+\omega^2(f,g^2)\right)^3 = p(a+b\sqrt{-27})/2,\\ \text{where } 4p = a^2+27b^2.\end{array}$$

The right hand side is an integer in the ring $\mathbf{Z}[\omega]$, and one can show that $\pi = (a+b\sqrt{-27})/2$ is a primary prime in $\mathbf{Z}[\omega]$ and that $p = \pi\overline{\pi}$. This is how Gauss first encountered $\mathbf{Z}[\omega]$ in connection with cubic residues. Notice also that if we set $\chi(a) = (a/\pi)_3$ and pick the primitive root g so that

$\chi(g) = \omega$, then

$$(f,1) + \omega(f,g) + \omega^2(f,g^2) = \sum_{a=1}^{p-1} \chi(a)\zeta_p^a. \tag{4.26}$$

This is an example of what we now call a cubic Gauss sum. See Ireland and Rosen [59, §§8.2–8.3] for the basic properties of Gauss sums and a modern treatment of (4.24) and (4.25).

The above discussion shows that Gauss was aware of cubic residues and $\mathbf{Z}[\omega]$ when he made his "Beautiful Observation" of 1805, and it's not surprising that two years later he was able to prove a version of cubic reciprocity [42, Vol. VIII, pp. 9–13]. The biquadratic case was harder, taking him until sometime around 1813 or 1814 to find a complete proof. We know this from a letter Gauss wrote Dirichlet in 1828, where Gauss mentions that he has possessed a proof of the "Main Theorem" for around 14 years [42, Vol. II, p. 516]. Exact dates are hard to come by, for most of the fragments Gauss left are undated, and it's not easy to match them up with his diary entries. For a fuller discussion of Gauss' work on biquadratic reciprocity, see Bachman [42, Vol. X.2.1, pp. 52–60] or Reiger [84].

Gauss' proofs of cubic and biquadratic reciprocity probably used Gauss sums similar to (4.26), and many modern proofs run along the same lines (see Ireland and Rosen [59, Chapter 9]). Gauss sums were first used in Gauss' sixth proof of quadratic reciprocity (see [42, Vol. II, pp. 55–59] or [41, pp. 501–505, German editions]). This is no accident, for as Gauss explained in 1818:

> From 1805 onwards I have investigated the theory of cubic and biquadratic residues... Theorems were found by induction... which had a wonderful analogy with the theorems for quadratic residues. On the other hand, for a long time all attempts at complete proofs have been futile. This was the motive for endeavoring to add yet more proofs to those already known for quadratic residues, in the hope that of the many different methods given, one or the other would contribute to the illumination of the related arguments [for cubic and biquadratic residues]. This hope was in no way in vain, for at last tireless labor has led to favorable success. Soon the fruit of this vigilance will be permitted to come to public light...

(see [42, Vol. II, p. 50] or [41, p. 497, German editions]). The irony is that Gauss never did publish his proofs, and it was left to Eisenstein and Jacobi to give us the first complete treatments of cubic and biquadratic reciprocity (see Collinson [22] or Smith [95, pp. 76–92] for more on the history of these reciprocity theorems).

We will conclude this section with some remarks about what happened after Gauss. Number theory was becoming a much larger area of mathematics, and the study of quadratic forms and reciprocity laws began to diverge. In the 1830s and 1840s, Dirichlet introduced L-series and began the analytic study of quadratic forms, and simultaneously, Eisenstein and Jacobi worked out not only cubic and biquadratic reciprocity, but also reciprocity for 5th, 8th and 12th powers. Kummer was also studying higher reciprocity, and he introduced his "ideal numbers" to make up for the lack of unique factorization in $\mathbf{Q}(e^{2\pi i/p})$. Both he and Eisenstein were able to prove generalized reciprocity laws using these "ideal numbers" (see Ireland and Rosen [59, Chapter 14] and Smith [95, pp. 93–126]). In 1871 Dedekind made the transition from "ideal numbers" to ideals in rings of algebraic integers, thereby laying the foundation for modern algebraic number theory and class field theory.

But reciprocity was not the only force leading to class field theory: there was also complex multiplication. Euler, Lagrange, Legendre and others studied transformations of the elliptic integrals

$$\int \frac{dx}{\sqrt{(1-x^2)(1-k^2x^2)}},$$

and they discovered that certain values of k, called singular moduli, gave elliptic integrals that could be transformed into complex multiples of themselves. This phenomenon came to be called *complex multiplication*. In working with complex multiplication, Abel observed that singular moduli and the roots of the corresponding transformation equations have remarkable algebraic properties. In modern terms, they generate *Abelian extensions* of $\mathbf{Q}(\sqrt{-n})$, i.e., Galois extensions of $\mathbf{Q}(\sqrt{-n})$ with Abelian Galois group. These topics will be discussed in more detail in Chapter Three.

Kronecker extended and completed Abel's work on complex multiplication, and in so doing he made the amazing conjecture that every Abelian extension of $\mathbf{Q}(\sqrt{-n})$ lies in one of the fields described above. Kronecker had earlier conjectured that every Abelian extension of $\mathbf{Q}$ lies in one of the cyclotomic fields $\mathbf{Q}(e^{2\pi i/n})$ (this is the famous Kronecker–Weber theorem, to be proved in §8). Abelian extensions may seem far removed from reciprocity theorems, but Kronecker also noticed relations between singular moduli and quadratic forms. For example, his results on complex multiplication by $\sqrt{-31}$ led to the following corollary which he was fond of quoting:

$$p = x^2 + 31y^2 \iff \begin{cases} (x^3 - 10x)^2 + 31(x^2 - 1)^2 \equiv 0 \bmod p \\ \text{has an integral solution} \end{cases}$$

(see [68, Vol. II, pp. 93 and 99–100, Vol. IV, pp. 123–129]). This is similar to what we just proved for $x^2 + 27y^2$ and $x^2 + 64y^2$ using cubic and biquadratic reciprocity. So something interesting is going on here.

We thus have two interrelated questions of interest:

(i) Is there a general reciprocity law that subsumes the known ones?

(ii) Is there a general method for describing all Abelian extensions of a number field?

The crowning acheivement of class field theory is that it solves both of these problems simultaneously: an Abelian extension L of a number field K is classified in terms of data intrinsic to K, and the key ingredient linking L to this data is the Artin reciprocity theorem. Complete statements of the theorems of class field theory will be given in Chapter Two, and in Chapter Three we will explain how complex multiplication is related to the class field theory of imaginary quadratic fields.

For a fuller account of the history of class field theory, see the article by W. and F. Ellison [32, §§III–IV] in Dieudonné's *Abrégé d'Histoire des Mathématiques 1700–1900*. Weil has a nice discussion of reciprocity and cyclotomic fields in [105] and [107], and Edwards describes Kummer's "ideal numbers" in [31, Chapter 4].

D. Exercises

4.1. Prove that $\mathbf{Z}[\omega]$ and $\mathbf{Z}[i]$ are subrings of the complex numbers and are closed under complex conjugation.

4.2. Let $\mathbf{Q}(\omega) = \{r + s\omega : r, s \in \mathbf{Q}\}$, and define the norm of $r + s\omega$ to be $N(r + s\omega) = (r + s\omega)\overline{(r + s\omega)}$.

(i) Show that $N(r + s\omega) = r^2 - rs + s^2$.

(ii) Show that $N(uv) = N(u)N(v)$ for $u, v \in \mathbf{Q}(\omega)$.

4.3. It is well-known that $R = \mathbf{C}[x, y]$ is a UFD (see Herstein [54, Corollary 2 to Theorem 3.11.1]). Prove that $I = \{f(x,y) \in R : f(0,0) = 0\}$ is an ideal of R which is not principal, so that R is not a PID. Hint: $x, y \in I$.

4.4. Given $\alpha \neq 0$ in a PID R, prove that α is irreducible $\Longleftrightarrow$ α is prime $\Longleftrightarrow$ αR is a prime ideal $\Longleftrightarrow$ αR is a maximal ideal.

4.5. Prove Lemma 4.5. Hint for (ii): use (i) and (2.4).

4.6. While $\mathbf{Z}[\omega]$ is a PID and a UFD, this exercise will show that the closely related ring $\mathbf{Z}[\sqrt{-3}]$ has neither property.

(a) Show that ± 1 are the only units of $\mathbf{Z}[\sqrt{-3}]$.

(b) Show that 2, $1+\sqrt{-3}$ and $1-\sqrt{-3}$ are nonassociate and irreducible in $\mathbf{Z}[\sqrt{-3}]$. Since $4 = 2\cdot 2 = (1+\sqrt{-3})(1-\sqrt{-3})$, these elements are not prime and thus $\mathbf{Z}[\sqrt{-3}]$ is not a UFD.

(c) Show that the ideal in $\mathbf{Z}[\sqrt{-3}]$ generated by 2 and $1+\sqrt{-3}$ is not principal. Thus $\mathbf{Z}[\sqrt{-3}]$ is not a PID.

4.7. This exercise is concerned with the proof of Proposition 4.7. Let p be a prime number.

(a) When $p \equiv 1 \bmod 3$, we showed that $p = \pi\overline{\pi}$, where π and $\overline{\pi}$ are prime in $\mathbf{Z}[\omega]$. Prove that π and $\overline{\pi}$ are nonassociate in $\mathbf{Z}[\omega]$.

(b) When $p \equiv 2 \bmod 3$, prove that p is prime in $\mathbf{Z}[\omega]$. Hint: show that p is irreducible. Note that by Lemma 2.5, the equation $p = N(\alpha)$ has no solutions.

4.8. Complete the proof of Lemma 4.8.

4.9. Let π be a prime of $\mathbf{Z}[\omega]$ not associate to $1-\omega$.

(a) Show that $3 \mid N(\pi)-1$.

(b) If any two of 1, ω and ω^2 are congruent modulo π, then show that $1 \equiv \omega \bmod \pi$, and explain why this contradicts our assumption on π. This proves that 1, ω and ω^2 are distinct modulo π.

4.10. Let π be prime in $\mathbf{Z}[\omega]$, and let $\alpha, \beta \in \mathbf{Z}[\omega]$ be not divisible by π. Verify the following properties of the Legendre symbol.

(a) $(\alpha\beta/\pi)_3 = (\alpha/\pi)_3(\beta/\pi)_3$.

(b) $(\alpha/\pi)_3 = (\beta/\pi)_3$ when $\alpha \equiv \beta \bmod \pi$.

4.11. Let π be prime in $\mathbf{Z}[\omega]$. Assuming that $(\mathbf{Z}[\omega]/\pi\mathbf{Z}[\omega])^*$ is cyclic, prove (4.11).

4.12. Let π be a prime of $\mathbf{Z}[\omega]$ which is not associate to $1-\omega$. Prove that exactly two of the six associates of π are primary.

4.13. Prove the top line of (4.13).

4.14. Use the hints in the text to prove that the congruence $x^3 \equiv a \bmod p$ is always solvable when p is a prime congruent to 2 modulo 3.

4.15. In this problem we will give an application of cubic reciprocity which is similar to Theorem 4.15. Let $p \equiv 1 \bmod 3$ be a prime.

(a) Use the proof of Theorem 4.15 to show that $4p$ can be written in the form $4p = a^2 + 27b^2$, where $a \equiv 1 \bmod 3$. Conclude that $\pi = (a + 3\sqrt{-3})/2$ is a primary prime of $\mathbf{Z}[\omega]$ and that $p = \pi\overline{\pi}$.

(b) Show that the supplementary laws (4.13) can be written

$$\left(\frac{\omega}{\pi}\right)_3 = \omega^{2(a+2)/3}$$

$$\left(\frac{1-\omega}{\pi}\right)_3 = \omega^{(a+2)/3+b}$$

where π is as in part (a).

(c) Use (b) to show that $(3/\pi)_3 = \omega^{2b}$.

(d) Use (c) and (4.14) to prove that for a prime p,

$$4p = x^2 + 243y^2 \iff \begin{cases} p \equiv 1 \bmod 3 \text{ and } 3 \text{ is a} \\ \text{cubic residue modulo } p. \end{cases}$$

Euler conjectured the result of (c) (in a slightly different form) in his *Tractatus* [33, Vol. V, pp. XXII and 250].

4.16. In this exercise we will discuss the properties of the Gaussian integers $\mathbf{Z}[i]$.

(a) Use the norm function to prove that $\mathbf{Z}[i]$ is Euclidean.

(b) Prove the analogs of Lemmas 4.5 and 4.6 for $\mathbf{Z}[i]$.

(c) Prove Proposition 4.18.

(d) Formulate and prove the analog of Lemma 4.8 for $\mathbf{Z}[i]$.

(e) Prove (4.19).

4.17. If π is a prime of $\mathbf{Z}[i]$ not associate to $1+i$, show that $4 \mid N(\pi)-1$ and that ± 1 and $\pm i$ are all distinct modulo π.

4.18. This exercise is devoted to the properties of the Legendre symbol $(\alpha/\pi)_4$, where π is prime in $\mathbf{Z}[i]$ and α is not divisible by π.

(a) Show that $\alpha^{(N(\pi)-1)/4}$ is congruent to a unique fourth root of unity modulo π. This shows that the Legendre symbol, as given in (4.20), is well-defined. Hint: use Exercise 4.17.

(b) Prove that the analogs of the properties given in Exercise 4.10 hold for $(\alpha/\pi)_4$.

(c) Prove that

$$\left(\frac{\alpha}{\pi}\right)_4 = 1 \iff x^4 \equiv \alpha \bmod \pi \text{ is solvable in } \mathbf{Z}[i].$$

4.19. In this exercise we will study the integer congruence $x^4 \equiv a \bmod p$, where $p \equiv 1 \bmod 4$ is prime and a is an integer not divisible by p.

(a) Write $p = \pi\overline{\pi}$ in $\mathbf{Z}[i]$. Then use (4.20) to show that $(a/\pi)_4^2 = (a/p)$, and conclude that $(a/\pi)_4 = \pm 1$ if and only if $(a/p) = 1$.

(b) Verify the partition of $(\mathbb{Z}/p\mathbb{Z})^*$ described in the discussion following (4.20).

4.20. Here we will study the congruence $x^4 \equiv a \bmod p$ when $p \equiv 3 \bmod 4$ is prime and a is an integer not divisible by p.

(a) Use (4.20) to show that $(a/p)_4 = 1$. Thus a is a fourth power modulo p in the ring $\mathbb{Z}[i]$.

(b) Show that a is the biquadratic residue of an *integer* modulo p if and only if $(a/p) = 1$. Hint: study the maps $\phi_k(x) = x^{2^k}$ on an Abelian group of order $2m$, m odd.

4.21. If a prime π of $\mathbb{Z}[i]$ is not associate to $1+i$, then show that a unique associate of π is primary.

4.22. Prove the top formula of (4.22).

4.23. Use the supplementary laws (4.22) to prove part (i) of Theorem 4.23.

4.24. Let $p \equiv 1 \bmod 4$ be prime, and write $p = a^2 + b^2$, where a is odd and b is even. The goal of this exercise is to present Dirichlet's elementary proof that $(2/\pi)_4 = i^{ab/2}$, where $\pi = a + bi$.

(a) Use quadratic reciprocity for the Jacobi symbol to prove that $(a/p) = 1$.

(b) Use $2p = (a+b)^2 + (a-b)^2$ and quadratic reciprocity to show that

$$\left(\frac{a+b}{p}\right) = (-1)^{((a+b)^2-1)/8}.$$

(c) Use (b) and (4.20) to show that

$$\left(\frac{a+b}{p}\right) = \left(\frac{i}{\pi}\right)_4 i^{ab/2}.$$

Hint: $-1 = i^2$.

(d) From $(a+b)^2 \equiv 2ab \bmod p$, deduce that

(i) $(a+b)^{(p-1)/2} \equiv (2ab)^{(p-1)/4} \bmod p$.

(ii) $(a+b/p) = (2ab/\pi)_4$.

(e) Show that $2ab \equiv 2a^2 i \bmod \pi$, and then use (a) and Exercise 4.19 to show that

$$\left(\frac{2ab}{\pi}\right)_4 = \left(\frac{2i}{\pi}\right)_4.$$

(f) Combine (c), (d) and (e) to show that $(2/\pi)_4 = i^{ab/4}$.

4.25. In this exercise we will study Gauss' statement of biquadratic reciprocity.

(a) If π is a primary prime of $\mathbb{Z}[i]$, then show that either $\pi \equiv 1 \bmod 4$ or $\pi \equiv 3+2i \bmod 4$.

(b) Let π and θ be distinct primary primes in $\mathbb{Z}[i]$. Show that biquadratic reciprocity is equivalent to the following two statements:

If either π or θ is congruent to 1 modulo 4,

then $(\pi/\theta)_4 = (\theta/\pi)_4$.

If π and θ are both congruent to $3+2i$ modulo 4,

then $(\pi/\theta)_4 = -(\theta/\pi)_4$.

This is how Gauss states biquadratic reciprocity in [42, Vol. II, §67, p. 138].

4.26. If 2 is a biquadratic residue modulo an odd prime p, prove that $p \equiv \pm 1 \bmod 8$.

4.27. In this exercise, we will present Gauss' proof that for a prime $p \equiv 1 \bmod 8$, the biquadratic character of 2 is determined by the decomposition $p = a^2 + 2b^2$. As usual, we write $p = \pi\bar{\pi}$ in $\mathbb{Z}[i]$.

(a) Show that $(-1/\pi)_4 = 1$ when $p \equiv 1 \bmod 8$.

(b) Use the properties of the Jacobi symbol to show that

$$\left(\frac{a}{p}\right) = (-1)^{(a^2-1)/8}.$$

(c) Use the Jacobi symbol to show that $(b/p) = 1$. Hint: write $b = 2^m c$, c odd, and first show that $(c/p) = 1$.

(d) Show that

$$\left(\frac{2}{\pi}\right)_4 = \left(\frac{-2b^2}{\pi}\right)_4 = \left(\frac{a^2}{\pi}\right)_4 = \left(\frac{a}{p}\right).$$

Hint: use Exercise 4.19.

Combining (c) and (d), we see that $(2/\pi)_4 = (-1)^{(a^2-1)/8}$, and Gauss' claim follows. If you read Gauss' original argument [42, Vol. II, §13], you'll appreciate how much the Jacobi symbol simplifies things.

4.28. Let (f,λ) and (f,μ) be periods, and write $(f,\mu) = \zeta^{\mu_1} + \cdots \zeta^{\mu_f}$. Then prove that

$$(f,\lambda)\cdot(f,\mu) = \sum_{j=1}^{f}(f,\lambda+\mu_j).$$

4.29. Let $p \equiv 1 \bmod 3$ be prime, and set $p-1 = 3f$. Let $(f,1)$, (f,g) and (f,g^2) be the periods as in the text. Recall that g is a primitive root modulo p. In this problem we will describe Gauss' proof of (4.24) (see [41, §358]). For $i,\ j \in \{0,1,2\}$, let (ij) be the number of pairs (m,n), $0 \le m,n \le f-1$, such that

$$1 + g^{3m+i} \equiv g^{3n+j} \bmod p.$$

(a) Show that the number of solutions modulo p of the equation $x^3 - y^3 \equiv 1 \bmod p$ is $N = 9(00) + 6$.

(b) Use Exercise 4.28 to show that

$$(f,1)\cdot(f,1) = f + (00)(f,1) + (01)(f,g) + (02)(f,g^2)$$
$$(f,1)\cdot(f,g) = (10)(f,1) + (11)(f,g) + (12)(f,g^2)$$

and conclude that $(00)+(01)+(02) = f-1$ and $(10)+(11)+(12) = f$. Hint: $(f,0) = f$ and $-1 = (-1)^3$.

(c) Show that $(10) = (22)$, $(11) = (20)$ and $(12) = (21)$. Hint: expand $(f,g)\cdot(f,1)$ and compare it to what you got in (b).

(d) Arguing as in (c), show that the 9 quantities (ij) reduce to three:

$$\alpha = (12) = (21) = (00) + 1$$
$$\beta = (01) = (10) = (22)$$
$$\gamma = (02) = (20) = (11).$$

(e) Note that $(f,1)\cdot(f,g)\cdot(f,g^2)$ is an integer. By expanding this quantity in terms of α, β and γ, show that

$$\alpha^2 + \beta^2 + \gamma^2 - \alpha = \alpha\beta + \beta\gamma + \alpha\gamma.$$

(f) Using (e), show that

$$(6\alpha - 3\beta - 3\gamma - 2)^2 + 27(\beta - \gamma)^2 = 12(\alpha + \beta + \gamma) - 4.$$

(g) Recall that $\alpha + \beta + \gamma = f$ (this was proved in (b)) and that $p - 1 = 3f$. Then use (e) to show that

$$4p = a^2 + 27b^2,$$

where $a = 6\alpha - 3\beta - 3\gamma - 2$ and $b = \beta - \gamma$.

(h) Let a be as in (g). Show that

$$a = 9\alpha - 3(\alpha + \beta + \gamma) - 2 = 9\alpha - p - 1.$$

Then use $\alpha = (00) + 1$ and (a) to conclude that

$$a = N - p + 2.$$

This proves (4.24).

In his first memoir on biquadratic residues [42, Vol. II, §§15–20, pp. 78–89], Gauss used the (ij)'s (without using Gauss sums) to determine the biquadratic character of 2.

CHAPTER TWO

CLASS FIELD THEORY

§5. THE HILBERT CLASS FIELD AND $p = x^2 + ny^2$

In Chapter One, we used elementary techniques to study the primes represented by $x^2 + ny^2$, $n > 0$. Genus theory told us when $p = x^2 + ny^2$ for a large but finite number of n's, and cubic and biquadratic reciprocity enabled us to treat two cases where genus theory failed. These methods are lovely but limited in scope. To solve $p = x^2 + ny^2$ when $n > 0$ is arbitrary, we will need class field theory, and this is the main task of Chapter Two. But rather than go directly to the general theorems of class field theory, in §5 we will first study the special case of the Hilbert class field. Theorem 5.1 below will use Artin Reciprocity for the Hilbert class field to solve our problem for infinitely many (but not all) $n > 0$. We will then study the case $p = x^2 + 14y^2$ in detail. This is a case where our previous methods failed, but once we determine the Hilbert class field of $\mathbb{Q}(\sqrt{-14})$, Theorem 5.1 will immediately give us a criterion for when $p = x^2 + 14y^2$.

The central notion of this section is the Hilbert class field of a number field K. We do not assume any previous acquaintance with this topic, for one of our goals is to introduce the reader to this more accessible part of class field theory. To see what the Hilbert class field has to do with the problem of representing primes by $x^2 + ny^2$, let's state the main theorem we intend to prove:

Theorem 5.1. *Let $n > 0$ be an integer satisfying the following condition:*

$$n \text{ squarefree}, \ n \not\equiv 3 \bmod 4. \tag{5.2}$$

Then there is a monic irreducible polynomial $f_n(x) \in \mathbb{Z}[x]$ of degree $h(-4n)$ such that if an odd prime p divides neither n nor the discriminant of $f_n(x)$, then

$$p = x^2 + ny^2 \iff \begin{cases} (-n/p) = 1 \text{ and } f_n(x) \equiv 0 \bmod p \\ \text{has an integer solution.} \end{cases}$$

Furthermore, $f_n(x)$ may be taken to be the minimal polynomial of a real algebraic integer α for which $L = K(\alpha)$ is the Hilbert class field of $K = \mathbb{Q}(\sqrt{-n})$.

While (5.2) does not give all integers $n > 0$, it gives infinitely many, so that Theorem 5.1 represents some real progress. In §9 we will use the full power of class field theory to prove a version of Theorem 5.1 that holds for *all* positive integers n.

A. Number Fields

We will review some basic facts from algebraic number theory, including Dedekind domains, factorization of ideals, and ramification. Most of the proofs will be omitted, though references will be given. Readers looking for a more complete treatment should consult Borevich and Shafarevich [8], Lang [72] or Marcus [77]. For an especially compact presentation of this material, see Ireland and Rosen [59, Chapter 12].

To begin, we define a *number field* K to be a subfield of the complex numbers $\mathbb{C}$ which has finite degree over $\mathbb{Q}$. The degree of K over $\mathbb{Q}$ is denoted $[K : \mathbb{Q}]$. Given such a field K, we let $\mathcal{O}_K$ denote the algebraic integers of K, i.e., the set of all $\alpha \in K$ which are roots of a monic integer polynomial. The basic structure of $\mathcal{O}_K$ is given in the following proposition:

Proposition 5.3. *Let K be a number field.*

(i) *$\mathcal{O}_K$ is a subring of $\mathbb{C}$ whose field of fractions is K.*

(ii) *$\mathcal{O}_K$ a free $\mathbb{Z}$-module of rank $[K : \mathbb{Q}]$.*

Proof. See Borevich and Shafarevich [8, §2.2] or Marcus [77, Corollaries to Theorems 2 and 9]. Q.E.D.

We will often call $\mathcal{O}_K$ the number ring of K. To begin our study of $\mathcal{O}_K$, we note that part (ii) of Proposition 5.3 has the following useful consequence concerning the ideals of $\mathcal{O}_K$:

Corollary 5.4. *If K is a number field and $\mathfrak{a}$ is a nonzero ideal of $\mathcal{O}_K$, then the quotient ring $\mathcal{O}_K/\mathfrak{a}$ is finite.*

Proof. See Exercise 5.1. Q.E.D.

Given a nonzero ideal $\mathfrak{a}$ of the number ring $\mathcal{O}_K$, its *norm* is defined to be $N(\mathfrak{a}) = |\mathcal{O}_K/\mathfrak{a}|$. Corollary 5.4 guarantees that $N(\mathfrak{a})$ is finite.

When we studied the rings $\mathbf{Z}[\omega]$ and $\mathbf{Z}[i]$ in §4, we used the fact that they were unique factorization domains. In general, the rings $\mathcal{O}_K$ are not UFDs, but they have another property which is almost as good: they are Dedekind domains. This means the following:

Theorem 5.5. *Let $\mathcal{O}_K$ be the ring of integers in a number field K. Then $\mathcal{O}_K$ is a Dedekind domain, which means that*

(i) *$\mathcal{O}_K$ is integrally closed in K, i.e., if $\alpha \in K$ satisfies a monic polynomial with coefficients in $\mathcal{O}_K$, then $\alpha \in \mathcal{O}_K$.*

(ii) *$\mathcal{O}_K$ is Noetherian, i.e., given any chain of ideals $\mathfrak{a}_1 \subset \mathfrak{a}_2 \subset \cdots$, there is an integer n such that $\mathfrak{a}_n = \mathfrak{a}_{n+1} = \cdots$.*

(iii) *Every nonzero prime ideal of $\mathcal{O}_K$ is maximal.*

Proof. The proof of (i) follows easily from the properties of algebraic integers (see Lang [72, §I.2] or Marcus [77, Exercise 4 to Chapter 2]), while (ii) and (iii) are straightforward consequences of Corollary 5.4 (see Exercise 5.1). Q.E.D.

The most important property of a Dedekind domain is that it has unique factorization at the level of ideals. More precisely:

Corollary 5.6. *If K is a number field, then any nonzero ideal $\mathfrak{a}$ in $\mathcal{O}_K$ can be written as a product*

$$\mathfrak{a} = \mathfrak{p}_1 \cdots \mathfrak{p}_r$$

of prime ideals, and the decomposition is unique up to order. Furthermore, the $\mathfrak{p}_i$'s are exactly the prime ideals of $\mathcal{O}_K$ containing $\mathfrak{a}$.

Proof. This corollary holds for any Dedekind domain. For a proof, see Lang [72, §I.6] or Marcus [77, Chapter 3, Theorem 16]. In Ireland and Rosen [59, §12.2] there is a nice proof (due to Hurwitz) that is special to the number field case. Q.E.D.

Prime ideals play an especially important role in algebraic number theory. We will often say "prime" rather than "nonzero prime ideal", and the terms "prime of K" and "nonzero prime ideal of $\mathcal{O}_K$" will be used interchangeably. Notice that when $\mathfrak{p}$ is a prime of K, the quotient ring $\mathcal{O}_K/\mathfrak{p}$ is a finite field by Corollary 5.4 and Theorem 5.5. This field is called the *residue field* of $\mathfrak{p}$.

Besides ideals of $\mathcal{O}_K$, we will also use *fractional ideals*, which are the nonzero finitely generated $\mathcal{O}_K$-submodules of K. The name "fractional ideal" comes from the fact that such an ideal can be written in the form $\alpha\mathfrak{a}$, where $\alpha \in K$ and $\mathfrak{a}$ is an ideal of $\mathcal{O}_K$ (see Exercise 5.2). Readers unfamiliar with fractional ideals should consult Marcus [77, Exercise 31 to Chapter 3]. The basic properties of fractional ideals are:

Proposition 5.7. *Let $\mathfrak{a}$ be a fractional $\mathcal{O}_K$-ideal.*

(i) *$\mathfrak{a}$ is invertible, i.e., there is a fractional $\mathcal{O}_K$-ideal $\mathfrak{b}$ such that $\mathfrak{a}\mathfrak{b} = \mathcal{O}_K$. The ideal $\mathfrak{b}$ will be denoted $\mathfrak{a}^{-1}$.*

(ii) *$\mathfrak{a}$ can be written uniquely as a product $\mathfrak{a} = \prod_{i=1}^{r} \mathfrak{p}_i^{r_i}$, $r_i \in \mathbf{Z}$, where the $\mathfrak{p}_i$'s are distinct prime ideals of $\mathcal{O}_K$.*

Proof. See Lang [72, §I.6] or Marcus [77, Exercise 31 to Chapter 3]. Q.E.D.

We will let I_K denote the set of all fractional ideals of K. I_K is closed under multiplication of ideals (see Exercise 5.2), and then part (i) of Proposition 5.7 shows that I_K is a group. The most important subgroup of I_K is the subgroup P_K of *principal fractional ideals*, i.e., those of the form $\alpha\mathcal{O}_K$ for some $\alpha \in K^*$. The quotient I_K/P_K is the *ideal class group* and is denoted by $C(\mathcal{O}_K)$. A basic fact is that $C(\mathcal{O}_K)$ is a finite group (see Borevich and Shafarevich [8, §3.7] or Marcus [77, Corollary 2 to Theorem 35]). In the case of imaginary quadratic fields, we will see in Theorem 5.30 that the ideal class group is closely related to the form class group defined in §3.

We will next introduce the idea of ramification, which is concerned with the behavior of primes in finite extensions. Suppose that K is a number field, and let L be a finite extension of K. If $\mathfrak{p}$ is a prime ideal of $\mathcal{O}_K$, then $\mathfrak{p}\mathcal{O}_K$ is an ideal of $\mathcal{O}_L$, and hence has a prime factorization

$$\mathfrak{p}\mathcal{O}_L = \mathfrak{P}_1^{e_1} \cdots \mathfrak{P}_g^{e_g}$$

where the $\mathfrak{P}_i$'s are the distinct primes of L containing $\mathfrak{p}$. The integer e_i, also written $e_{\mathfrak{P}_i|\mathfrak{p}}$, is called the *ramification index* of $\mathfrak{p}$ in $\mathfrak{P}_i$. Each prime $\mathfrak{P}_i$ containing $\mathfrak{p}$ also gives a residue field extension $\mathcal{O}_K/\mathfrak{p} \subset \mathcal{O}_L/\mathfrak{P}_i$, and its degree, written f_i or $f_{\mathfrak{P}_i|\mathfrak{p}}$, is the *inertial degree* of $\mathfrak{p}$ in $\mathfrak{P}_i$. The basic relation between the e_i's and f_i's is given by

Theorem 5.8. *Let $K \subset L$ be number fields, and let $\mathfrak{p}$ be a prime of K. If e_i (resp. f_i), $i = 1,\ldots,g$ are the ramification indices (resp. inertial degrees)*

defined above, then

$$\sum_{i=1}^{g} e_i f_i = [L:K].$$

Proof. See Borevich and Shafarevich [8, §3.5] or Marcus [77, Theorem 21]. Q.E.D.

In the above situation, we say that a prime $\mathfrak{p}$ of K *ramifies* in L if any of the ramification indices e_i are greater than 1. It can be proved that only a finite number of primes of K ramify in L (see Lang [72, §III.2] or Marcus [77, Corollary 3 to Theorem 24]).

Most of the extensions $K \subset L$ we will deal with will be Galois extensions, and in this case the above description can be simplified as follows:

Theorem 5.9. *Let $K \subset L$ be a Galois extension, and let $\mathfrak{p}$ be prime in K.*

(i) *The Galois group* $\mathrm{Gal}(L/K)$ *acts transitively on the primes of L containing $\mathfrak{p}$, i.e., if $\mathfrak{P}$ and $\mathfrak{P}'$ are primes of L containing $\mathfrak{p}$, then there is $\sigma \in \mathrm{Gal}(L/K)$ such that $\sigma(\mathfrak{P}) = \mathfrak{P}'$.*

(ii) *The primes $\mathfrak{P}_1, \ldots, \mathfrak{P}_g$ of L containing $\mathfrak{p}$ all have the same ramification index e and the same inertial degree f, and the formula of Theorem 5.8 becomes*

$$efg = [L:K].$$

Proof. For a proof of (i), see Lang [72, §I.7] or Marcus [77, Theorem 23]. The proof of (ii) follows easily from (i) and is left to the reader (see Exercise 5.3). Q.E.D.

Given a Galois extension $K \subset L$, an ideal $\mathfrak{p}$ of K ramifies if $e > 1$, and is unramified if $e = 1$. If $\mathfrak{p}$ satisfies the stronger condition $e = f = 1$, we say that $\mathfrak{p}$ *splits completely* in L. Such a prime is unramified, and in addition $\mathfrak{p}\mathcal{O}_L$ is the product of $[L:K]$ distinct primes, the maximum number allowed by Theorem 5.9. In §8 we will show that L is determined uniquely by the primes of K that split completely in L.

We will also need some facts concerning decomposition and inertia groups. Let $K \subset L$ be Galois, and let $\mathfrak{P}$ be a prime of L. Then the *decomposition group* and *inertia group* of $\mathfrak{P}$ are defined by

$$D_{\mathfrak{P}} = \{\sigma \in \mathrm{Gal}(L/K) : \sigma(\mathfrak{P}) = \mathfrak{P}\}$$
$$I_{\mathfrak{P}} = \{\sigma \in \mathrm{Gal}(L/K) : \sigma(\alpha) \equiv \alpha \bmod \mathfrak{P} \text{ for all } \alpha \in \mathcal{O}_L\}.$$

It is easy to show that $I_{\mathfrak{P}} \subset D_{\mathfrak{P}}$ and that an element $\sigma \in D_{\mathfrak{P}}$ induces an automorphism $\tilde{\sigma}$ of $\mathcal{O}_L/\mathfrak{P}$ which is the identity on $\mathcal{O}_K/\mathfrak{p}$, $\mathfrak{p} = \mathfrak{P} \cap \mathcal{O}_K$ (see Exercise 5.4). If $\widetilde{G}$ denotes the Galois group of $\mathcal{O}_K/\mathfrak{p} \subset \mathcal{O}_L/\mathfrak{P}$, it follows that $\tilde{\sigma} \in \widetilde{G}$. Thus the map $\sigma \mapsto \tilde{\sigma}$ defines a homomorphism $D_{\mathfrak{P}} \to \widetilde{G}$ whose kernel is exactly the inertia group $I_{\mathfrak{P}}$ (see Exercise 5.4). Then we have:

Proposition 5.10. *Let $D_{\mathfrak{P}}$, $I_{\mathfrak{P}}$ and $\widetilde{G}$ be as above.*

(i) *The homomorphism $D_{\mathfrak{P}} \to \widetilde{G}$ is surjective. Thus $D_{\mathfrak{P}}/I_{\mathfrak{P}} \simeq \widetilde{G}$.*

(ii) $|I_{\mathfrak{P}}| = e_{\mathfrak{P}|\mathfrak{p}}$ *and* $|D_{\mathfrak{P}}| = e_{\mathfrak{P}|\mathfrak{p}} f_{\mathfrak{P}|\mathfrak{p}}$.

Proof. See Lang [72, §I.7] or Marcus [77, Theorem 28]. Q.E.D.

The following proposition will help us decide when a prime is unramified or split completely in a Galois extension:

Proposition 5.11. *Let $K \subset L$ be a Galois extension, where $L = K(\alpha)$ for some $\alpha \in \mathcal{O}_L$. Let $f(x)$ be the monic minimal polynomial of α over K, so that $f(x) \in \mathcal{O}_K[x]$. If $\mathfrak{p}$ is prime in $\mathcal{O}_K$ and $f(x)$ is separable modulo $\mathfrak{p}$, then*

(i) *$\mathfrak{p}$ is unramified in L.*

(ii) *If $f(x) \equiv f_1(x) \cdots f_g(x) \bmod \mathfrak{p}$, where the $f_i(x)$ are distinct and irreducible modulo $\mathfrak{p}$, then $\mathfrak{P}_i = \mathfrak{p}\mathcal{O}_L + f_i(\alpha)\mathcal{O}_L$ is a prime ideal of $\mathcal{O}_L$, $\mathfrak{P}_i \neq \mathfrak{P}_j$ for $i \neq j$, and*

$$\mathfrak{p}\mathcal{O}_L = \mathfrak{P}_1 \cdots \mathfrak{P}_g.$$

Furthermore, all of the $f_i(x)$ have the same degree, which is the inertial degree f.

(iii) *$\mathfrak{p}$ splits completely in L if and only if $f(x) \equiv 0 \bmod \mathfrak{p}$ has a solution in $\mathcal{O}_K$.*

Proof. Note that (i) and (iii) are immediate consequences of (ii) (see Exercise 5.5). To prove (ii), note that $f(x)$ separable modulo $\mathfrak{p}$ implies that

$$f(x) \equiv f_1(x) \cdots f_g(x) \bmod \mathfrak{p},$$

where the $f_i(x)$ are distinct and irreducible modulo $\mathfrak{p}$. The fact that the above congruence governs the splitting of $\mathfrak{p}$ in $\mathcal{O}_L$ is a general fact that holds for arbitrary finite extensions (see Marcus [77, Theorem 27]). However, the decomposition group from Proposition 5.10 makes the proof in the Galois case especially easy. See Exercise 5.6. Q.E.D.

B. Quadratic Fields

To better understand the theory just sketched, let's apply apply it to the case of quadratic number fields. Such a field can be written uniquely in the form $K = \mathbf{Q}(\sqrt{N})$, where $N \neq 0, 1$ is a squarefree integer. The basic invariant of K is its *discriminant* d_K, which is defined to be

$$(5.12) \qquad d_K = \begin{cases} N & \text{if } N \equiv 1 \bmod 4 \\ 4N & \text{otherwise.} \end{cases}$$

Note that $d_K \equiv 0, 1 \bmod 4$ and $K = \mathbf{Q}(\sqrt{d_K})$, so that a quadratic field is determined by its discriminant.

The next step is to describe the integers $\mathcal{O}_K$ of K. Writing $K = \mathbf{Q}(\sqrt{N})$, N squarefree, one can show that

$$(5.13) \qquad \mathcal{O}_K = \begin{cases} \mathbf{Z}[\sqrt{N}], & N \not\equiv 1 \bmod 4 \\ \mathbf{Z}\left[\dfrac{1+\sqrt{N}}{2}\right], & N \equiv 1 \bmod 4 \end{cases}$$

(see Exercise 5.7 or Marcus [77, Corollary 2 to Theorem 1]). Hence the rings $\mathbf{Z}[\omega]$ and $\mathbf{Z}[i]$ from §4 are the full rings of integers in their respective fields. Using the discriminant, this description of $\mathcal{O}_K$ may be written more elegantly as follows:

$$(5.14) \qquad \mathcal{O}_K = \mathbf{Z}\left[\frac{d_K + \sqrt{d_K}}{2}\right]$$

(see Exercise 5.7).

We can now explain the restriction (5.2) made on n in Theorem 5.1. Namely, given $n > 0$, let K be the imaginary quadratic field $\mathbf{Q}(\sqrt{-n})$. Then (5.12) and (5.13) imply that

$$(5.15) \qquad d_K = -4n \iff \mathcal{O}_K = \mathbf{Z}[\sqrt{-n}] \iff n \text{ satisfies (5.2)}$$

(see Exercise 5.8). Thus the condition (5.2) on n is equivalent to $\mathbf{Z}[\sqrt{-n}]$ being the full ring of integers in K. For other n's, we will see in §7 that $\mathbf{Z}[\sqrt{-n}]$ is no longer a Dedekind domain but still has a lot of interesting structure.

We next want to discuss the arithmetic of a quadratic field K. As in §4, this means describing units and primes, the difference being that "prime" now means "prime ideal". Let's first consider units. Quadratic fields come in two flavors, real ($d_K > 0$) and imaginary ($d_K < 0$), and the units $\mathcal{O}_K^*$ behave quite differently in the two cases. In the imaginary case, there are only finitely many units. In §4 we computed $\mathcal{O}_K^*$ for $K = \mathbf{Q}(\sqrt{-3})$ or $\mathbf{Q}(i)$, and for all other imaginary quadratic fields it turns out that $\mathcal{O}_K^* = \{\pm 1\}$ (see Exercise 5.9). On the other hand, real quadratic fields always have infinitely many units, and determining them is related to Pell's equation and continued fractions (see Borevich and Shafarevich [8, §2.7]).

Before describing the primes of $\mathcal{O}_K$, we will need one useful bit of notation: if $D \equiv 0, 1 \bmod 4$, then the *Kronecker symbol* $(D/2)$ is defined by

$$\left(\frac{D}{2}\right) = \begin{cases} 0 & \text{if } D \equiv 0 \bmod 4 \\ 1 & \text{if } D \equiv 1 \bmod 8 \\ -1 & \text{if } D \equiv 5 \bmod 8. \end{cases}$$

We will most often apply this when $D = d_K$ is the discriminant of a quadratic field K. The following proposition tells us about the primes of quadratic fields:

Proposition 5.16. *Let K be a quadratic field of discriminant d_K, and let the nontrivial automorphism of K be denoted $\alpha \mapsto \alpha'$. Let p be prime in $\mathbb{Z}$.*

(i) *If $(d_K/p) = 0$ (i.e., $p \mid d_K$), then $p\mathcal{O}_K = \mathfrak{p}^2$ for some prime ideal $\mathfrak{p}$ of $\mathcal{O}_K$.*

(ii) *If $(d_K/p) = 1$, then $p\mathcal{O}_K = \mathfrak{p}\mathfrak{p}'$, where $\mathfrak{p} \neq \mathfrak{p}'$ are prime in $\mathcal{O}_K$.*

(iii) *If $(d_K/p) = -1$, then $p\mathcal{O}_K$ is prime in $\mathcal{O}_K$.*

Furthermore, the primes in (i)–(iii) *above give all nonzero primes of $\mathcal{O}_K$.*

Proof. To prove (i), suppose that p is an odd prime dividing d_K, and let $\mathfrak{p}$ be the ideal

$$\mathfrak{p} = p\mathcal{O}_K + \sqrt{d_K}\mathcal{O}_K.$$

Squaring, one obtains

$$\mathfrak{p}^2 = p^2\mathcal{O}_K + p\sqrt{d_K}\mathcal{O}_K + d_K\mathcal{O}_K.$$

However, d_K is squarefree (except for a possible factor of 4) and p is an odd divisor, so that $\gcd(p^2, d_K) = p$. It follows easily that $\mathfrak{p}^2 = p\mathcal{O}_K$, and then the relation $efg = [K : \mathbb{Q}] = 2$ from Theorem 5.9 implies that $\mathfrak{p}$ is a prime ideal. The case when $p = 2$ is similar and is left as part of Exercise 5.10.

Let's next prove (ii) and (iii) for an odd prime p not dividing d_K. The key tool will be Proposition 5.11. Note that $f(x) = x^2 - d_K$ is the minimal polynomial of the primitive element $\sqrt{d_K}$ of K over $\mathbb{Q}$, and since $p \nmid d_K$, $f(x)$ is separable modulo p. Then Proposition 5.11 shows that p is unramified in K.

If $(d_K/p) = 1$, then the congruence $x^2 \equiv d_K \bmod p$ has a solution, and consequently p splits completely in K by part (iii) of Proposition 5.11, i.e., $p\mathcal{O}_K = \mathfrak{p}_1\mathfrak{p}_2$ for distinct primes $\mathfrak{p}_1$ and $\mathfrak{p}_2$ of $\mathcal{O}_K$. Since $\mathrm{Gal}(K/\mathbb{Q})$ acts transitively on the primes of K containing p (Theorem 5.9), we must have $\mathfrak{p}_1' = \mathfrak{p}_2$, and it follows that $p\mathcal{O}_K$ factors as claimed. If $(d_K/p) = -1$, then $f(x) = x^2 - d_K$ is irreducible modulo p, and hence by part (ii) of Proposition 5.11, $p\mathcal{O}_K$ is prime in K.

The proof of (ii) and (iii) for $p = 2$ is similar and is left as an exercise (see Exercise 5.10). It remains to prove that the prime ideals listed so far are *all* nonzero primes in $\mathcal{O}_K$. The argument is analagous to what we did in Proposition 4.7, and the details are left to the reader (see Exercise 5.10). Q.E.D.

From this proposition, we get the following immediate corollary which tells us how primes of $\mathbf{Z}$ behave in a quadratic extension:

Corollary 5.17. *Let K be a quadratic field of discriminant d_K, and let p be an integer prime. Then:*

(i) *p ramifies in K if and only if p divides d_K.*

(ii) *p splits completely in K if and only if $(d_K/p) = 1$.* Q.E.D.

C. The Hilbert Class Field

The Hilbert class field of a number field K is defined in terms of the unramified Abelian extensions of K. To see what these terms mean, we begin with the "Abelian" part. This is easy, for an extension $K \subset L$ is *Abelian* if it is Galois and $\text{Gal}(L/K)$ is an Abelian group. But we aren't quite ready to define "unramified", for we first need to discuss the ramification of infinite primes.

Prime ideals of $\mathcal{O}_K$ are often called *finite primes* to distinguish them from the *infinite primes*, which are determined by the embeddings of K into $\mathbf{C}$. A *real infinite prime* is an embedding $\sigma : K \to \mathbf{R}$, while a *complex infinite prime* is a pair of complex conjugate embeddings $\sigma, \overline{\sigma} : K \to \mathbf{C}$, $\sigma \neq \overline{\sigma}$. Given an extension $K \subset L$, an infinite prime σ of K *ramifies* in L provided that σ is real but it has an extension to L which is complex. For example, the infinite prime of $\mathbf{Q}$ is unramified in $\mathbf{Q}(\sqrt{2})$ but ramified in $\mathbf{Q}(\sqrt{-2})$.

An extension $K \subset L$ is *unramified* if it is unramified at *all* primes, finite or infinite. While this is a very strong restriction, it can still happen that a given field has unramified extensions of arbitrarily high degree (an example is $K = \mathbf{Q}(\sqrt{-2 \cdot 3 \cdot 5 \cdot 7 \cdot 11 \cdot 13})$, a consequence of the work of Golod and Shafarevich on class field towers—see Roquette [85]). But if we ask for unramified *Abelian* extensions, a much nicer picture emerges. In §8 we will use class field theory to prove the following result:

Theorem 5.18. *Given a number field K, there is a finite Galois extension L of K such that:*

(i) *L is an unramified Abelian extension of K.*

(ii) *Any unramified Abelian extension of K lies in L.* Q.E.D.

The field L of Theorem 5.18 is called the *Hilbert class field* of K. It is the maximal unramified Abelian extension of K and is clearly unique.

To unlock the full power of the Hilbert class field L of K, we will use the Artin symbol to link L to the ideal structure of $\mathcal{O}_K$. The following lemma is needed to define the Artin symbol:

Lemma 5.19. *Let $K \subset L$ be a Galois extension, and let $\mathfrak{p}$ be a prime of $\mathcal{O}_K$ which is unramified in L. If $\mathfrak{P}$ is a prime of $\mathcal{O}_L$ containing $\mathfrak{p}$, then there is a unique element $\sigma \in \mathrm{Gal}(L/K)$ such that for all $\alpha \in \mathcal{O}_L$,*

$$\sigma(\alpha) \equiv \alpha^{N(\mathfrak{p})} \bmod \mathfrak{P},$$

where $N(\mathfrak{p}) = |\mathcal{O}_K/\mathfrak{p}|$ is the norm of $\mathfrak{p}$.

Proof. As in Proposition 5.10, let $D_{\mathfrak{P}}$ and $I_{\mathfrak{P}}$ be the decomposition and inertia groups of $\mathfrak{P}$. Recall that $\sigma \in D_{\mathfrak{P}}$ induces an element $\tilde{\sigma} \in \widetilde{G}$, where $\widetilde{G}$ is the Galois group of $\mathcal{O}_L/\mathfrak{P}$ over $\mathcal{O}_K/\mathfrak{p}$. Since $\mathfrak{p}$ is unramified in L, part (ii) of Proposition 5.10 tells us that $|I_{\mathfrak{P}}| = e_{\mathfrak{P}|\mathfrak{p}} = 1$, and then the first part of the proposition implies that $\sigma \mapsto \tilde{\sigma}$ defines an isomorphism

$$D_{\mathfrak{P}} \xrightarrow{\sim} \widetilde{G}.$$

The structure of the Galois group $\widetilde{G}$ is well known: if $\mathcal{O}_K/\mathfrak{p}$ has q elements, then $\widetilde{G}$ is a cyclic group with canonical generator given by the *Frobenius automorphism* $x \mapsto x^q$ (see Hasse [50, pp. 40–41]). Thus there is a *unique* $\sigma \in D_{\mathfrak{P}}$ which maps to the Frobenius element. Since $q = N(\mathfrak{p})$ by definition, σ satisfies our desired condition

$$\sigma(\alpha) \equiv \alpha^{N(\mathfrak{p})} \bmod \mathfrak{P} \quad \text{for all } \alpha \in \mathcal{O}_L.$$

To prove uniqueness, note that any σ satisfying this condition must lie in $D_{\mathfrak{P}}$, and then we are done. Q.E.D.

The unique element σ of Lemma 5.19 is called the *Artin symbol* and is denoted $((L/K)/\mathfrak{P})$ since it depends on the prime $\mathfrak{P}$ of L. Its crucial property is that for any $\alpha \in \mathcal{O}_L$, we have

$$\left(\frac{L/K}{\mathfrak{P}}\right)(\alpha) \equiv \alpha^{N(\mathfrak{p})} \bmod \mathfrak{P}, \tag{5.20}$$

where $\mathfrak{p} = \mathfrak{P} \cap \mathcal{O}_K$. The Artin symbol $((L/K)/\mathfrak{P})$ has the following useful properties:

Corollary 5.21. *Let $K \subset L$ be a Galois extension, and let $\mathfrak{p}$ be an unramified prime of K. Given a prime $\mathfrak{P}$ of L containing $\mathfrak{p}$, we have:*

(i) *If $\sigma \in \mathrm{Gal}(L/K)$, then*

$$\left(\frac{L/K}{\sigma(\mathfrak{P})}\right) = \sigma\left(\frac{L/K}{\mathfrak{P}}\right)\sigma^{-1}.$$

(ii) *The order of $((L/K)/\mathfrak{P})$ is the inertial degree $f = f_{\mathfrak{P}|\mathfrak{p}}$.*

(iii) *$\mathfrak{p}$ splits completely in L if and only if $((L/K)/\mathfrak{P}) = 1$.*

Proof. The proof of (i) is a direct consequence of the uniqueness of the Artin symbol. The details are left to the reader (see Exercise 5.12).

To prove (ii), recall from the proof of Lemma 5.19 that since $\mathfrak{p}$ is unramified, the decomposition group $D_{\mathfrak{P}}$ is isomorphic to the Galois group of the finite extension $\mathcal{O}_K/\mathfrak{p} \subset \mathcal{O}_L/\mathfrak{P}$ whose degree is the inertial degree f. By definition, the Artin symbol maps to a generator of the Galois group, so that the Artin symbol has order f as desired.

To prove (iii), recall that $\mathfrak{p}$ splits completely in L if and only if $e = f = 1$. Since we're already assuming that $e = 1$, (iii) follows immediately from (ii). Q.E.D.

When $K \subset L$ is an Abelian extension, the Artin symbol $((L/K)/\mathfrak{P})$ depends only on the underlying prime $\mathfrak{p} = \mathfrak{P} \cap \mathcal{O}_K$. To see this, let $\mathfrak{P}'$ be another prime containing $\mathfrak{p}$. We've seen that $\mathfrak{P}' = \sigma(\mathfrak{P})$ for some $\sigma \in \mathrm{Gal}(L/K)$. Then Corollary 5.21 implies that

$$\left(\frac{L/K}{\mathfrak{P}'}\right) = \left(\frac{L/K}{\sigma(\mathfrak{P})}\right) = \sigma\left(\frac{L/K}{\mathfrak{P}}\right)\sigma^{-1} = \left(\frac{L/K}{\mathfrak{P}}\right)$$

since $\mathrm{Gal}(L/K)$ is Abelian. It follows that whenever $K \subset L$ is Abelian, the Artin symbol can be written as $((L/K)/\mathfrak{p})$.

To see the relevance of the Artin symbol to reciprocity, let's work out an example. Let $K = \mathbf{Q}(\sqrt{-3})$ and $L = K(\sqrt[3]{2})$. Since $\mathcal{O}_K$ is the ring $\mathbf{Z}[\omega]$ of §4, it's a PID, and consequently a prime ideal $\mathfrak{p}$ can be written as $\pi\mathbf{Z}[\omega]$, where π is prime in $\mathbf{Z}[\omega]$. If π doesn't divide 6, it follows from Proposition 5.11 that π is unramified in L (see part (a) of Exercise 5.14). Since $\mathrm{Gal}(L/K) \simeq \mathbf{Z}/3\mathbf{Z}$ is Abelian, we see that $((L/K)/\pi)$ is defined. To determine which automorphism it is, we need only evaluate it on $\sqrt[3]{2}$. The answer is very nice:

$$\left(\frac{L/K}{\pi}\right)(\sqrt[3]{2}) = \left(\frac{2}{\pi}\right)_3 \sqrt[3]{2}. \tag{5.22}$$

So the Artin symbol generalizes the Legendre symbol! To prove this, let $\mathfrak{P}$ be a prime of $\mathcal{O}_L$ containing π. Then, by (5.20),

$$\begin{aligned}\left(\frac{L/K}{\pi}\right)(\sqrt[3]{2}) &\equiv \left(\sqrt[3]{2}\right)^{N(\pi)} \bmod \mathfrak{P}\\ &\equiv 2^{(N(\pi)-1)/3}\cdot\sqrt[3]{2} \bmod \mathfrak{P}.\end{aligned}$$

However, we know from (4.10) that

$$2^{(N(\pi)-1)/3} \equiv \left(\frac{2}{\pi}\right)_3 \bmod \pi,$$

and then $\pi \in \mathfrak{P}$ implies

$$\left(\frac{L/K}{\pi}\right)(\sqrt[3]{2}) \equiv \left(\frac{2}{\pi}\right)_3 \sqrt[3]{2} \bmod \mathfrak{P}.$$

Since $((L/K)/\pi)(\sqrt[3]{2})$ equals $\sqrt[3]{2}$ times a cube root of unity (which are distinct modulo $\mathfrak{P}$—see part (a) of Exercise 5.13), (5.22) is proved. In Exercise 5.14, we will generalize (5.22) to the case of the nth power Legendre symbol.

When $K \subset L$ is an unramified Abelian extension, things are especially nice because $((L/K)/\mathfrak{p})$ is defined for *all* primes $\mathfrak{p}$ of $\mathcal{O}_K$. To exploit this, let I_K be the set of all fractional ideals of $\mathcal{O}_K$. As we saw in Proposition 5.7, any fractional ideal $\mathfrak{a} \in I_K$ has a prime factorization

$$\mathfrak{a} = \prod_{i=1}^{r} \mathfrak{p}_i^{r_i}, \qquad r_i \in \mathbf{Z},$$

and then we define the Artin symbol $((L/K)/\mathfrak{a})$ to be the product

$$\left(\frac{L/K}{\mathfrak{a}}\right) = \prod_{i=1}^{r} \left(\frac{L/K}{\mathfrak{p}_i}\right)^{r_i}.$$

The Artin symbol thus defines a homomorphism, called the *Artin map*,

$$\left(\frac{L/K}{\cdot}\right) : I_K \longrightarrow \mathrm{Gal}(L/K).$$

Notice that when $K \subset L$ is ramified, the Artin map is not defined on all of I_K. This is one reason why the general theorems of class field theory are complicated to state.

The *Artin reciprocity theorem for the Hilbert class field* relates the Hilbert class field to the ideal class group $C(\mathcal{O}_K)$ as follows:

Theorem 5.23. *If L is the Hilbert class field of a number field K, then the Artin map*

$$\left(\frac{L/K}{\cdot}\right) : I_K \longrightarrow \mathrm{Gal}(L/K)$$

is surjective, and its kernel is exactly the subgroup P_K of principal fractional ideals. Thus the Artin map induces an isomorphism

$$C(\mathcal{O}_K) \xrightarrow{\sim} \mathrm{Gal}(L/K). \qquad \text{Q.E.D.}$$

This theorem will follow from the results of §8. The appearance of the class group $C(\mathcal{O}_K)$ explains why L is called a "class field".

If we apply Galois theory to Theorems 5.18 and 5.23, we get the following classification of unramified Abelian extensions of K (see Exercise 5.17):

Corollary 5.24. *Given a number field K, there is a one-to-one correspondence between unramified Abelian extensions M of K and subgroups H of the ideal class group $C(\mathcal{O}_K)$. Furthermore, if the extension $K \subset M$ corresponds to the subgroup $H \subset C(\mathcal{O}_K)$, then the Artin map induces an isomorphism*

$$C(\mathcal{O}_K)/H \xrightarrow{\sim} \mathrm{Gal}(M/K). \qquad \text{Q.E.D.}$$

This corollary is *class field theory for unramified Abelian extensions*, and it illustrates one of the main themes of class field theory: a certain class of extensions of K (unramified Abelian extensions) are classified in terms of data intrinsic to K (subgroups of the ideal class group). The theorems we encounter in §8 will follow the same format.

Theorem 5.23 also allows us to characterize the primes of K which split completely in the Hilbert class field:

Corollary 5.25. *Let L be the Hilbert class field of a number field K, and let $\mathfrak{p}$ be a prime ideal of K. Then*

$$\mathfrak{p} \textit{ splits completely in } L \iff \mathfrak{p} \textit{ is a principal ideal.}$$

Proof. By Corollary 5.21, we know that $\mathfrak{p}$ splits completely if and only if $((L/K)/\mathfrak{p}) = 1$. Since the Artin map induces an isomorphism $C(\mathcal{O}_K) \simeq \mathrm{Gal}(L/K)$, we see that $((L/K)/\mathfrak{p}) = 1$ if and only if $\mathfrak{p}$ determines the trivial class of $C(\mathcal{O}_K)$. By the definition of the ideal class group, this means that $\mathfrak{p}$ is principal, and the corollary is proved. Q.E.D.

In §8, we will see that the Hilbert class field is characterized by the property that the primes that split completely are exactly the principal prime ideals.

D. Solution of $p = x^2 + ny^2$ for infinitely many n

Now that we know about the Hilbert class field, we can prove Theorem 5.1:

Proof of Theorem 5.1. The first step is to relate $p = x^2 + ny^2$ to the behavior of p in the Hilbert class field L. This result is sufficiently interesting to be a theorem in its own right:

Theorem 5.26. *Let L be the Hilbert class field of $K = \mathbb{Q}(\sqrt{-n})$. Assume that n satisfies (5.2), so that $\mathcal{O}_K = \mathbb{Z}[\sqrt{-n}]$. If p is an odd prime not dividing n, then*

$$p = x^2 + ny^2 \iff p \text{ splits completely in } L.$$

Proof. Since n satisfies (5.2), we know from (5.15) that $d_K = -4n$ and $\mathcal{O}_K = \mathbb{Z}[\sqrt{-n}]$. Let p be an odd prime not dividing n. Then $p \nmid d_K$, so that p is unramified in K by Corollary 5.17. We will prove the following equivalences:

$$\begin{aligned} p = x^2 + ny^2 &\iff p\mathcal{O}_K = \mathfrak{p}\bar{\mathfrak{p}},\ \mathfrak{p} \neq \bar{\mathfrak{p}}, \text{ and } \mathfrak{p} \text{ is principal in } \mathcal{O}_K \\ &\iff p\mathcal{O}_K = \mathfrak{p}\bar{\mathfrak{p}},\ \mathfrak{p} \neq \bar{\mathfrak{p}}, \text{ and } \mathfrak{p} \text{ splits completely in } L \\ &\iff p \text{ splits completely in } L, \end{aligned} \tag{5.27}$$

and Theorem 5.26 will follow.

To prove the first equivalence, suppose that $p = x^2 + ny^2 = (x + \sqrt{-n}y) \times (x - \sqrt{-n}y)$. Setting $\mathfrak{p} = (x + \sqrt{-n}y)\mathcal{O}_K$, then $p\mathcal{O}_K = \mathfrak{p}\bar{\mathfrak{p}}$ must be the prime factorization of $p\mathcal{O}_K$ in $\mathcal{O}_K$. Note that $\mathfrak{p} \neq \bar{\mathfrak{p}}$ since p is unramified in K. Conversely, suppose that $p\mathcal{O}_K = \mathfrak{p}\bar{\mathfrak{p}}$, where $\mathfrak{p}$ is principal. Since $\mathcal{O}_K = \mathbb{Z}[\sqrt{-n}]$, we can write $\mathfrak{p} = (x + \sqrt{-n}y)\mathcal{O}_K$. This implies that $p\mathcal{O}_K = (x^2 + ny^2)\mathcal{O}_K$, and it follows that $p = x^2 + ny^2$.

The second equivalence follows immediately from Corollary 5.25. To prove the final equivalence, we will use the following lemma:

Lemma 5.28. *Let L be the Hilbert class field of an imaginary quadratic field K, and let τ denote complex conjugation. Then $\tau(L) = L$, and consequently L is Galois over $\mathbb{Q}$.*

Proof. It is easy to see that $\tau(L)$ is an unramified Abelian extension of $\tau(K) = K$. Since L is the maximal such extension, we have $\tau(L) \subset L$, and then $\tau(L) = L$ since they have the same degree over K. Hence $\tau \in \mathrm{Gal}(L/\mathbb{Q})$, which implies that L is Galois over $\mathbb{Q}$ (see Exercise 5.19). Q.E.D.

To finish the proof of (5.27), note that the condition

$$p\mathcal{O}_K = \mathfrak{p}\bar{\mathfrak{p}},\ \mathfrak{p} \neq \bar{\mathfrak{p}}, \text{ and } \mathfrak{p} \text{ splits completely in } L$$

says that p splits completely in K and that some prime of K containing p splits completely in L. Since L is Galois over $\mathbf{Q}$, this is easily seen to be equivalent to p splitting completely in L (see Exercise 5.18), and Theorem 5.26 is proved. Q.E.D.

The next step in the proof of Theorem 5.1 is to give a more elementary way of saying that p splits completely in L. We have the following criterion:

Proposition 5.29. *Let K be an imaginary quadratic field, and let L be a finite extension of K which is Galois over $\mathbf{Q}$. Then:*

(i) *There is a real algebraic integer α such that $L = K(\alpha)$.*

(ii) *Given α as in* (i), *let $f(x) \in \mathbf{Z}[x]$ denote its monic minimal polynomial. If p is a prime not dividing the discriminant of $f(x)$, then*

$$p \text{ splits completely in } L \iff \begin{cases} (d_K/p) = 1 \text{ and } f(x) \equiv 0 \bmod p \\ \text{has an integer solution.} \end{cases}$$

Proof. By Lemma 5.28, L is Galois over $\mathbf{Q}$, and thus $[L \cap \mathbf{R} : \mathbf{Q}] = [L : K]$ since $L \cap \mathbf{R}$ is the fixed field of complex conjugation. This implies that for $\alpha \in L \cap \mathbf{R}$,

$$L \cap \mathbf{R} = \mathbf{Q}(\alpha) \iff L = K(\alpha)$$

(see Exercise 5.19). Hence, if $\alpha \in \mathcal{O}_L \cap \mathbf{R}$ satisfies $L \cap \mathbf{R} = \mathbf{Q}(\alpha)$, then α is a real integral primitive element of L over K, and (i) is proved. Furthermore, given such an α, let $f(x)$ be its monic minimal polynomial over $\mathbf{Q}$. Then $f(x) \in \mathbf{Z}[x]$, and since $[L \cap \mathbf{R} : \mathbf{Q}] = [L : K]$, $f(x)$ is also the minimal polynomial of α over K.

To prove the final part of (ii), let p be a prime not dividing the discriminant of $f(x)$. This tells us that $f(x)$ is separable modulo p. By Corollary 5.17 we have

$$p\mathcal{O}_K = \mathfrak{p}\bar{\mathfrak{p}},\ \mathfrak{p} \neq \bar{\mathfrak{p}} \iff \left(\frac{d_K}{p}\right) = 1.$$

We may assume that p splits completely in K, so that $\mathbf{Z}/p\mathbf{Z} \simeq \mathcal{O}_K/\mathfrak{p}$. Since $f(x)$ is separable over $\mathbf{Z}/p\mathbf{Z}$, it is separable over $\mathcal{O}_K/\mathfrak{p}$, and then Proposition 5.11 shows that

$$\begin{aligned} \mathfrak{p} \text{ splits completely in } L &\iff f(x) \equiv 0 \bmod \mathfrak{p} \text{ is solvable in } \mathcal{O}_K \\ &\iff f(x) \equiv 0 \bmod p \text{ is solvable in } \mathbf{Z}, \end{aligned}$$

where the last equivalence again uses $\mathbf{Z}/p\mathbf{Z} \simeq \mathcal{O}_K/\mathfrak{p}$. The proposition now follows from the last equivalence of (5.27). Q.E.D.

We can now prove the main equivalence of Theorem 5.1. Since the Hilbert class field L of $K = \mathbf{Q}(\sqrt{-n})$ is Galois over $\mathbf{Q}$, Proposition 5.29 implies that there is a real algebraic integer α which is a primitive element of L over K. Let $f_n(x)$ be the monic minimal polynomial of α, and let p be an odd prime dividing neither n nor the discriminant of $f_n(x)$, then Theorem 5.26 and Proposition 5.29 imply that

$$p = x^2 + ny^2 \iff p \text{ splits completely in } L$$
$$\iff \begin{cases} (-n/p) = 1 \text{ and } f_n(x) \equiv 0 \bmod p \\ \text{has an integer solution.} \end{cases}$$

In the second equivalence, recall that n satisfies (5.2), so that $d_K = -4n$, and hence $(d_K/p) = (-n/p)$.

It remains to show that the degree of $f_n(x)$ is the class number $h(-4n)$. Using Galois theory and Theorem 5.23, it follows that $f_n(x)$ has degree

$$[L:K] = |\mathrm{Gal}(L/K)| = |C(\mathcal{O}_K)|.$$

In Theorem 5.30 below we will see that when $d_K < 0$, there is a natural isomorphism

$$C(\mathcal{O}_K) \simeq C(d_K)$$

between the ideal class group $C(\mathcal{O}_K)$ and the form class group $C(d_K)$ from §3. Since $d_K = -4n$ in our case, we have $|C(\mathcal{O}_K)| = |C(-4n)| = h(-4n)$, which completes the proof of Theorem 5.1. Q.E.D.

The polynomial $f_n(x)$ of Theorem 5.1 is not unique—there are lots of primitive elements. However, we can at least predict its degree in advance by computing the class number $h(-4n)$. In §8 we will see that knowing $f_n(x)$ is *equivalent* to knowing the Hilbert class field.

We have now answered our basic question of when $p = x^2 + ny^2$, at least for those n satisfying (5.2). Notice that quadratic forms have almost completely disappeared! We used $x^2 + ny^2$ in Theorem 5.26, but otherwise all of the action took place using *ideals* rather than *forms*. This is typical of what happens in modern algebraic number theory—ideals are the dominant language. At the same time, we don't want to waste the work done on quadratic forms in §§2–3. So can we translate quadratic forms into ideals? In §7 we will study this question in detail. The full story is somewhat complicated, but the case of negative field discriminants rather nice: here, the form class group $C(d_K)$ from §3 is isomorphic to the ideal class group $C(\mathcal{O}_K)$. More precisely, we get the following theorem, which is a special case of the results of §7:

Theorem 5.30. *Let K be an imaginary quadratic field K of discriminant $d_K < 0$. Then:*

(i) *If* $f(x,y) = ax^2 + bxy + cy^2$ *is a primitive positive definite quadratic form of discriminant* d_K, *then*

$$[a,(-b+\sqrt{d_K})/2] = \{ma + n(-b+\sqrt{d_K})/2 : m,n \in \mathbf{Z}\}$$

is an ideal of $\mathcal{O}_K$.

(ii) *The map sending* $f(x,y)$ *to* $[a,(-b+\sqrt{d_K})/2]$ *induces an isomorphism between the form class group* $C(d_K)$ *of* §3 *and the ideal class group* $C(\mathcal{O}_K)$. *Hence the order of* $C(\mathcal{O}_K)$ *is the class number* $h(d_K)$. Q.E.D.

If we combine Theorems 5.30 and 5.23, we see that the Galois group $\text{Gal}(L/K)$ of the Hilbert class field of an imaginary quadratic field K is canonically isomorphic to the form class group $C(d_K)$. Thus the "class" in "Hilbert class field" refers to Gauss' classes of properly equivalent quadratic forms.

This theorem allows us to compute ideal class groups using what we know about quadratic forms. For example, consider the quadratic field $K = \mathbf{Q}(\sqrt{-14})$ of discriminant -56. In §2 we saw that the reduced forms of discriminant -56 are $x^2 + 14y^2$, $2x^2 + 7y^2$ and $3x^2 \pm 2xy + 5y^2$. The form class group $C(-56)$ is thus cyclic of order 4 since only $x^2 + 14y^2$ and $2x^2 + 7y^2$ give classes of order ≤ 2. Then, using Theorem 5.30, we see that the ideal class group $C(\mathcal{O}_K)$ is isomorphic to $\mathbf{Z}/4\mathbf{Z}$, and furthermore, ideal class representatives are given by $[1,\sqrt{-14}] = \mathcal{O}_K$, $[2,\sqrt{-14}]$ and $[3, 1 \pm \sqrt{-14}]$. See Exercises 5.20–5.22 for some other applications of Theorem 5.30.

The final task of §5 is to work out an explicit example of Theorem 5.1. We will discuss the case $p = x^2 + 14y^2$, which was left unresolved at the end of §3. Of course, we know from Theorem 5.1 that there is *some* polynomial $f_{14}(x)$ such that

$$p = x^2 + 14y^2 \iff \begin{cases} (-14/p) = 1 \text{ and } f_{14}(x) \equiv 0 \bmod p \\ \text{has an integer solution,} \end{cases}$$

but so far all we know about $f_{14}(x)$ is that it has degree 4 since $h(-56) = 4$. This illustrates one weakness of Theorem 5.1: it tells us that $f_{14}(x)$ exists, but doesn't tell us how to find it. To determine $f_{14}(x)$, we need to know the Hilbert class field of $\mathbf{Q}(\sqrt{-14})$. The answer is as follows:

Proposition 5.31. *The Hilbert class field of* $K = \mathbf{Q}(\sqrt{-14})$ *is* $L = K(\alpha)$, *where* $\alpha = \sqrt{2\sqrt{2}-1}$.

Proof. Since $h(-56) = 4$, the Hilbert class field has degree 4 over K. Then $L = K(\alpha)$ will be the Hilbert class field once we show that $K \subset L$ is an unramified Abelian extension of degree 4. It's easy to see that $K \subset L$ is Abelian of degree 4, so that we need only show that it is unramified. Furthermore, since K is imaginary quadratic, the infinite primes are automatically unramified.

Note that $\alpha^2 = 2\sqrt{2} - 1$, so that $\sqrt{2} \in L$. If we let $K_1 = K(\sqrt{2})$, then we have the extensions

$$K \subset K_1 \subset L,$$

and it suffices to show that $K \subset K_1$ and $K_1 \subset L$ are unramified (see Exercise 5.15). Since each of these extensions is obtained by adjoining a square root ($K_1 = K(\sqrt{2})$ and $L = K_1(\sqrt{\mu})$, $\mu = 2\sqrt{2} - 1$), let's first prove a general lemma about this situation:

Lemma 5.32. *Let $L = K(\sqrt{u})$ be a quadratic extension with $u \in \mathcal{O}_K$, and let $\mathfrak{p}$ be prime in $\mathcal{O}_K$.*

(i) *If $2u \notin \mathfrak{p}$, then $\mathfrak{p}$ is unramified in L.*

(ii) *If $2 \in \mathfrak{p}$, $u \notin \mathfrak{p}$ and $u = b^2 - 4c$ for some $b, c \in \mathcal{O}_K$, then $\mathfrak{p}$ is unramified in L.*

Proof. (i) Since the discriminant of $x^2 - u$ is $4u \notin \mathfrak{p}$, $x^2 - u$ is separable modulo $\mathfrak{p}$. Thus $\mathfrak{p}$ is unramified by Proposition 5.11.

(ii) Note that $L = K(\beta)$, where $\beta = (-b + \sqrt{u})/2$ is a root of $x^2 + bx + c$. The discriminant is $b^2 - 4c = u \notin \mathfrak{p}$, so again $\mathfrak{p}$ is unramified by Proposition 5.11. Q.E.D.

Now we can prove Proposition 5.31. To study $K \subset K_1$, let $\mathfrak{p}$ be prime in $\mathcal{O}_K$. Since $K_1 = K(\sqrt{2})$, part (i) of Lemma 5.32 implies that $\mathfrak{p}$ is unramified whenever $2 \notin \mathfrak{p}$. It remains to study the case $2 \in \mathfrak{p}$. Since $\sqrt{-14} \in K$ and $\sqrt{2} \in K_1$, we also have $\sqrt{-7} \in K_1$, i.e., $K_1 = K(\sqrt{-7})$. Since $-7 \notin \mathfrak{p}$ and $-7 = 1^2 - 4 \cdot 2$, $\mathfrak{p}$ is unramified by part (ii) of Lemma 5.32.

The extension $K_1 \subset L$ is almost as easy. We know that $L = K_1(\sqrt{\mu})$, $\mu = 2\sqrt{2} - 1$. Let $\mu' = -2\sqrt{2} - 1$. Since $\sqrt{\mu\mu'} = \sqrt{-7} \in K_1$, it follows that $\sqrt{\mu'} \in L$, and in fact

$$L = K_1(\sqrt{\mu}) = K_1(\sqrt{\mu'}).$$

Now let $\mathfrak{p}$ be prime in K_1. If $2 \notin \mathfrak{p}$, then $\mu + \mu' = -2$ shows that $\mu \notin \mathfrak{p}$ or $\mu' \notin \mathfrak{p}$, and $\mathfrak{p}$ is unramified by part (i) of Lemma 5.32. If $2 \in \mathfrak{p}$, then $\mu \notin \mathfrak{p}$ since $\mu = 2\sqrt{2} - 1$. We also have $\mu = (1 + \sqrt{2})^2 - 4$, and then part (ii) of Lemma 5.32 shows that $\mathfrak{p}$ is unramified. Q.E.D.

We can now characterize when a prime p is represented by $x^2 + 14y^2$:

Theorem 5.33. *If $p \neq 7$ is an odd prime, then*

$$p = x^2 + 14y^2 \iff \begin{cases} (-14/p) = 1 \textit{ and } (x^2+1)^2 \equiv 8 \bmod p \\ \textit{has an integer solution.} \end{cases}$$

Proof. Since $\alpha = \sqrt{2\sqrt{2}-1}$ is a real integral primitive element of the Hilbert class field of $K = \mathbf{Q}(\sqrt{-14})$, its minimal polynomial $x^4 + 2x^2 - 7 = (x^2+1)^2 - 8$ can be chosen to be the polynomial $f_{14}(x)$ of Theorem 5.1. Its discriminant is $-2^{14} \cdot 7$ (see Exercise 5.24), so that the only excluded primes are 2 and 7. Then Theorem 5.33 follows immediately from Theorem 5.1. Q.E.D.

These methods can be used to compute the Hilbert class field in other cases (see Herz [56]). For example, in Exercise 5.25, we will see that the Hilbert class field of $K = \mathbf{Q}(\sqrt{-17})$ is $L = K(\alpha)$, where $\alpha = \sqrt{(1+\sqrt{17})/2}$. This gives us an explicit criterion for a prime to be of the form $x^2 + 17y^2$ (see Exercise 5.26).

One unsatisfactory aspect of these examples is that they don't explain how the primitive element α of the Hilbert class field was found. In general, the Hilbert class field is difficult to describe explicitly, though this can be done for class numbers ≤ 4 (see Herz [56]). In §6 we will use genus theory to discover the above primitive elements when $K = \mathbf{Q}(\sqrt{-14})$ or $\mathbf{Q}(\sqrt{-17})$, and in Chapter Three we will use complex multiplication to give a general method for finding the Hilbert class field of any imaginary quadratic field.

E. Exercises

5.1. Let $\mathcal{O}_K$ be the algebraic integers in a number field K.

(a) Show that a nonzero ideal $\mathfrak{a}$ of $\mathcal{O}_K$ contains a nonzero integer m. Hint: if $\alpha \neq 0$ is in $\mathfrak{a}$, let $x^n + a_1 x^{n-1} + \cdots + a_n$ be its minimal polynomial. Show that $m = a_n$ is what we want.

(b) Show that $\mathcal{O}_K/\mathfrak{a}$ is finite whenever $\mathfrak{a}$ is a nonzero ideal of $\mathcal{O}_K$. Hint: if m is the integer from (a), consider the surjection $\mathcal{O}_K/m\mathcal{O}_K \to \mathcal{O}_K/\mathfrak{a}$. Use part (ii) of Proposition 5.3 to compute the order of $\mathcal{O}_K/m\mathcal{O}_K$.

(c) Use (b) to show that every nonzero ideal of $\mathcal{O}_K$ is a free $\mathbf{Z}$-module of rank $[K:\mathbf{Q}]$.

(d) If we have ideals $\mathfrak{a}_1 \subset \mathfrak{a}_2 \subset \cdots$, show that there is an integer n such that $\mathfrak{a}_n = \mathfrak{a}_{n+1} = \cdots$. Hint: consider the surjections $\mathcal{O}_K/\mathfrak{a}_1 \to \mathcal{O}_K/\mathfrak{a}_2 \to \cdots$, and use (b).

(e) Use (b) to show that a nonzero prime ideal of $\mathcal{O}_K$ is maximal.

5.2. We will study the elementary properties of fractional ideals in a number field K. Recall that $\mathfrak{a} \subset K$ is a fractional ideal if, under ordinary addition and multiplication, it is a finitely generated $\mathcal{O}_K$-module.

(a) Show that $\mathfrak{a}$ is a fractional ideal if and only if $\mathfrak{a} = \alpha\mathfrak{b}$, where $\alpha \in K$ and $\mathfrak{b}$ is an ideal of $\mathcal{O}_K$. Hint: write each generator of $\mathfrak{a}$ in the form α/β, $\alpha, \beta \in \mathcal{O}_K$. Going the other way, use part (c) of Exercise 5.1 to show that $\alpha\mathfrak{b}$ is a finitely generated $\mathcal{O}_K$-module.

(b) Show that a nonzero fractional ideal is a free $\mathbf{Z}$-module of rank $[K:\mathbf{Q}]$. Hint: use (a) and part (c) of Exercise 5.1.

(c) Show that the product of two fractional ideals is a fractional ideal.

5.3. Let $K \subset L$ be a Galois extension, and let $\mathfrak{p} \subset \mathfrak{P}$ be prime ideals of K and L respectively.

(a) If $\sigma \in \mathrm{Gal}(L/K)$, then prove that $e_{\sigma(\mathfrak{P})|\mathfrak{p}} = e_{\mathfrak{P}|\mathfrak{p}}$ and $f_{\sigma(\mathfrak{P})|\mathfrak{p}} = f_{\mathfrak{P}|\mathfrak{p}}$.

(b) Prove part (ii) of Theorem 5.9.

5.4. Let $K \subset L$ be a Galois extension, and let $\mathfrak{P}$ be prime in L. Then we have the decomposition group $D_{\mathfrak{P}} = \{\sigma \in \mathrm{Gal}(L/K) : \sigma(\mathfrak{P}) = \mathfrak{P}\}$ and the inertia group $I_{\mathfrak{P}} = \{\sigma \in \mathrm{Gal}(L/K) : \sigma(\alpha) \equiv \alpha \bmod \mathfrak{P}$ for all $\alpha \in \mathcal{O}_L\}$.

(a) Show that $I_{\mathfrak{P}} \subset D_{\mathfrak{P}}$.

(b) Show that $\sigma \in D_{\mathfrak{P}}$ induces an automorphism $\tilde{\sigma}$ of $\mathcal{O}_L/\mathfrak{P}$ which is the identity on $\mathcal{O}_K/\mathfrak{p}$, $\mathfrak{p} = \mathfrak{P} \cap \mathcal{O}_K$.

(c) Let $\sigma \in D_{\mathfrak{P}}$. Then show that $\sigma \in I_{\mathfrak{P}}$ if and only if the automorphism $\tilde{\sigma}$ from (b) is the identity.

5.5. In Proposition 5.11, prove that parts (i) and (iii) are consequences of part (ii).

5.6. In this exercise, we will prove part (ii) of Proposition 5.11. Let $\mathfrak{P}$ be a prime of $\mathcal{O}_L$ containing $\mathfrak{p}$, and let $D_{\mathfrak{P}} = \{\sigma \in \mathrm{Gal}(L/K) : \sigma(\mathfrak{P}) = \mathfrak{P}\}$ be the decomposition group. In Proposition 5.10 we observed that the order of $D_{\mathfrak{P}}$ is ef, where $e = e_{\mathfrak{P}|\mathfrak{p}}$ and $f = f_{\mathfrak{P}|\mathfrak{p}}$.

(a) Since $f(x) \equiv f_1(x) \cdots f_g(x) \bmod \mathfrak{p}$, show that $f_i(\alpha) \in \mathfrak{P}$ for some i. We can assume that $f_1(\alpha) \in \mathfrak{P}$.

(b) Using $f = [\mathcal{O}_L/\mathfrak{P} : \mathcal{O}_K/\mathfrak{p}]$, prove that $f \geq \deg(f_1(x))$.

(c) Since $f_1(\sigma(\alpha)) \in \mathfrak{P}$ for all $\sigma \in D_{\mathfrak{P}}$, show that $\deg(f_1(x)) \geq |D_{\mathfrak{P}}| = ef$. Hint: this is where separability is used.

(d) From (b) and (c) conclude that $e = 1$ and $f = \deg(f_1(x))$. Thus $\mathfrak{p}$ is unramified in L.

(e) Show that $\mathfrak{p}\mathcal{O}_L = \mathfrak{P}_1 \cdots \mathfrak{P}_g$ where $\mathfrak{P}_i$ is prime in $\mathcal{O}_L$ and $f_i(\alpha) \in \mathfrak{P}_i$. This shows that all of the $f_i(x)$'s have the same degree.

(f) Show that $\mathfrak{P}_i$ is generated by $\mathfrak{p}$ and $f_i(\alpha)$. Hint: let $I_i = \mathfrak{p}\mathcal{O}_L + f_i(\alpha)\mathcal{O}_K$. Use $I_i \subset \mathfrak{P}_i$ and $I_1 \cdots I_g \subset \mathfrak{p}\mathcal{O}_L$ to show $I_i = \mathfrak{P}_i$.

5.7. In this problem we will determine the integers in the quadratic field $K = \mathbf{Q}(\sqrt{N})$, where N is squarefree. Let $\alpha \mapsto \alpha'$ denote the nontrivial automorphism of K.

(a) Given $\alpha = r + s\sqrt{N} \in K$, define the *trace* and *norm* of α to be

$$T(\alpha) = \alpha + \alpha' = 2r$$
$$N(\alpha) = \alpha\alpha' = r^2 - s^2 N.$$

Then prove that for $\alpha, \beta \in K$,

$$T(\alpha + \beta) = T(\alpha) + T(\beta)$$
$$N(\alpha\beta) = N(\alpha)N(\beta).$$

(b) Given $\alpha \in K$, prove that $\alpha \in \mathcal{O}_K$ if and only if $T(\alpha), N(\alpha) \in \mathbf{Z}$.

(c) Use (b) to prove the description of $\mathcal{O}_K$ given in (5.13).

(d) Prove the description of $\mathcal{O}_K$ given in (5.14).

5.8. Use (5.12) and (5.13) to prove (5.15).

5.9. In this exercise we will study the units in an imaginary quadratic field K. Let $N(\alpha)$ be the norm of $\alpha \in K$ from Exercise 5.7.

(a) Prove that $\alpha \in \mathcal{O}_K$ is a unit if and only if $N(\alpha) = 1$.

(b) Show that $\mathcal{O}_K^* = \{\pm 1\}$ unless $K = \mathbf{Q}(i)$ or $\mathbf{Q}(\omega)$, in which case $\mathcal{O}_K^* = \{\pm 1, \pm i\}$ or $\{\pm 1, \pm\omega, \pm\omega^2\}$ respectively. Hint: use (a) and (5.13). Exercises 4.5 and 4.16 will also be useful.

5.10. Let K be a quadratic field of discriminant d_K, and let the nontrivial automorphism of K be $(a + b\sqrt{d_K})' = a - b\sqrt{d_K}$. We want to complete the description of the prime ideals $\mathfrak{p}$ of $\mathcal{O}_K$ begun in Proposition 5.16. Our basic tools will be Proposition 5.11 and the formula $efg = 2$ from Theorem 5.9.

(a) If $2 \mid d_K$, then show that $2\mathcal{O}_K = \mathfrak{p}^2$, $\mathfrak{p} = \mathfrak{p}'$ prime. Hint: write $d_K = 4N$ and set

$$\mathfrak{p} = \begin{cases} 2\mathcal{O}_K + (1+\sqrt{N})\mathcal{O}_K, & N \text{ odd} \\ 2\mathcal{O}_K + \sqrt{N}\mathcal{O}_K, & N \text{ even.} \end{cases}$$

(b) If $2 \nmid d_K$, then show that

$$d_K \equiv 1 \bmod 8 \iff 2\mathcal{O}_K = \mathfrak{p}\mathfrak{p}',\ \mathfrak{p} \neq \mathfrak{p}' \text{ prime}$$
$$d_K \equiv 5 \bmod 8 \iff 2\mathcal{O}_K \text{ is prime in } \mathcal{O}_K.$$

Hint: apply Proposition 5.11 to $K = \mathbf{Q}(\alpha)$, $\alpha = (1 + \sqrt{d_K})/2$.

(c) Show that the ideals described in parts (i)–(iii) of Proposition 5.16 give all prime ideals of $\mathcal{O}_K$. Hint: use norms to prove that any prime ideal $\mathfrak{p}$ contains a nonzero integer m. Thus $\mathfrak{p} \mid m\mathcal{O}_K$, and we are done by unique factorization.

Notice how these results generalize the descriptions given in Propositions 4.7 and 4.18 of the primes in $\mathbf{Z}[\omega]$ and $\mathbf{Z}[i]$.

5.11. This problem will study the norm of a prime $\mathfrak{p}$ in a number field K. Recall that the norm $N(\mathfrak{p})$ is defined by $N(\mathfrak{p}) = |\mathcal{O}_K/\mathfrak{p}|$. Let p be the unique prime of $\mathbf{Z}$ contained in $\mathfrak{p}$.

(i) Show that $N(\mathfrak{p}) = p^f$, where f is the inertial degree of $\mathfrak{p}$ over p.

(ii) Now assume that $\mathfrak{p}$ is prime in a quadratic field K. Show that

$$p \mid d_K : N(\mathfrak{p}) = p$$

$$p \nmid d_K : N(\mathfrak{p}) = \begin{cases} p, & p \text{ splits completely in } K \\ p^2, & p\mathcal{O}_K \text{ is prime in } \mathcal{O}_K. \end{cases}$$

Hint: use $efg = 2$.

5.12. This exercise is concerned with the Artin symbol $((L/K)/\mathfrak{P})$.

(a) Prove part (i) of Corollary 5.21.

(b) Let $K \subset L$ be a Galois extension and let $\mathfrak{p}$ be a prime of K unramified in L. Prove that the set $\{((L/K)/\mathfrak{P}) : \mathfrak{P} \text{ is a prime of } L \text{ containing } \mathfrak{p}\}$ is a conjugacy class of $\mathrm{Gal}(L/K)$. This conjugacy class is defined to be the Artin symbol $((L/K)/\mathfrak{p})$ of $\mathfrak{p}$.

5.13. Assume that the number field K contains a primitive nth root of unity ζ. In this problem we will discuss a generalization of the Legendre symbol. Let $a \in \mathcal{O}_K$ and let $\mathfrak{p}$ be a prime ideal of $\mathcal{O}_K$ such that $na \notin \mathfrak{p}$.

(a) Prove that $1, \zeta, \ldots, \zeta^{n-1}$ are distinct modulo $\mathfrak{p}$. Hint: show that $x^n - 1$ is separable modulo $\mathfrak{p}$.

(b) Use (a) to prove that $n \mid N(\mathfrak{p}) - 1$.

(c) Show that $a^{(N(\mathfrak{p})-1)/n}$ is congruent to a unique nth root of unity modulo $\mathfrak{p}$. This allows us to define the nth power Legendre symbol $(a/\mathfrak{p})_n$ to be the unique nth root of unity such that

$$a^{(N(\mathfrak{p})-1)/n} \equiv \left(\frac{a}{\mathfrak{p}}\right)_n \bmod \mathfrak{p}.$$

(d) Prove that $(a/\mathfrak{p})_n = 1$ if and only if a is an nth power residue modulo $\mathfrak{p}$.

5.14. Let K, n, a and $\mathfrak{p}$ be as in the previous exercise, and let $L = K(\sqrt[n]{a})$. Note that L is an Abelian extension of K. In this problem we will relate the Legendre symbol $(a/\mathfrak{p})_n$ to the Artin symbol $((L/K)/\mathfrak{p})$.

(a) Show that $\mathfrak{p}$ is unramified in L. Hint: show that $x^n - a$ is separable modulo $\mathfrak{p}$ and use Proposition 5.11.

(b) Generalize the argument of (5.22) to show that

$$\left(\frac{L/K}{\mathfrak{p}}\right)(\sqrt[n]{a}) = \left(\frac{a}{\mathfrak{p}}\right)_n \sqrt[n]{a}.$$

5.15. Suppose that $K \subset M \subset L$ are number fields.

(a) Let $\mathfrak{p}$ be prime in $\mathcal{O}_K$, and assume that $\mathfrak{p} \subset \mathfrak{P} \subset \mathfrak{P}'$, where $\mathfrak{P}$ (resp. $\mathfrak{P}'$) is prime in $\mathcal{O}_M$ (resp. $\mathcal{O}_L$). Then show that $e_{\mathfrak{P}'|\mathfrak{p}} = e_{\mathfrak{P}'|\mathfrak{P}} e_{\mathfrak{P}|\mathfrak{p}}$.

(b) Prove that a prime $\mathfrak{p}$ of $\mathcal{O}_K$ is unramified in L if and only if $\mathfrak{p}$ is unramified in M and every prime of $\mathcal{O}_M$ lying over $\mathfrak{p}$ is unramified in L.

(c) Prove that L is an unramified extension of K if and only if L is unramified over M and M is unramified over K.

5.16. Let $K \subset L$ be an unramified Abelian extension, and assume that $K \subset M \subset L$. By the previous exercise, $K \subset M$ is unramified, and it is clearly Abelian. We thus have Artin maps

$$\left(\frac{L/K}{\cdot}\right) : I_K \longrightarrow \mathrm{Gal}(L/K)$$

$$\left(\frac{M/K}{\cdot}\right) : I_K \longrightarrow \mathrm{Gal}(M/K)$$

and we also have the restriction map $r : \mathrm{Gal}(L/K) \to \mathrm{Gal}(M/K)$. Then use Lemma 5.19 to prove that

$$\left(\frac{M/K}{\cdot}\right) = r \circ \left(\frac{L/K}{\cdot}\right).$$

5.17. Prove Corollary 5.24. Hint: besides Galois theory and Theorems 5.18 and 5.23, you will also need Exercises 5.15 and 5.16.

5.18. If $K \subset M \subset L$, where L and M are Galois over K, then prove that a prime $\mathfrak{p}$ of $\mathcal{O}_K$ splits completely in L if and only if it splits completely in M and some prime of $\mathcal{O}_M$ containing $\mathfrak{p}$ splits completely in L.

5.19. Let K be an imaginary quadratic field, and let $K \subset L$ be a Galois extension. As usual, τ will denote complex conjugation.

(a) Show that L is Galois over $\mathbb{Q}$ if and only if $\tau(L) = L$.

(b) If L is Galois over $\mathbf{Q}$, then prove that

(i) $[L \cap \mathbf{R} : \mathbf{Q}] = [L : K]$.

(ii) For $\alpha \in L \cap \mathbf{R}$, $L \cap \mathbf{R} = \mathbf{Q}(\alpha) \iff L = K(\alpha)$.

5.20. Show that $\mathbf{Z}[(1+\sqrt{-19})/2]$ is a UFD. Hint: every PID is a UFD (see Ireland and Rosen [59, §1.3] or Marcus [77, pp. 255–256]). Thus, by Theorem 5.30, it suffices to show that $h(-19) = 1$.

5.21. In this exercise we will study the ring $\mathbf{Z}[\sqrt{-2}]$.

(a) Use Theorem 5.30 to show that $\mathbf{Z}[\sqrt{-2}]$ is a UFD.

(b) Show that $\sqrt{-2}$ is a prime in $\mathbf{Z}[\sqrt{-2}]$.

(c) If $ab = u^3$ in $\mathbf{Z}[\sqrt{-2}]$ and a and b are relatively prime, then prove that a and b are cubes in $\mathbf{Z}[\sqrt{-2}]$.

5.22. We can now give a second proof of Fermat's theorem that $(x,y) = (3,\pm 5)$ are the only integer solutions of the equation $x^3 = y^2 + 2$.

(a) If $x^3 = y^2 + 2$, show that $y + \sqrt{-2}$ and $y - \sqrt{-2}$ are relatively prime in $\mathbf{Z}[\sqrt{-2}]$. Hint: use part (b) of Exercise 5.21.

(b) Use part (c) of Exercise 5.21 to show that $(x,y) = (3,\pm 5)$.

This argument is due to Euler [33, Vol. I, Chapter XII, §§191–193], though he assumed (without proof) that Exercise 5.21 was true.

5.23. If $D \equiv 1 \bmod 4$ is negative and squarefree, prove a version of Theorems 5.1 and 5.26 for primes of the form $x^2 + xy + ((1-D)/4)y^2$.

5.24. Prove that the discriminant of $x^4 + bx^2 + c$ equals $2^4c(b^2 - 4c)^2$. Hint: write down the roots explicitly.

5.25. Let $K = \mathbf{Q}(\sqrt{-17})$.

(a) Show that $C(\mathcal{O}_K) \simeq \mathbf{Z}/4\mathbf{Z}$.

(b) Show that the Hilbert class field of K is $L = K(\alpha)$, where $\alpha = \sqrt{(1+\sqrt{17})/2}$. Hint: use the methods of Proposition 5.31. The only tricky part concerns primes of $K(\sqrt{17})$ which contain 2. Setting $u = (1+\sqrt{17})/2$ and $u' = (1-\sqrt{17})/2$, note that u and u' satisfy $x = x^2 - 4$.

5.26. Prove an analog of Theorem 5.33 for primes of the form $x^2 + 17y^2$.

§6. THE HILBERT CLASS FIELD AND GENUS THEORY

In Chapter One we studied the genus theory of primitive positive definite quadratic forms, and our main result (Theorem 3.15) was that for a fixed discriminant D:

(i) There are $2^{\mu-1}$ genera, where μ is the number defined in Proposition 3.11.

(ii) The principal genus consists of squares of classes.

In this section, we will use Artin reciprocity for the Hilbert class field of an imaginary quadratic field K to prove (i) and (ii) when D is the discriminant d_K of K. This result is less general than what we proved in §3, but the proof is such a nice application of the Hilbert class field that we couldn't resist including it. Readers more interested in $p = x^2 + ny^2$ may skip to §7 without loss of continuity.

The key to the class field theory interpretation of genus theory is the concept of the *genus field*. Given an imaginary quadratic field K of discriminant d_K, Theorem 5.30 tells us that the form class group $C(d_K)$ is isomorphic to the ideal class group $C(\mathcal{O}_K)$. The principal genus is a subgroup of $C(d_K)$ and hence maps to a subgroup of $C(\mathcal{O}_K)$. By Corollary 5.24, this subgroup determines an unramified Abelian extension of K which is called the *genus field* of K. Theorem 6.1 below will describe the genus field explicitly and show that the characters used in Gauss' definition of genus appear in the Artin map of the genus field. This will take a fair amount of work, but once done, (i) and (ii) above will follow easily by Artin Reciprocity. We will then discuss how the genus field can help in the harder problem of determining the Hilbert class field.

A. Genus Theory for Field Discriminants

Here is the main result of this section:

Theorem 6.1. *Let K be an imaginary quadratic field of discriminant d_K. Let μ be the number of primes dividing d_K, and let $p_1, \ldots, p_r$ be the odd primes dividing d_K (so that $\mu = r$ or $r+1$ according to whether $d_K \equiv 0$ or $1 \bmod 4$). Set $p_i^* = (-1)^{(p_i-1)/2} p_i$. Then:*

(i) *The genus field of K is the maximal unramified extension of K which is an Abelian extension of $\mathbb{Q}$.*

(ii) *The genus field of K is $K(\sqrt{p_1^*}, \ldots, \sqrt{p_r^*})$.*

(iii) *The number of genera of primitive positive definite forms of discriminant d_K is $2^{\mu-1}$.*

(iv) *The principal genus of primitive positive definite forms of discriminant d_K consists of squares of classes.*

Proof. First, note that for field discriminants d_K, the number μ defined in the statement of the theorem agrees with the one defined in Proposition 3.11 (see Exercise 6.1). Note also that (iii) and (iv) of the theorem are the facts about genus theory that we want to prove.

To start the proof, let L be the Hilbert class field of K, and let M be the unramified Abelian extension of K corresponding to the subgroup $C(\mathcal{O}_K)^2 \subset C(\mathcal{O}_K)$ via Corollary 5.24. We claim that

(6.2) $\quad M$ is the maximal unramified extension of K Abelian over $\mathbb{Q}$.

To prove this, consider an unramified extension $\tilde{M}$ of K which is Abelian over $\mathbb{Q}$. Then $\tilde{M}$ is also Abelian over K, so that $M \subset L$, and we thus have the following diagram of fields:

$$\begin{array}{c} L \\ | \\ \tilde{M} \\ | \\ K \\ | \\ \mathbb{Q} \end{array} \quad \text{Abelian} \tag{6.3}$$

We want the maximal such $\tilde{M}$. Since L is Galois over $\mathbb{Q}$ (see Lemma 5.28), we can interpret (6.3) via Galois theory. Let $G = \mathrm{Gal}(L/\mathbb{Q})$. Then $\tilde{M}$ being Abelian over $\mathbb{Q}$ is equivalent to $[G,G] \subset \mathrm{Gal}(L/\tilde{M})$, where $[G,G]$ is the commutator subgroup of G (see Exercise 6.2). Note also that $[G,G] \subset \mathrm{Gal}(L/K)$ since the latter has index two in G. Thus $\tilde{M}$ satisfies (6.3) if and only if

$$[G,G] \subset \mathrm{Gal}(L/\tilde{M}) \subset \mathrm{Gal}(L/K).$$

It follows by Galois theory that the maximal unramified extension of K Abelian over $\mathbb{Q}$ is the one that corresponds to $[G,G]$. By Theorem 5.23, $\mathrm{Gal}(L/K)$ can be identified with $C(\mathcal{O}_K)$ via the Artin map. If we can show that $[G,G] \subset \mathrm{Gal}(L/K)$ maps to $C(\mathcal{O}_K)^2 \subset C(\mathcal{O}_K)$, then (6.2) will follow.

We first compute $G = \mathrm{Gal}(L/\mathbb{Q})$. We have a short exact sequence

$$1 \longrightarrow \mathrm{Gal}(L/K) \longrightarrow G \longrightarrow \mathrm{Gal}(K/\mathbb{Q}) \longrightarrow 1$$

which splits because complex conjugation τ is in G by Lemma 5.28. Thus G is the semidirect product $\mathrm{Gal}(L/K) \rtimes (\mathbb{Z}/2\mathbb{Z})$, where $\mathbb{Z}/2\mathbb{Z}$ acts by conjugation by τ.

Under the isomorphism $\mathrm{Gal}(L/K) \simeq C(\mathcal{O}_K)$, conjugation by τ operates on $C(\mathcal{O}_K)$ by sending an ideal to its conjugate. To see this, let $\mathfrak{p}$ be a prime ideal of $\mathcal{O}_K$. Then the uniqueness part of Lemma 5.19 shows that

$$\tau\left(\frac{L/K}{\mathfrak{p}}\right)\tau^{-1} = \left(\frac{L/K}{\tau(\mathfrak{p})}\right) \tag{6.4}$$

(see Exercise 6.3), and our claim follows. However, for any ideal $\mathfrak{a}$ of $\mathcal{O}_K$, we will prove in Lemma 7.14 that the product $\mathfrak{a}\bar{\mathfrak{a}}$ is always a principal ideal, and it follows that the class of $\bar{\mathfrak{a}}$ is the inverse of the class of $\mathfrak{a}$ in $C(\mathcal{O}_K)$. Hence G may be identified with the semidirect product $C(\mathcal{O}_K) \rtimes (\mathbf{Z}/2\mathbf{Z})$, where $\mathbf{Z}/2\mathbf{Z}$ acts by sending an element of $C(\mathcal{O}_K)$ to its inverse.

It is now easy to show that $[G,G] = C(\mathcal{O}_K)^2$. First, note that $C(\mathcal{O}_K)^2$ is normal in G (any subgroup of $C(\mathcal{O}_K)$ is, which has unexpected consequences—see Exercise 6.4), and since $\mathbf{Z}/2\mathbf{Z}$ acts trivially on $C(\mathcal{O}_K)/C(\mathcal{O}_K)^2$ (every element is its own inverse), we have

$$(6.5) \qquad \begin{aligned} G/C(\mathcal{O}_K)^2 &\simeq (C(\mathcal{O}_K) \rtimes (\mathbf{Z}/2\mathbf{Z}))/C(\mathcal{O}_K)^2 \\ &\simeq (C(\mathcal{O}_K)/C(\mathcal{O}_K)^2) \times (\mathbf{Z}/2\mathbf{Z}), \end{aligned}$$

so that $G/C(\mathcal{O}_K)^2$ is Abelian (see Exercise 6.5). It follows that $[G,G] \subset C(\mathcal{O}_K)^2$. To prove the opposite inclusion, note that for any $a \in C(\mathcal{O}_K)$, we have $(a,1) \in C(\mathcal{O}_K) \rtimes (\mathbf{Z}/2\mathbf{Z})$, and then

$$(a,1)(1,\tau)(a,1)^{-1}(1,\tau)^{-1} = (a^2,1),$$

where τ is the nontrivial element of $\mathbf{Z}/2\mathbf{Z}$. This proves that $[G,G] = C(\mathcal{O}_K)^2$, and (6.2) is proved.

We will next show that

$$(6.6) \qquad M = K(\sqrt{p_1^*},\ldots,\sqrt{p_r^*}),$$

where p_i^*'s are as in the statement of the theorem. We begin with two preliminary lemmas. The first concerns some general facts about ramification and the Artin symbol:

Lemma 6.7. *Let L and M be Abelian extensions of a number field K, and let $\mathfrak{p}$ be prime in $\mathcal{O}_K$.*

(i) *$\mathfrak{p}$ is unramified in LM if and only if it is unramified in both L and M.*

(ii) *If $\mathfrak{p}$ is unramified in LM, then under the natural injection*

$$\mathrm{Gal}(LM/K) \longrightarrow \mathrm{Gal}(L/K) \times \mathrm{Gal}(M/K),$$

the Artin symbol $((LM/K)/\mathfrak{p})$ maps to $(((L/K)/\mathfrak{p}),((M/K)/\mathfrak{p}))$.

Proof. See Exercise 6.6, or, for a more general version of these facts, Marcus [77, Exercises 10–11, pp. 117–118]. Q.E.D.

The second lemma tells us when a quadratic extension $K \subset K(\sqrt{a})$, $a \in \mathbf{Z}$, is unramified:

Lemma 6.8. *Let K be a quadratic field of discriminant d_K, and let $K(\sqrt{a})$ be a quadratic extension where $a \in \mathbf{Z}$. Then $K \subset K(\sqrt{a})$ is unramified if and only if a can be chosen so that $a \mid d_K$ and $a \equiv 1 \bmod 4$.*

Proof. For the most part, the proof is a straightforward application of the techniques used in the proof of Proposition 5.31. See Exercises 6.7, 6.8 and 6.9 for the details. Q.E.D.

We can now prove (6.6). Let $M^* = K(\sqrt{p_1^*}, \ldots, \sqrt{p_r^*})$. Since p_i^* divides d_K and satisfies $p_i^* \equiv 1 \bmod 4$, $K \subset K(\sqrt{p_i^*})$ is unramified by Lemma 6.8, and consequently $K \subset M^*$ is unramified by Lemma 6.7. But $M^* = \mathbf{Q}(\sqrt{d_K}, \sqrt{p_1^*}, \ldots, \sqrt{p_r^*})$ is clearly Abelian over $\mathbf{Q}$, so that $M^* \subset M$ by the maximality of M.

To prove the opposite inclusion, we first study $\mathrm{Gal}(M/\mathbf{Q})$. Since $\mathbf{Q} \subset M \subset L$ corresponds to $G \supset C_K^2 \supset \{1\}$ under the Galois correspondence, we have

$$\mathrm{Gal}(M/\mathbf{Q}) \simeq \mathrm{Gal}(L/\mathbf{Q})/\mathrm{Gal}(L/M) = G/C(\mathcal{O}_K)^2,$$

so that by (6.5), $\mathrm{Gal}(M/\mathbf{Q}) \simeq (\mathbf{Z}/2\mathbf{Z})^m$ for some m. Then Galois theory shows that $M = \mathbf{Q}(\sqrt{a_1}, \ldots, \sqrt{a_m})$ where $a_1, \ldots, a_m \in \mathbf{Z}$ (see Exercise 6.10). Thus M is the compositum of quadratic extensions $K \subset K(\sqrt{a_i})$, $a_i \in \mathbf{Z}$, and by Lemma 6.7, each of these is unramified.

It thus suffices to show that M^* contains all unramified extensions $K \subset K(\sqrt{a})$, $a \in \mathbf{Z}$. By Lemma 6.8, we may assume that $a \equiv 1 \bmod 4$ and that $a \mid d_K$. It follows that a must be of the form $p_{i_1}^* \cdots p_{i_s}^*$, $1 \le i_1 < \cdots < i_s \le r$, so that $K(\sqrt{a})$ is clearly contained in M^*. This completes the proof of (6.6).

We will next show that $[M:\mathbf{Q}] = 2^\mu$. Note that $M = \mathbf{Q}(\sqrt{d_K}, \sqrt{p_1^*}, \ldots, \sqrt{p_r^*})$. When $d_K \equiv 1 \bmod 4$, we have $d_K = p_1^* \cdots p_r^*$, so that $[M:\mathbf{Q}] = 2^r = 2^\mu$ since $\mu = r$ in this case. When $d_K \equiv 0 \bmod 4$, we can write $d_K = -4n$, $n > 0$, and then we have

$$M = \begin{cases} \mathbf{Q}(i, \sqrt{p_1^*}, \ldots, \sqrt{p_r^*}), & n \equiv 3 \bmod 4 \\ \mathbf{Q}(\sqrt{2}, \sqrt{p_1^*}, \ldots, \sqrt{p_r^*}), & n \equiv 6 \bmod 8 \\ \mathbf{Q}(\sqrt{-2}, \sqrt{p_1^*}, \ldots, \sqrt{p_r^*}), & n \equiv 2 \bmod 8 \end{cases} \tag{6.9}$$

(see Exercise 6.11). Thus $[M:\mathbf{Q}] = 2^{r+1} = 2^\mu$. Since $[C(\mathcal{O}_K): C(\mathcal{O}_K)^2]$ equals half of $[G: C(\mathcal{O}_K)^2] = [M:\mathbf{Q}] = 2^\mu$, we have proved that

$$[C(\mathcal{O}_K): C(\mathcal{O}_K)^2] = 2^{\mu-1}. \tag{6.10}$$

We can now compute the Artin map $((M/K)/\cdot): I_K \to \mathrm{Gal}(M/K)$. If we set $K_i = K(\sqrt{p_i^*})$, then M is the compositum $K_1 \cdots K_r$, and we have a natural injection

$$\text{(6.11)} \qquad \mathrm{Gal}(M/K) \longrightarrow \prod_{i=1}^{r} \mathrm{Gal}(K_i/K).$$

Furthermore, we may identify $\mathrm{Gal}(K_i/K)$ with $\{\pm 1\}$, so that composing the Artin map with (6.11) gives us a homomorphism

$$\Phi_K : I_K \longrightarrow \{\pm 1\}^r.$$

We claim that if $\mathfrak{a}$ is an ideal of $\mathcal{O}_K$ prime to $2d_K$, then $\Phi_K(\mathfrak{a})$ can be computed in terms of Legendre symbols as follows:

$$\text{(6.12)} \qquad \Phi_K(\mathfrak{a}) = \left(\left(\frac{N(\mathfrak{a})}{p_1} \right), \dots, \left(\frac{N(\mathfrak{a})}{p_r} \right) \right),$$

where $N(\mathfrak{a}) = |\mathcal{O}_K/\mathfrak{a}|$ is the norm of $\mathfrak{a}$.

To prove (6.12), we will need one basic fact about norms: if $\mathfrak{a}$ and $\mathfrak{b}$ are ideals of $\mathcal{O}_K$, then $N(\mathfrak{a}\mathfrak{b}) = N(\mathfrak{a})N(\mathfrak{b})$ (see Lemma 7.14 or Marcus [77, Theorem 22]). It follows that both sides of (6.12) are multiplicative in $\mathfrak{a}$, so that we may assume that $\mathfrak{a}$ is a prime ideal $\mathfrak{p}$ of $\mathcal{O}_K$. Then Lemma 6.7, applied to (6.11), shows that $((M/K)/\mathfrak{p})$ maps to the r-tuple

$$\left(\left(\frac{K_1/K}{\mathfrak{p}} \right), \dots, \left(\frac{K_r/K}{\mathfrak{p}} \right) \right).$$

If we can show that

$$\text{(6.13)} \qquad \left(\frac{K_i/K}{\mathfrak{p}} \right)(\sqrt{p_i^*}) = \left(\frac{N(\mathfrak{p})}{p_i} \right) \sqrt{p_i^*},$$

then (6.12) will follow immediately.

To prove (6.13), let $\mathfrak{P}$ be a prime of $\mathcal{O}_{K_i}$ containing $\mathfrak{p}$, and set $\sigma = ((K_i/K)/\mathfrak{p})$. By Lemma 5.19 we see that

$$\text{(6.14)} \qquad \sigma\left(\sqrt{p_i^*}\right) \equiv \left(\sqrt{p_i^*}\right)^{N(\mathfrak{p})} \equiv (p_i^*)^{(N(\mathfrak{p})-1)/2}\sqrt{p_i^*} \bmod \mathfrak{P}.$$

Since K is a quadratic field, it follows that $N(\mathfrak{p}) = p$ or p^2 (see Exercise 5.11), and thus here are two cases to consider.

If $N(\mathfrak{p}) = p$, then we know that

$$(p_i^*)^{(p-1)/2} \equiv \left(\frac{p_i^*}{p} \right) \bmod p.$$

Since $p \in \mathfrak{P}$ and $(p_i^*/p) = (p/p_i)$ by quadratic reciprocity, we see that (6.14) reduces to

$$\sigma\left(\sqrt{p_i^*}\right) \equiv \left(\frac{p}{p_i} \right) \sqrt{p_i^*} \equiv \left(\frac{N(\mathfrak{p})}{p_i} \right) \sqrt{p_i^*} \bmod \mathfrak{P},$$

and we are done. If $N(\mathfrak{p}) = p^2$, then by Fermat's Little Theorem,

$$(p_i^*)^{(p^2-1)/2} \equiv \left((p_i^*)^{(p+1)/2}\right)^{p-1} \equiv 1 \bmod p,$$

so that (6.14) becomes

$$\sigma\left(\sqrt{p_i^*}\right) \equiv \sqrt{p_i^*} \equiv \left(\frac{N(\mathfrak{p})}{p_i}\right)\sqrt{p_i^*} \bmod \mathfrak{P},$$

and (6.13) is proved. This proves (6.12).

For the rest of the proof, we will assume that $d_K = -4n$, $n > 0$ (see Exercise 6.12 for the case $d_K \equiv 1 \bmod 4$). Here, it is easily checked that the map (6.11) is an isomorphism, and then Artin reciprocity (Corollary 5.24) for $K \subset M$ means that the map $\Phi_K : I_K \to \{\pm 1\}^r$ of (6.12) induces an isomorphism

$$A : C(\mathcal{O}_K)/C(\mathcal{O}_K)^2 \xrightarrow{\sim} \{\pm 1\}^r,$$

where the A stands for Artin.

It's now time to bring in quadratic forms. Let $C(d_K)$ be the class group of primitive positive definite forms of discriminant $d_K = -4n$, and let P be the principal genus. Recall from the proof of Theorem 3.15 that we have the $\mu = r+1$ assigned characters $\chi_0, \chi_1, \ldots, \chi_r$, where χ_0 is one of δ, ϵ or $\delta\epsilon$, and $\chi_i(a) = (a/p_i)$ for $i = 1, \ldots, r$. In Lemma 3.20, we proved that if $f(x,y)$ represents a number a prime to $4n$, then the genus of $f(x,y)$ is determined by the $(r+1)$-tuple $(\chi_0(a), \chi_1(a), \ldots, \chi_r(a))$. Thus we have an injective map

$$G : C(d_K)/P \longrightarrow \{\pm 1\}^{r+1},$$

where the G stands for Gauss.

To relate the two maps A and G, we will use the isomorphism $C(d_K) \simeq C(\mathcal{O}_K)$ of Theorem 5.30. Since $C(d_K)^2 \subset P$, we get the following diagram:

$$\begin{array}{ccccc} C(d_K)/C(d_K)^2 & \xrightarrow{\alpha} & C(d_K)/P & \xrightarrow{G} & \{\pm 1\}^{r+1} \\ \downarrow \wr & & & & \downarrow \pi \\ C(\mathcal{O}_K)/C(\mathcal{O}_K)^2 & & \xrightarrow{A} & & \{\pm 1\}^r \end{array} \tag{6.15}$$

where $\alpha : C(d_K)/C(d_K)^2 \to C(d_K)/P$ is the natural surjection and π is the projection onto the last r factors.

We claim that this diagram commutes, which means that Gauss's definition of genus is amazingly close to the Artin map of the genus field. (The full story of the relation is worked out in Exercise 6.13.)

To prove that (6.15) commutes, let $f(x,y) = ax^2 + 2bxy + cy^2$ be a form of discriminant $-4n$. We can assume that a is relatively prime to $4n$. Then, in (6.15), if we first go across and then down, we see that the class of $f(x,y)$ maps to

$$(6.16)\qquad (\chi_1(a),\ldots,\chi_r(a)) = \left(\left(\frac{a}{p_1}\right),\ldots,\left(\frac{a}{p_r}\right)\right).$$

Let's see what happens when we go the other way. By Theorem 5.30, $f(x,y)$ corresponds to the ideal $\mathfrak{a} = [a, b+\sqrt{-n}]$ of $\mathcal{O}_K$. However, it is easy to see that the natural map

$$(6.17)\qquad \mathbf{Z}/a\mathbf{Z} \longrightarrow \mathcal{O}_K/\mathfrak{a}$$

is an isomorphism (see Exercise 6.14). Thus $\mathfrak{a}$ has norm $N(\mathfrak{a}) = a$, and our description of the Artin map from (6.12) shows that $f(x,y)$ maps to

$$(6.18)\qquad \left(\left(\frac{N(\mathfrak{a})}{p_1}\right),\ldots,\left(\frac{N(\mathfrak{a})}{p_r}\right)\right) = \left(\left(\frac{a}{p_1}\right),\ldots,\left(\frac{a}{p_r}\right)\right).$$

Comparing (6.16) and (6.18), we see that (6.15) commutes as claimed.

Now everything is easy to prove. If we go down and across in (6.15), the resulting map is injective. By commutivity, it follows that $\alpha : C(d_K)/C(d_K)^2 \to C(d_K)/P$ must be injective, which proves that $C(d_K)^2 = P$, and part (iv) of the theorem is done. The number of genera is thus $[C(d_K): P] = [C(d_K): C(d_K)^2] = [C(\mathcal{O}_K): C(\mathcal{O}_K)^2] = 2^{\mu-1}$ (the last equality is (6.10)), and (iii) follows. Finally, since $P = C(d_K)^2$ corresponds to $C(\mathcal{O}_K)^2$, we see that M is the genus field of K, and then (i) and (ii) follow from (6.2) and (6.6). Theorem 6.1 is proved. Q.E.D.

B. Applications to the Hilbert Class Field

Theorem 6.1 makes it easy to compute the genus field. So let's see if the genus field can help us find the Hilbert class field, which in general is more difficult to compute. The nicest case is when the genus field *equals* the Hilbert class field, which happens for field discriminants where every genus consists of a single class (see Exercise 6.15). In particular, if $d_K = -4n$, then this means that n is one of Euler's convenient numbers (see Proposition 3.24). Of the 65 convenient numbers on Gauss' list in §3, 35 satisfy the additional condition that $d_K = -4n$ (see Exercise 6.15), so that we can determine lots of Hilbert class fields. For example, when $K = \mathbf{Q}(\sqrt{-5})$, Theorem 6.1 tells us that the Hilbert class field is $K(\sqrt{5}) = K(i)$. Other examples are just as easy to work out (see Exercise 6.16).

The more typical situation is when the Hilbert class field is strictly bigger than the genus field. It turns out that the genus field can still provide us with useful information about the Hilbert class field. Let's consider the case

$K = \mathbf{Q}(\sqrt{-14})$. Here, the genus field is $M = K(\sqrt{-7}) = K(\sqrt{2})$ by Theorem 6.1. Since the class number is 4, we know that the Hilbert class field is a quadratic extension of M, so that $L = M(\sqrt{u})$ for some $u \in M$. This is already useful information, but we can do better. In Theorem 5.1, we saw the importance of a *real* primitive element of the Hilbert class field. So let's intersect everything with the real numbers. This gives us the quadratic extension $M \cap \mathbf{R} \subset L \cap \mathbf{R}$. Since $M = K(\sqrt{2}) = \mathbf{Q}(\sqrt{-14}, \sqrt{2})$, it follows that $M \cap \mathbf{R} = \mathbf{Q}(\sqrt{2})$. Thus we can write $L \cap \mathbf{R} = \mathbf{Q}(\sqrt{2}, \sqrt{u})$, where $u > 0$ is in $\mathbf{Q}(\sqrt{2})$, and from this it is easy to prove that

$$L = K(\sqrt{u}), \qquad u = a + b\sqrt{2} > 0, \quad a, b \in \mathbf{Z}$$

(see Exercise 6.17). Hence genus theory explains the form of the primitive element $\alpha = \sqrt{2\sqrt{2} - 1}$ of Proposition 5.31. In Exercise 6.18, we will continue this discussion and show how one can take $u = a + b\sqrt{2}$ and discover the precise form $u = 2\sqrt{2} - 1$ of the primitive element of the Hilbert class field.

It's interesting to compare this discussion of $x^2 + 14y^2$ to what we did in §3. The genus theory developed in §3 told us when p was represented by $x^2 + 14y^2$ or $2x^2 + 7y^2$, but this partial information didn't help in deciding when $p = x^2 + 14y^2$. In contrast, the genus theory of Theorem 6.1 determines the genus field, which helps us understand the Hilbert class field. The field-theoretic approach seems to have more useful information.

This ends our discussion of genus theory, but it by no means exhausts the topic. For more complete treatments of genus theory from the point of view of class field theory, see Hasse [51], Janusz [62, §VI.3] and Cohn's two books [19, Chapters 14 and 18] and [21, Chapter 8]. Genus theory can also be studied by standard methods of algebraic number theory, with no reference to class field theory. Both Cohn [20, Chapter XIII] and Hasse [50, §§26.8 and 29.3] use the Hilbert symbol in their discussion of genera. For a more elementary approach, see Zagier [111]. Genus theory can also be generalized in several ways. It is possible to define the genus field of an arbitrary number field (see Ishida [60]), and in another direction, one can formulate genus theory from the point of view of algebraic groups and Tamagawa numbers (Ono [82] has a nice introduction to this subject). For a survey of all these aspects of genus theory, see Frei [39].

C. Exercises

6.1. Let d_K be the discriminant of a quadratic field. When considering forms of discriminant d_K, show that the number μ from Proposition 3.11 is just the number of primes dividing d_K.

6.2. Suppose that we have fields $K \subset M \subset L$, where L is Galois over K with group $G = \mathrm{Gal}(L/K)$. Prove that M is Abelian over K if and only if $[G,G] \subset \mathrm{Gal}(L/M)$.

6.3. Prove statement (6.4).

6.4. If K is an imaginary quadratic field and M is an unramified Abelian extension of K, then prove that M is Galois over $\mathbf{Q}$. Hint: use the description of $\mathrm{Gal}(L/\mathbf{Q})$, where L is the Hilbert class field of K.

6.5. Prove statement (6.5).

6.6. In this problem we will prove Lemma 6.7. Let $\mathfrak{p}$ be a prime of $\mathcal{O}_K$.

(a) If $\mathfrak{p}$ is unramified in LM, then use Exercise 5.15 to show that it's unramified in both L and M.

(b) Prove the converse of (a). Hint: assume not. Then use the facts about the decomposition group from Proposition 5.10 to find $\sigma \in \mathrm{Gal}(LM/K)$, $\sigma \neq 1$, such that $\sigma(\alpha) \equiv \alpha \bmod \mathfrak{P}$ for all $\alpha \in \mathcal{O}_{LM}$ (and $\mathfrak{P}$ is a prime of $\mathcal{O}_{LM}$ containing $\mathfrak{p}$). Argue that $\sigma_{|L}$ and $\sigma_{|M}$ are the identity. Note that (a) and (b) prove part (i) of Lemma 6.7.

(c) Use Exercise 5.16 to prove part (ii) of Lemma 6.7.

(d) With the same hypothesis as Lemma 6.7, show that $\mathfrak{p}$ splits completely in LM if and only if it splits completely in both L and M. In Exercise 8.14 we will see that this result can be proved without assuming that L and M are Galois over K.

6.7. Let $K = \mathbf{Q}(i, \sqrt{2m})$, where $m \in \mathbf{Z}$ is odd and squarefree.

(a) Let $\alpha = (1+i)\sqrt{2m}/2$. Show that $\alpha^2 = im$, and conclude that $\alpha \in \mathcal{O}_K$. (It turns out that 1, i, $\sqrt{2m}$ and α form an integral basis of $\mathcal{O}_K$—see Marcus [77, Exercise 42 to Chapter 2].)

(b) Let $\mathfrak{P}$ be the ideal of $\mathcal{O}_K$ generated by $1+i$ and $1+\alpha$. Show that $2\mathcal{O}_K = \mathfrak{P}^4$, and conclude that $\mathfrak{P}$ is prime. Hint: compute $\mathfrak{P}^2$.

6.8. Let K be a quadratic field. We want to show that if $K \subset K(i)$ is unramified, then $d_K \equiv 12 \bmod 16$.

(a) Show that $K \subset K(i)$ is ramified when $d_K \equiv 1 \bmod 4$. Hint: consider the diagram of fields

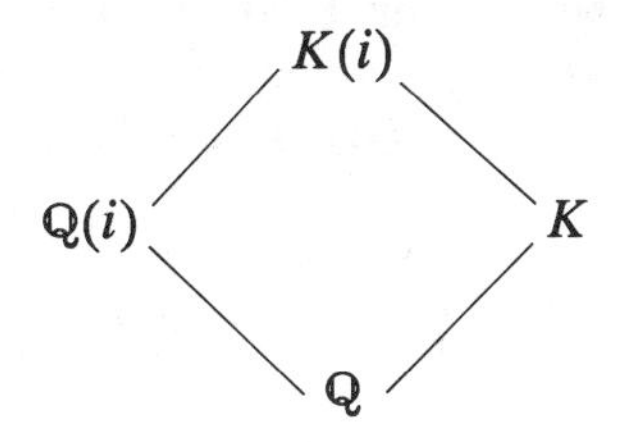

If $K \subset K(i)$ is unramified, show that 2 is unramified in $K(i)$. But 2 ramifies in $\mathbb{Q}(i)$. Exercise 5.15 will be useful.

(b) Show that the extension is ramified when $d_K \equiv 0 \bmod 8$. Hint: if it's unramified, show that the ramification index of 2 in $K(i)$ is at most 2. Then use Exercise 6.7.

Since an even discriminant is of the form $4N$, where $N \equiv 2,3 \bmod 4$, it follows from (a) and (b) that $d_K \equiv 12 \bmod 16$ when $K \subset K(i)$ is unramified.

6.9. In this exercise we will prove Lemma 6.8.

(a) Prove that $K \subset K(\sqrt{a})$ is unramified when $a \mid d_K$ and $a \equiv 1 \bmod 4$. Hint: when $2 \notin \mathfrak{p}$, note that $d_K = ab$, where $K(\sqrt{a}) = K(\sqrt{b})$.

(b) Assume that $K \subset K(\sqrt{a})$ is unramified. Show that $a \mid d_K$ and consequently that a may be chosen to be odd. Hint: if p is a prime such that $p \mid a$, $p \nmid d_K$, then analyze p in the fields

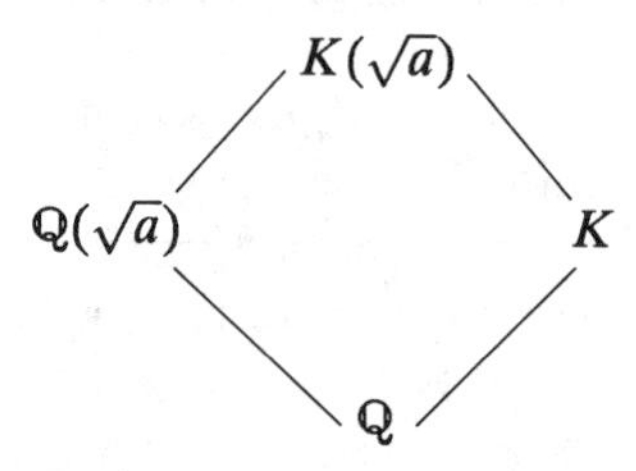

(c) Let $K \subset K(\sqrt{a})$ be unramified, where $a \mid d_K$ is odd.

(i) If $a \equiv 3 \bmod 4$, show that $d_K \equiv 12 \bmod 16$. Hint: apply (b) to $-a$, and then use Exercise 6.7.

(ii) If $d_K \equiv 12 \bmod 16$, show that $K(\sqrt{a}) = K(\sqrt{b})$, where $b \mid d_K$ and $b \equiv 1 \bmod 4$. Hint: factor d_K.

Lemma 6.8 follows easily from (a)–(c).

6.10. If M is a Galois extension of $\mathbb{Q}$ and $\mathrm{Gal}(M/\mathbb{Q}) \simeq (\mathbb{Z}/2\mathbb{Z})^m$, then show that $M = \mathbb{Q}(\sqrt{a_1}, \ldots, \sqrt{a_m})$, $a_i \in \mathbb{Z}$ squarefree.

6.11. Prove the description of the genus field M given in (6.9).

6.12. Complete the proof of Theorem 6.1 when $d_K \equiv 1 \bmod 4$, $d_K < 0$.

6.13. Let K be an imaginary quadratic field of discriminant $-4n$. The description of the genus field M given in (6.9) gives us an isomorphism

$$\mathrm{Gal}(M/\mathbb{Q}) \xrightarrow{\sim} \{\pm 1\}^{\mu}.$$

However, we also have maps

$$C(-4n) \longrightarrow C(\mathcal{O}_K) \longrightarrow \mathrm{Gal}(M/K).$$

If we combine these with the natural inclusion $\mathrm{Gal}(M/K) \subset \mathrm{Gal}(M/\mathbb{Q})$, then we get a map

$$C(-4n) \longrightarrow \{\pm 1\}^{\mu}.$$

Show that this map is exactly what Gauss used in his definition of genus. Hint: it's fun to see the characters ϵ, δ and $\epsilon\delta$ from §3 reappear. For example, when n is odd, the key step is to show that

$$\left(\frac{M/K}{\mathfrak{a}}\right)(i) = \epsilon(N(\mathfrak{a}))i$$

for ideals $\mathfrak{a}$ prime to $4n$. The proof is similar to the proof of (6.13).

6.14. Prove that the map (6.17) is an isomorphism.

6.15. In this exercise we will study when the genus field equals the Hilbert class field.

(a) Prove that the genus field of an imaginary quadratic field K equals its Hilbert class field if and only if for primitive positive forms of discriminant d_K, there is only one class per genus.

(b) Of Gauss' list of 65 convenient numbers n in §3, which satisfy the condition (5.2) that guarantees that $-4n$ is a field discriminant? This gives us a list of fields where we know the Hilbert class field.

6.16. Compute the Hilbert class fields of the fields $\mathbb{Q}(\sqrt{-6})$, $\mathbb{Q}(\sqrt{-10})$ and $\mathbb{Q}(\sqrt{-35})$.

6.17. Let $K = \mathbb{Q}(\sqrt{-14})$, and let L be the Hilbert class field of K. The genus field M of K is $K(\sqrt{-7}) = K(\sqrt{2})$, so that L is a degree 2 extension of M. Use the hints in the text to show that $L = K(\sqrt{u})$, where $u = a + b\sqrt{2} > 0$, $a, b \in \mathbb{Z}$.

6.18. In this exercise we will discover a primitive element for the Hilbert class field L of $K = \mathbb{Q}(\sqrt{-14})$. From the previous exercise, we know that $L = K(\sqrt{u})$, where $u = a + b\sqrt{2} > 0$, $a, b \in \mathbb{Z}$. Let $u' = a - b\sqrt{2}$.

(a) Show that $\mathrm{Gal}(L/\mathbb{Q})$ is the dihedral group $\langle \sigma, \tau : \sigma^4 = 1,\ \tau^2 = 1,\ \sigma\tau = \tau\sigma^3 \rangle$ of order 8, where $\sigma(\sqrt{u}) = \sqrt{u'}$ and τ is complex conjugation. Conclude that $\sigma^2(\sqrt{u}) = -\sqrt{u}$ and $\tau(\sqrt{u'}) = -\sqrt{u'}$.

(b) Show that $\mathbb{Q}(\sqrt{-7})$ is the fixed field of σ^2 and τ.

(c) Show that $\sqrt{uu'}$ is fixed by σ^2 and $\sigma\tau$, and then using τ, conclude that $\sqrt{uu'} = m\sqrt{-7}$, $m \in \mathbb{Z}$.

(d) Let N be the norm function on $\mathbb{Q}(\sqrt{2})$, and let $\pi = 2\sqrt{2} - 1$. Note that $N(\pi) = -7$. Show that $u = \pi\alpha$, where $N(\alpha) = m^2$. Hint: $\mathbb{Z}[\sqrt{2}]$ is a UFD. You may have to switch u and u'.

(e) Assume that u has no square factors in $\mathbf{Z}[\sqrt{2}]$. Then show that $u = \epsilon\pi n$, where ϵ is a unit and n is a squarefree integer prime to 14. Hint: use Proposition 5.16 to describe the primes in $\mathbf{Z}[\sqrt{2}]$.

(f) Show that n must be ± 1. Hint: note that $u\mathcal{O}_L = \pi\mathcal{O}_L \cdot n\mathcal{O}_L$ is a square, and conclude that any prime dividing n ramifies in L.

(g) Thus $u = \epsilon\pi$ by (f). All units of $\mathbf{Z}[\sqrt{2}]$ are of the form $\pm(\sqrt{2}-1)^m$ (see Hasse [50, pp. 554–556]). Since $N(u) = -7$ and $N(\sqrt{2}-1) = -1$, we can assume $u = \pi$ since $u > 0$.

This proves that $\sqrt{u} = \sqrt{\pi} = \sqrt{2\sqrt{2}-1}$ is the desired primitive element.

6.19. Adapt Exercises 6.17 and 6.18 to discover a primitive element for the Hilbert class field of $\mathbf{Q}(\sqrt{-17})$. Hint: see Exercise 5.25. You may assume that the integers in $\mathbf{Q}(\sqrt{17})$ are a UFD and that the units are all of the form $\pm(4+\sqrt{17})^m$, $m \in \mathbf{Z}$ (see Borevich and Shafarevich [8, p. 422]). This method will lead most naturally to $u = 4 + \sqrt{17}$, which is related to our earlier choice $(1+\sqrt{17})/2$ via

$$(4+\sqrt{17})\cdot(1+\sqrt{17})/2 = ((5+\sqrt{17})/2)^2.$$

This problem may also be done without using the fact that $\mathbf{Q}(\sqrt{17})$ has class number 1 (see Herz [56]).

6.20. Let $K = \mathbf{Q}(\sqrt{-55})$.

(a) Show that $C(\mathcal{O}_K) \simeq \mathbf{Z}/4\mathbf{Z}$.

(b) Determine the Hilbert class field of K. Hint: use the methods of Proposition 5.31. Exercises 6.17 and 6.18 will show you what to look for.

(c) Prove an analog of Theorem 5.33 for primes of the form $x^2 + 55y^2$.

§7. ORDERS IN IMAGINARY QUADRATIC FIELDS

In §5, we solved our basic question of $p = x^2 + ny^2$ for those n's where $\mathbf{Z}[\sqrt{-n}]$ is the full ring of integers $\mathcal{O}_K$ in $K = \mathbf{Q}(\sqrt{-n})$ (see (5.15)). This holds for infinitely many n's, but it also leaves out infinitely many. The full story of what happens for these other n's will be told in §9, and we will see that the answer involves the ring $\mathbf{Z}[\sqrt{-n}]$. Such a ring is an example of an *order* in an imaginary quadratic field, which brings us to the main topic of §7.

We begin this section with a study of orders in a quadratic fields K. Unlike $\mathcal{O}_K$, an order $\mathcal{O}$ is usually *not* a Dedekind domain, so that the ideal theory of $\mathcal{O}$ is more complicated. This will lead us to restrict the class of ideals under consideration. In the case of imaginary quadratic fields, there is a nice relation between ideals in orders and quadratic forms. In particular, an order $\mathcal{O}$ has an ideal class group $C(\mathcal{O})$, and we will show that for any discriminant $D < 0$, the form class group $C(D)$ from §3 is naturally isomorphic to $C(\mathcal{O})$ for a suitable order $\mathcal{O}$. Then, to prepare the way for class field theory, we will show how to translate ideals for an order $\mathcal{O}$ in K into terms of the maximal order $\mathcal{O}_K$. The section will conclude with a discussion of class numbers.

A. Orders in Quadratic Fields

An *order* $\mathcal{O}$ in a quadratic field K is a subset $\mathcal{O} \subset K$ such that

(i) $\mathcal{O}$ is a subring of K containing 1.
(ii) $\mathcal{O}$ is a finitely generated $\mathbf{Z}$-module.
(iii) $\mathcal{O}$ contains a $\mathbf{Q}$-basis of K.

Since $\mathcal{O}$ is clearly torsion-free, (ii) and (iii) are equivalent to $\mathcal{O}$ being a free $\mathbf{Z}$-module of rank 2 (see Exercise 7.1). Note also that by (iii), K is the field of fractions of $\mathcal{O}$.

The ring $\mathcal{O}_K$ of integers in K is always an order in K—this follows from the description (5.13) of $\mathcal{O}_K$ given in §5. More importantly, (i) and (ii) above imply that for *any* order $\mathcal{O}$ of K, we have $\mathcal{O} \subset \mathcal{O}_K$ (see Exercise 7.2), so that $\mathcal{O}_K$ is the *maximal order* of K.

To describe orders in quadratic fields more explicitly, first note that by (5.14), the maximal order $\mathcal{O}_K$ can be written as follows:

$$(7.1) \qquad \mathcal{O}_K = [1, w_K], \qquad w_K = \frac{d_K + \sqrt{d_K}}{2},$$

where d_K is the discriminant of K. We can now describe all orders in quadratic fields:

Lemma 7.2. *Let $\mathcal{O}$ be an order in a quadratic field K of discriminant d_K. Then $\mathcal{O}$ has finite index in $\mathcal{O}_K$, and if we set $f = [\mathcal{O}_K : \mathcal{O}]$, then*

$$\mathcal{O} = \mathbf{Z} + f\mathcal{O}_K = [1, f w_K],$$

where w_K is as in (7.1).

Proof. Since $\mathcal{O}$ and $\mathcal{O}_K$ are free $\mathbf{Z}$-modules of rank 2, it follows that $[\mathcal{O}_K : \mathcal{O}] < \infty$. Setting $f = [\mathcal{O}_K : \mathcal{O}]$, we have $f\mathcal{O}_K \subset \mathcal{O}$, and then $\mathbf{Z} + f\mathcal{O}_K \subset \mathcal{O}$ follows. However, (7.1) implies $\mathbf{Z} + f\mathcal{O}_K = [1, f w_K]$, so that to prove the lemma, we need only show that $[1, f w_K]$ has index f in $\mathcal{O}_K = [1, w_K]$. This last fact is obvious, and we are done. Q.E.D.

Given an order $\mathcal{O}$ in a quadratic field K, the index $f = [\mathcal{O}_K : \mathcal{O}]$ is called the *conductor* of the order. Another important invariant of $\mathcal{O}$ is its *discriminant*, which is defined as follows. Let $\alpha \mapsto \alpha'$ be the nontrivial automorphism of K, and suppose that $\mathcal{O} = [\alpha, \beta]$. Then the *discriminant* of $\mathcal{O}$ is the number

$$D = \left(\det\begin{pmatrix} \alpha & \beta \\ \alpha' & \beta' \end{pmatrix}\right)^2.$$

The discriminant is independent of the integral basis used, and if we compute D using the basis $\mathcal{O} = [1, f w_K]$ from Lemma 7.2, then we obtain the formula

$$D = f^2 d_K. \tag{7.3}$$

Thus the discriminant satisfies $D \equiv 0,1 \bmod 4$. From (7.3) we also see that $K = \mathbf{Q}(\sqrt{D})$, so that K is real or imaginary according to whether $D > 0$ or $D < 0$. In fact, one can show that D determines $\mathcal{O}$ uniquely and that any nonsquare integer $D \equiv 0,1 \bmod 4$ is the discriminant of an order in a quadratic field. See Exercise 7.3 for proofs of these elementary facts. Note that by (7.3), the discriminant of the maximal order $\mathcal{O}_K$ is d_K, which agrees with the definition given in §5.

For an example of an order, consider $\mathbf{Z}[\sqrt{-n}] \subset K = \mathbf{Q}(\sqrt{-n})$. The discriminant of $\mathbf{Z}[\sqrt{-4n}]$ is easily computed to be $-4n$, and then (7.3) shows that

$$-4n = f^2 d_K.$$

This makes it easy to compute the conductor of $\mathbf{Z}[\sqrt{-n}]$. This order will be used in §9 when we give the general solution of $p = x^2 + ny^2$.

Now let's study the ideals of an order $\mathcal{O}$. If $\mathfrak{a}$ is a nonzero ideal of $\mathcal{O}$, then the proof of Corollary 5.4 adapts easily to show that $\mathcal{O}/\mathfrak{a}$ is finite (see Exercise 7.4). Thus we can define the *norm* of $\mathfrak{a}$ to be $N(\mathfrak{a}) = |\mathcal{O}/\mathfrak{a}|$. Furthermore, as in the proof of Theorem 5.5, it follows that $\mathcal{O}$ is Noetherian and that every nonzero prime ideal of $\mathcal{O}$ is maximal (see Exercise 7.4). However, it is equally obvious that if the conductor f of $\mathcal{O}$ is greater than 1, then $\mathcal{O}$ is *not* integrally closed in K, so that $\mathcal{O}$ is *not* a Dedekind domain when $f > 1$. Thus we may not assume that the ideals of $\mathcal{O}$ have unique factorization.

To remedy this situation, we will introduce the concept of a *proper ideal* of an order. Namely, given any ideal $\mathfrak{a}$ of $\mathcal{O}$, notice that

$$\mathcal{O} \subset \{\beta \in K : \beta\mathfrak{a} \subset \mathfrak{a}\}$$

since $\mathfrak{a}$ is an ideal of $\mathcal{O}$. However, equality need not occur. For example, if $\mathcal{O} = \mathbf{Z}[\sqrt{-3}]$ is the order of conductor 2 in $K = \mathbf{Q}(\sqrt{-3})$, and $\mathfrak{a}$ is the ideal of $\mathcal{O}$ generated by 2 and $1 + \sqrt{-3}$, then one sees easily that

$$\mathcal{O} \neq \{\beta \in K : \beta\mathfrak{a} \subset \mathfrak{a}\} = \mathcal{O}_K$$

(see Exercise 7.5). In general, we say that an ideal $\mathfrak{a}$ of $\mathcal{O}$ is *proper* whenever equality holds, i.e., when

$$\mathcal{O} = \{\beta \in K : \beta\mathfrak{a} \subset \mathfrak{a}\}.$$

For example, principal ideals are always proper, and for the maximal order, *all* ideals are proper (see Exercise 7.6).

We can also extend this terminology to fractional ideals. A *fractional ideal* of $\mathcal{O}$ is a subset of K which is a nonzero finitely generated $\mathcal{O}$-module. One can show that every fractional ideal is of the form $\alpha\mathfrak{a}$, where $\alpha \in K^*$ and $\mathfrak{a}$ is an $\mathcal{O}$-ideal (see Exercise 7.7). Then a fractional $\mathcal{O}$-ideal $\mathfrak{b}$ is *proper* provided that

$$\mathcal{O} = \{\beta \in K : \beta\mathfrak{b} \subset \mathfrak{b}\}.$$

Once we have fractional ideals, we can also talk about invertible ideals: a fractional $\mathcal{O}$-ideal $\mathfrak{a}$ is *invertible* if there is another fractional $\mathcal{O}$-ideal $\mathfrak{b}$ such that $\mathfrak{a}\mathfrak{b} = \mathcal{O}$. Note that principal fractional ideals (those of the form $\alpha\mathcal{O}$, $\alpha \in K^*$) are obviously invertible. The basic result is that for orders in quadratic fields, the notions of proper and invertible coincide:

Proposition 7.4. *Let $\mathcal{O}$ be an order in a quadratic field K, and let $\mathfrak{a}$ be a fractional $\mathcal{O}$-ideal. Then $\mathfrak{a}$ is proper if and only if $\mathfrak{a}$ is invertible.*

Proof. If $\mathfrak{a}$ is invertible, then $\mathfrak{a}\mathfrak{b} = \mathcal{O}$ for some fractional $\mathcal{O}$-ideal $\mathfrak{b}$. If $\beta \in K$ and $\beta\mathfrak{a} \subset \mathfrak{a}$, then we have

$$\beta\mathcal{O} = \beta(\mathfrak{a}\mathfrak{b}) = (\beta\mathfrak{a})\mathfrak{b} \subset \mathfrak{a}\mathfrak{b} = \mathcal{O},$$

and $\beta \in \mathcal{O}$ follows, proving that $\mathfrak{a}$ is proper.

To argue the other way, we will need the following lemma:

Lemma 7.5. *Let $K = \mathbb{Q}(\tau)$ be a quadratic field, and let $ax^2 + bx + c$ be the minimal polynomial of τ, where a, b and c are relatively prime integers. Then $[1,\tau]$ is a proper ideal for the order $[1,a\tau]$ of K.*

Proof. First, $[1,a\tau]$ is an order since $a\tau$ is an algebraic integer. Then, given $\beta \in K$, note that $\beta[1,\tau] \subset [1,\tau]$ is equivalent to

$$\begin{aligned} &\beta \cdot 1 \in [1,\tau] \\ &\beta \cdot \tau \in [1,\tau]. \end{aligned}$$

The first line says $\beta = m + n\tau$, $m,n \in \mathbf{Z}$. To understand the second, note that

$$\begin{aligned} \beta\tau = m\tau + n\tau^2 &= m\tau + \frac{n}{a}(-b\tau - c) \\ &= \frac{-cn}{a} + \left(\frac{-bn}{a} + m\right)\tau. \end{aligned}$$

Since $\gcd(a,b,c) = 1$, we see that $\beta\tau \in [1,\tau]$ if and only if $a \mid n$. It follows that

$$\{\beta \in K : \beta[1,\tau] \subset [1,\tau]\} = [1,a\tau],$$

which proves the lemma. Q.E.D.

Now we can prove that proper ideals are invertible. First note that $\mathfrak{a}$ is a $\mathbf{Z}$-module of rank 2 (see Exercise 7.8), so that $\mathfrak{a} = [\alpha,\beta]$ for some $\alpha,\beta \in K$. Then $\mathfrak{a} = \alpha[1,\tau]$, where $\tau = \beta/\alpha$. If $ax^2 + bx + c$, $\gcd(a,b,c) = 1$, is the minimal polynomial of τ, then Lemma 7.5 implies that $\mathcal{O} = [1,a\tau]$. Let $\beta \mapsto \beta'$ denote the nontrivial automorphism of K. Since τ' is the other root of $ax^2 + bx + c$, using Lemma 7.5 again shows that $\mathfrak{a}' = \alpha'[1,\tau']$ is a fractional ideal for $[1,a\tau] = [1,a\tau'] = \mathcal{O}$. We claim that

$$\mathfrak{a}\mathfrak{a}' = \frac{N(\alpha)}{a}\mathcal{O}. \tag{7.6}$$

To see why, note that

$$a\mathfrak{a}\mathfrak{a}' = a\alpha\alpha'[1,\tau][1,\tau'] = N(\alpha)[a,a\tau,a\tau',a\tau\tau'].$$

Since $\tau + \tau' = -b/a$ and $\tau\tau' = c/a$, this becomes

$$a\mathfrak{a}\mathfrak{a}' = N(\alpha)[a,a\tau,-b,c] = N(\alpha)[1,a\tau] = N(\alpha)\mathcal{O}$$

since $\gcd(a,b,c) = 1$. This implies (7.6), which in turn proves that $\mathfrak{a}$ is invertible. Q.E.D.

Unfortunately, Proposition 7.4 is not strong enough to prove unique factorization for proper ideals (see Exercise 7.9 for a counterexample). Later we will see that unique factorization holds for a slightly smaller class of ideals, those prime to the conductor.

Given an order $\mathcal{O}$, let $I(\mathcal{O})$ denote the set of proper fractional $\mathcal{O}$-ideals. By Proposition 7.4, $I(\mathcal{O})$ is a group under multiplication: the crucial issues are closure and the existence of inverses, both of which follow from the invertibility of proper ideals (see Exercise 7.10). The principal $\mathcal{O}$-ideals give a subgroup $P(\mathcal{O}) \subset I(\mathcal{O})$, and thus we can form the quotient

$$C(\mathcal{O}) = I(\mathcal{O})/P(\mathcal{O}),$$

which is the *ideal class group* of the order $\mathcal{O}$. When $\mathcal{O}$ is the maximal order $\mathcal{O}_K$, $I(\mathcal{O}_K)$ and $P(\mathcal{O}_K)$ will be denoted I_K and P_K, respectively. This is the notation used in §5, and in general we will reserve the subscript K exclusively for the maximal order. Note that the above definition of $C(\mathcal{O}_K)$ agrees with the one given in §5.

B. Orders and Quadratic Forms

We can relate the ideal class group $C(\mathcal{O})$ to the form class group $C(D)$ defined in §3 as follows:

Theorem 7.7. *Let $\mathcal{O}$ be the order of discriminant D in an imaginary quadratic field K.*

(i) *If $f(x,y) = ax^2 + bxy + cy^2$ is a primitive positive definite quadratic form of discriminant D, then $[a, (-b+\sqrt{D})/2]$ is a proper ideal of $\mathcal{O}$.*

(ii) *The map sending $f(x,y)$ to $[a, (-b+\sqrt{D})/2]$ induces an isomorphism between the form class group $C(D)$ and the ideal class group $C(\mathcal{O})$. Hence the order of $C(\mathcal{O})$ is the class number $h(D)$.*

(iii) *A positive integer m is represented by a form $f(x,y)$ if and only if m is the norm $N(\mathfrak{a})$ of some ideal $\mathfrak{a}$ in the corresponding ideal class in $C(\mathcal{O})$ (recall that $N(\mathfrak{a}) = |\mathcal{O}/\mathfrak{a}|$).*

Remark. Because of the isomorphism $C(D) \simeq C(\mathcal{O})$, we will sometimes write the class number as $h(\mathcal{O})$ instead of $h(D)$.

Proof. Let $f(x,y) = ax^2 + bxy + cy^2$ be a primitive positive definite form of discriminant $D < 0$. The roots of $f(x,1) = ax^2 + bx + c$ are complex, so that there is a unique $\tau \in \mathfrak{h}$ ($\mathfrak{h}$ is the upper half plane) such that $f(\tau, 1) = 0$. We call τ the *root* of $f(x,y)$. Since $a > 0$, it follows that $\tau = (-b + \sqrt{D})/2a$. Thus

$$[a, (-b+\sqrt{D})/2] = [a, a\tau] = a[1, \tau].$$

Note also that $\tau \in K$.

To prove (i), note that by Lemma 7.5, $a[1,\tau]$ is a proper ideal for the order $[1, a\tau]$. However, if f is the conductor of $\mathcal{O}$, then $D = f^2 d_K$ by (7.3), and thus

$$\begin{aligned} a\tau &= \frac{-b+\sqrt{D}}{2} = \frac{-b+f\sqrt{d_K}}{2} \\ &= -\frac{b+fd_K}{2} + f\left(\frac{d_K+\sqrt{d_K}}{2}\right) = -\frac{b+fd_K}{2} + fw_K. \end{aligned}$$

Since $D = b^2 - 4ac$, fd_K and b have the same parity, so that $(b + fd_K)/2 \in \mathbf{Z}$. It follows that $[1, a\tau] = [1, fw_K]$, so that $[1, a\tau] = \mathcal{O}$ by Lemma 7.2. This proves that $a[1,\tau]$ is a proper $\mathcal{O}$-ideal.

To prove (ii), let $f(x,y)$ and $g(x,y)$ be forms of discriminant D, and let τ and τ' be their respective roots. We will prove the following equivalences:

$$
\begin{aligned}
& f(x,y), g(x,y) \text{ are properly equivalent} \\
(7.8) \qquad & \iff \tau' = \frac{p\tau+q}{r\tau+s}, \ \begin{pmatrix} p & q \\ r & s \end{pmatrix} \in \mathrm{SL}(2,\mathbf{Z}) \\
& \iff [1,\tau] = \lambda[1,\tau'], \ \lambda \in K^*.
\end{aligned}
$$

To see why this is true, assume that $f(x,y) = g(px+qy, rx+sy)$, where $\left(\begin{smallmatrix} p & q \\ r & s \end{smallmatrix}\right) \in \mathrm{SL}(2,\mathbf{Z})$. Then

$$
(7.9) \qquad 0 = f(\tau,1) = g(p\tau+q, r\tau+s) = (r\tau+t)^2 g\left(\frac{p\tau+q}{r\tau+s}, 1\right),
$$

so that $g((p\tau+q)/(r\tau+s),1) = 0$. However, an easy computation shows that

$$
(7.10) \qquad \mathrm{Im}\left(\frac{p\tau+q}{r\tau+s}\right) = \det\begin{pmatrix} p & q \\ r & s \end{pmatrix} |r\tau+s|^{-2}\mathrm{Im}(\tau)
$$

(see Exercise 7.11). This implies $(p\tau+q)/(r\tau+s) \in \mathfrak{h}$, and thus $\tau' = (p\tau+q)/(r\tau+s)$ by the uniqueness of the root τ'. Conversely, if $\tau' = (p\tau+q)/(r\tau+s)$, then (7.9) shows that $f(x,y)$ and $g(px+qy, rx+sy)$ have the same root, and it follows easily that they must be equal (see Exercise 7.12). This proves the first equivalence of (7.8).

Next, if $\tau' = (p\tau+q)/(r\tau+s)$, let $\lambda = r\tau+s \in K^*$. Then

$$
\begin{aligned}
\lambda[1,\tau'] &= (r\tau+s)\left[1, \frac{p\tau+q}{r\tau+s}\right] \\
&= [r\tau+s, p\tau+q] = [1,\tau]
\end{aligned}
$$

since $\left(\begin{smallmatrix} p & q \\ r & s \end{smallmatrix}\right) \in \mathrm{SL}(2,\mathbf{Z})$. Conversely, if $[1,\tau] = \lambda[1,\tau']$ for some $\lambda \in K^*$, then $[1,\tau] = [\lambda, \lambda\tau']$, which implies

$$
\begin{aligned}
\lambda\tau' &= p\tau+q \\
\lambda &= r\tau+s
\end{aligned}
$$

for some $\left(\begin{smallmatrix} p & q \\ r & s \end{smallmatrix}\right) \in \mathrm{GL}(2,\mathbf{Z})$. This gives us

$$
\tau' = \frac{p\tau+q}{r\tau+s},
$$

and then (7.10) shows that $\left(\begin{smallmatrix} p & q \\ r & s \end{smallmatrix}\right) \in \mathrm{SL}(2,\mathbf{Z})$ since τ and τ' are both in $\mathfrak{h}$. This completes the proof of (7.8).

Using (7.8), one easily sees that the map sending $f(x,y)$ to $a[1,\tau]$ induces an *injection*

$$
C(D) \longrightarrow C(\mathcal{O}).
$$

To show that the map is surjective, let $\mathfrak{a}$ be a fractional $\mathcal{O}$-ideal. As in the proof of Proposition 7.4, we can write $\mathfrak{a} = [\alpha, \beta]$ for some $\alpha, \beta \in K$. Switching α and β if necessary, we can assume that $\tau = \beta/\alpha$ lies in $\mathfrak{h}$. Let $ax^2 + bx + c$ be the minimal polynomial of τ. We may assume that $\gcd(a, b, c) = 1$ and $a > 0$. Then $f(x, y) = ax^2 + bxy + cy^2$ is positive definite of discriminant D (see Exercise 7.12), and $f(x, y)$ maps to $a[1, \tau]$. This ideal lies in the class of $\mathfrak{a} = [\alpha, \beta] = \alpha[1, \tau]$ in $C(\mathcal{O})$, and surjectivity is proved.

We thus have a bijection of sets

$$C(D) \longrightarrow C(\mathcal{O}). \tag{7.11}$$

We next want to see what happens to the group structure, but we first need to review the formulas for Dirichlet composition from §3. Given two primitive positive definite forms $f(x, y) = ax^2 + bxy + cy^2$ and $g(x, y) = a'x^2 + b'xy + c'y^2$ of discriminant D, suppose that $\gcd(a, a', (b + b')/2) = 1$. Then the *Dirichlet composition* of $f(x, y)$ and $g(x, y)$ was defined to be the form

$$F(x, y) = aa'x^2 + Bxy + \frac{B^2 - D}{4aa'}y^2,$$

where B is the unique number modulo $2aa'$ such that

$$\begin{aligned} B &\equiv b \bmod 2a \\ B &\equiv b' \bmod 2a' \\ B^2 &\equiv D \bmod 4aa' \end{aligned} \tag{7.12}$$

(see Lemma 3.2 and (3.7)). In Theorem 3.9 we asserted that Dirichlet composition made $C(D)$ into an Abelian group, but the proof given in §3 was not complete. So our first task is to use the bijection (7.11) to finish the proof of Theorem 3.9.

Given $f(x, y)$, $g(x, y)$ and $F(x, y)$ as above, we get three proper ideals of $\mathcal{O}$:

$$[a, (-b + f\sqrt{d_K})/2], \ [a', (-b' + f\sqrt{d_K})/2] \text{ and } [aa', (-B + f\sqrt{d_K})/2].$$

If we set $\Delta = (-B + f\sqrt{d_K})/2$ and use the top two lines of (7.12), then these ideals can be written as

$$[a, \Delta], \ [a', \Delta] \text{ and } [aa', \Delta].$$

We claim that

$$[a, \Delta][a', \Delta] = [aa', \Delta]. \tag{7.13}$$

To see this, note that $\Delta^2 \equiv -B\Delta \bmod aa'$ by the last line of (7.12). Thus

$$[a, \Delta][a', \Delta] = [aa', a\Delta, a'\Delta, \Delta^2] = [aa', a\Delta, a'\Delta, -B\Delta].$$

However, from (7.12) one easily proves that $\gcd(a,a',B) = 1$ (see Exercise 7.13), and then (7.13) follows immediately.

By (7.11) and (7.13), we see that the Dirichlet composition of $f(x,y)$ and $g(x,y)$ corresponds to the product of their corresponding ideal classes, which proves that Dirichlet composition induces a well-defined binary operation on $C(D)$. Furthermore, since the product of ideals makes $C(\mathcal{O})$ into a group, it follows immediately that $C(D)$ is a group under Dirichlet composition. This completes the proof of Theorem 3.9, and it is now obvious that (7.11) is an isomorphism of groups.

Before we can prove part (iii) of the theorem, we need to learn more about the norm $N(\mathfrak{a}) = |\mathcal{O}/\mathfrak{a}|$ of a proper $\mathcal{O}$-ideal $\mathfrak{a}$. The basic properties of $N(\mathfrak{a})$ are:

Lemma 7.14. *Let $\mathcal{O}$ be an order in an imaginary quadratic field. Then:*

(i) $N(\alpha\mathcal{O}) = N(\alpha)$ *for* $\alpha \in \mathcal{O}$, $\alpha \neq 0$.

(ii) $N(\mathfrak{a}\mathfrak{b}) = N(\mathfrak{a})N(\mathfrak{b})$ *for proper $\mathcal{O}$-ideals $\mathfrak{a}$ and $\mathfrak{b}$.*

(iii) $\mathfrak{a}\overline{\mathfrak{a}} = N(\mathfrak{a})\mathcal{O}$ *for a proper $\mathcal{O}$-ideal $\mathfrak{a}$.*

Proof. The proof of (i) is covered in Exercises 7.14 and 7.15. We will next prove a special case of (ii): if $\alpha \neq 0$ in $\mathcal{O}$, we claim that

$$N(\alpha\mathfrak{a}) = N(\alpha)N(\mathfrak{a}). \tag{7.15}$$

To prove this, note that the inclusions $\alpha\mathfrak{a} \subset \alpha\mathcal{O} \subset \mathcal{O}$ give us the short exact sequence

$$0 \to \alpha\mathcal{O}/\alpha\mathfrak{a} \to \mathcal{O}/\alpha\mathfrak{a} \to \mathcal{O}/\alpha\mathcal{O} \to 0,$$

which implies that $|\mathcal{O}/\alpha\mathfrak{a}| = |\mathcal{O}/\alpha\mathcal{O}||\alpha\mathcal{O}/\alpha\mathfrak{a}|$. Since multiplication by α induces an isomorphism $\mathcal{O}/\mathfrak{a} \xrightarrow{\sim} \alpha\mathcal{O}/\alpha\mathfrak{a}$, we get $N(\alpha\mathfrak{a}) = N(\alpha\mathcal{O})N(\mathfrak{a})$, and then (7.15) follows from (i).

Before proving (ii) and (iii), we need study $N(\mathfrak{a})$. If we write $\mathfrak{a}$ in the form $\mathfrak{a} = \alpha[1,\tau]$, then Lemma 7.5 implies that $\mathcal{O} = [1,a\tau]$. Since $[a,a\tau]$ obviously has index a in $[1,a\tau]$, we obtain

$$N(a[1,\tau]) = a.$$

Then $a \cdot \mathfrak{a} = \alpha \cdot a[1,\tau]$ and (7.15) imply that

$$N(\mathfrak{a}) = \frac{N(\alpha)}{a}. \tag{7.16}$$

Now (iii) follows immediately by combining (7.16) with the equation

$$\mathfrak{a}\overline{\mathfrak{a}} = \frac{N(\alpha)}{a}\mathcal{O}$$

proved in (7.6). Turning to (ii), note that (iii) implies that

$$N(\mathfrak{ab})\mathcal{O} = \mathfrak{ab} \cdot \overline{\mathfrak{ab}} = \mathfrak{a}\overline{\mathfrak{a}} \cdot \mathfrak{b}\overline{\mathfrak{b}} = N(\mathfrak{a})\mathcal{O} \cdot N(\mathfrak{b})\mathcal{O} = N(\mathfrak{a})N(\mathfrak{b})\mathcal{O},$$

and then $N(\mathfrak{ab}) = N(\mathfrak{a})N(\mathfrak{b})$ follows. Q.E.D.

A useful consequence of this lemma is that if $\mathfrak{a}$ is a proper $\mathcal{O}$-ideal, then $\overline{\mathfrak{a}}$ gives the inverse of $\mathfrak{a}$ in $C(\mathcal{O})$. This follows immediately from $\mathfrak{a}\overline{\mathfrak{a}} = N(\mathfrak{a})\mathcal{O}$. In Exercise 7.16 we will use the isomorphism $C(D) \simeq C(\mathcal{O})$ to give a second proof of this fact.

We can now prove part (iii) of the theorem. If m is represented by $f(x,y)$, then $m = d^2a$, where a is properly represented by $f(x,y)$. We may assume that $f(x,y) = ax^2 + bxy + cy^2$. Then $f(x,y)$ maps to $\mathfrak{a} = a[1,\tau]$, so that $N(\mathfrak{a}) = a$ by (7.16). It follows that $N(d\mathfrak{a}) = d^2a = m$, so that m is the norm of an ideal in the class of $\mathfrak{a}$.

Conversely, assume that $N(\mathfrak{a}) = m$. We know that $\mathfrak{a} = \alpha[1,\tau]$, where $\mathrm{Im}(\tau) > 0$ and $a\tau^2 + b\tau + c = 0$, $\gcd(a,b,c) = 1$ and $a > 0$. Then $f(x,y) = ax^2 + bxy + cy^2$ maps to the class of $\mathfrak{a}$, so that we need only show that $f(x,y)$ represents m.

By (7.16), we know that

$$m = N(\mathfrak{a}) = \frac{N(\alpha)}{a}.$$

However, $\alpha[1,\tau] = \mathfrak{a} \subset \mathcal{O} = [1, a\tau]$, so that $\alpha = p + qa\tau$ and $\alpha\tau = r + sa\tau$ for some integers $p,q,r,s \in \mathbf{Z}$. Thus $(p + qa\tau)\tau = r + sa\tau$, and since $a\tau^2 = -b\tau - c$, comparing coefficients shows that $p = as + bq$. Hence

$$\begin{aligned} m = \frac{N(\alpha)}{a} &= \frac{1}{a}(p^2 - bpq + acq^2) \\ &= \frac{1}{a}\left((as + bq)^2 - b(as + bq)q + acq^2\right) \\ &= \frac{1}{a}(a^2s^2 + absq + acq^2) \\ &= as^2 + bsq + cq^2 = f(s,q). \end{aligned}$$

This proves (iii) and completes the proof of Theorem 7.7. Q.E.D.

Notice that Theorem 5.30 is an immediate corollary of Theorem 7.7.

The map $f(x,y) \mapsto \mathfrak{a} = [a, (-b + \sqrt{D})/2]$ of Theorem 7.7 has a natural inverse which is defined as follows. If $\mathfrak{a} = [\alpha, \beta]$ is a proper $\mathcal{O}$-ideal with $\mathrm{Im}(\beta/\alpha) > 0$, then

$$f(x,y) = \frac{N(\alpha x - \beta y)}{N(\mathfrak{a})}$$

is a positive definite form of discriminant D. On the level of classes, this map is the inverse to the map of Theorem 7.7 (see Exercise 7.17).

Theorem 7.7 allows us to translate what we know about quadratic forms into facts about ideal classes. Here is an example that will be useful later on:

Corollary 7.17. *Let $\mathcal{O}$ be an order in an imaginary quadratic field. Given a nonzero integer M, then every ideal class in $C(\mathcal{O})$ contains a proper $\mathcal{O}$-ideal whose norm is relatively prime to M.*

Proof. In Lemma 2.25 we learned that any primitive form represents numbers relatively prime to M, and the corollary then follows from part (iii) of Theorem 7.7. Q.E.D.

The reader may wonder if Theorem 7.7 holds for real quadratic fields. Simple examples show that this isn't true in general. For instance, when $K = \mathbf{Q}(\sqrt{3})$, the maximal order $\mathcal{O}_K = \mathbf{Z}[\sqrt{3}]$ is a UFD, which implies that $C(\mathcal{O}_K) \simeq \{1\}$. Yet the forms $\pm(x^2 - 3y^2)$ of discriminant $d_K = 12$ are not properly equivalent, so that $C(d_K) \not\simeq \{1\}$ (see Exercise 7.18 for the details). In order to make a version of Theorem 7.7 that holds for real quadratic fields, we need to change the notion of equivalence. In Exercises 7.19–7.24 we will explore two ways of doing this:

1. Change the notion of equivalence of ideals. Instead of using all principal ideals $P(\mathcal{O})$, use only $P^+(\mathcal{O})$, which consists of all principal ideals $\alpha\mathcal{O}$ where $N(\alpha) > 0$. The quotient $I(\mathcal{O})/P^+(\mathcal{O})$ is the *narrow* (or *strict*) *ideal class group* and is denoted by $C^+(\mathcal{O})$. In Exercise 7.21 we will construct a natural isomorphism $C(D) \simeq C^+(\mathcal{O})$ which holds for any orders in any quadratic field K. We also have $C^+(\mathcal{O}) = C(\mathcal{O})$ when K is imaginary, and the same is true when K is real and $\mathcal{O}$ has a unit ϵ with $N(\epsilon) = -1$. If K has no such unit, then $|C^+(\mathcal{O})| = 2|C(\mathcal{O})|$.
2. Change the notion of equivalence of forms. Instead of using proper equivalence, use the notion of *signed equivalence*, where $f(x,y)$ and $g(x,y)$ are *signed equivalent* if there is a matrix $\left(\begin{smallmatrix} p & q \\ r & s \end{smallmatrix}\right) \in \mathrm{GL}(2,\mathbf{Z})$ such that
$$f(x,y) = \det\begin{pmatrix} p & q \\ r & s \end{pmatrix} g(px+qy, rx+sy).$$
The set of signed equivalence classes is denoted $C_s(D)$, and in Exercise 7.22 we will see that there is a natural isomorphism $C_s(D) \simeq C(\mathcal{O})$. The criteria for when $C_s(D) = C(D)$ are the same as above.

For other treatments of the relation between forms and ideals, see Borevich and Shafarevich [8, Chapter 2, §7.5], Cohn [19, §§14.A–C] and Zagier [111, §§8 and 10].

C. Ideals Prime to the Conductor

The theory described so far does not interact well with the usual formulation of class field theory. The reason is that class field theory is always stated in terms of the maximal order $\mathcal{O}_K$. So given an order $\mathcal{O}$ in a quadratic field K, we will need to translate proper $\mathcal{O}$-ideals into terms of $\mathcal{O}_K$-ideals. This is difficult to do directly, but becomes much easier once we study $\mathcal{O}$-ideals prime to the conductor.

Given an order $\mathcal{O}$ of conductor f, we say that a nonzero $\mathcal{O}$-ideal $\mathfrak{a}$ is *prime to* f provided that $\mathfrak{a} + f\mathcal{O} = \mathcal{O}$. The following lemma gives the basic properties of $\mathcal{O}$-ideals prime to the conductor:

Lemma 7.18. *Let $\mathcal{O}$ be an order of conductor f.*

(i) *An $\mathcal{O}$-ideal $\mathfrak{a}$ is prime to f if and only if its norm $N(\mathfrak{a})$ is relatively prime to f.*

(ii) *Every $\mathcal{O}$-ideal prime to f is proper.*

Proof. To prove (i), let $m_f : \mathcal{O}/\mathfrak{a} \to \mathcal{O}/\mathfrak{a}$ be multiplication by f. Then

$$\mathfrak{a} + f\mathcal{O} = \mathcal{O} \iff m_f \text{ is surjective} \iff m_f \text{ is an isomorphism.}$$

By the structure theorem for finite Abelian groups, m_f is an isomorphism if and only if f is relatively prime to the order $N(\mathfrak{a})$ of $\mathcal{O}/\mathfrak{a}$, and (i) is proved.

To show that an $\mathcal{O}$-ideal $\mathfrak{a}$ prime to f is proper, let $\beta \in K$ satisfy $\beta\mathfrak{a} \subset \mathfrak{a}$. Then β is certainly in $\mathcal{O}_K$, and we thus have

$$\beta\mathcal{O} = \beta(\mathfrak{a} + f\mathcal{O}) = \beta\mathfrak{a} + \beta f\mathcal{O} \subset \mathfrak{a} + f\mathcal{O}_K.$$

However, $f\mathcal{O}_K \subset \mathcal{O}$, which proves that $\beta\mathcal{O} \subset \mathcal{O}$. Thus $\beta \in \mathcal{O}$, which proves that $\mathfrak{a}$ is proper. Q.E.D.

It follows that $\mathcal{O}$-ideals prime to f lie naturally in $I(\mathcal{O})$ and are closed under multiplication (since $N(\mathfrak{a}\mathfrak{b}) = N(\mathfrak{a})N(\mathfrak{b})$ will also be prime to f). The subgroup of fractional ideals they generate is denoted $I(\mathcal{O},f) \subset I(\mathcal{O})$, and inside of $I(\mathcal{O},f)$ we have the subgroup $P(\mathcal{O},f)$ generated by the principal ideals $\alpha\mathcal{O}$ where $\alpha \in \mathcal{O}$ has norm $N(\alpha)$ prime to f. We can then describe $C(\mathcal{O})$ in terms of $I(\mathcal{O},f)$ and $P(\mathcal{O},f)$ as follows:

Proposition 7.19. *The inclusion $I(\mathcal{O},f) \subset I(\mathcal{O})$ induces an isomorphism*

$$I(\mathcal{O},f)/P(\mathcal{O},f) \simeq I(\mathcal{O})/P(\mathcal{O}) = C(\mathcal{O}).$$

Proof. The map $I(\mathcal{O},f) \to C(\mathcal{O})$ is surjective by Corollary 7.17 (any ideal class in $C(\mathcal{O})$ contains an $\mathcal{O}$-ideal prime to f), and the kernel is $I(\mathcal{O},f) \cap$

$P(\mathcal{O})$. This obviously contains $P(\mathcal{O},f)$, but the inclusion $I(\mathcal{O},f)\cap P(\mathcal{O})\subset P(\mathcal{O},f)$ needs proof. An element of $I(\mathcal{O},f)\cap P(\mathcal{O})$ is a fractional ideal $\alpha\mathcal{O}=\mathfrak{a}\mathfrak{b}^{-1}$, where $\alpha\in K$ and $\mathfrak{a}$ and $\mathfrak{b}$ are $\mathcal{O}$-ideals prime to f. Let $m=N(\mathfrak{b})$. Then $m\mathcal{O}=N(\mathfrak{b})\mathcal{O}=\mathfrak{b}\overline{\mathfrak{b}}$, so that $m\mathfrak{b}^{-1}=\overline{\mathfrak{b}}$. Hence

$$m\alpha\mathcal{O}=\mathfrak{a}\cdot m\mathfrak{b}^{-1}=\mathfrak{a}\overline{\mathfrak{b}}\subset\mathcal{O},$$

which proves that $m\alpha\mathcal{O}\in P(\mathcal{O},f)$. Then $\alpha\mathcal{O}=m\alpha\mathcal{O}\cdot(m\mathcal{O})^{-1}$ is also in $P(\mathcal{O},f)$, and the proposition is proved. Q.E.D.

For any order $\mathcal{O}$, ideals prime to the conductor relate nicely to ideals for the maximal order $\mathcal{O}_K$. Before we can explain this, we need a definition: given a positive integer m, an $\mathcal{O}_K$-ideal $\mathfrak{a}$ is *prime to* m provided that $\mathfrak{a}+m\mathcal{O}_K=\mathcal{O}_K$. As in Lemma 7.18, this is equivalent to $\gcd(N(\mathfrak{a}),m)=1$. Thus, inside of the group of fractional $\mathcal{O}_K$-ideals I_K, we have the subgroup $I_K(m)\subset I_K$ generated by $\mathcal{O}_K$-ideals prime to m.

Proposition 7.20. *Let $\mathcal{O}$ be an order of conductor f in an imaginary quadratic field K.*

(i) *If $\mathfrak{a}$ is an $\mathcal{O}_K$-ideal prime to f, then $\mathfrak{a}\cap\mathcal{O}$ is an $\mathcal{O}$-ideal prime to f of the same norm.*

(ii) *If $\mathfrak{a}$ is an $\mathcal{O}$-ideal prime to f, then $\mathfrak{a}\mathcal{O}_K$ is an $\mathcal{O}_K$-ideal prime to f of the same norm.*

(iii) *The map $\mathfrak{a}\mapsto\mathfrak{a}\cap\mathcal{O}$ induces an isomorphism $I_K(f)\xrightarrow{\sim}I(\mathcal{O},f)$, and the inverse of this map is given by $\mathfrak{a}\mapsto\mathfrak{a}\mathcal{O}_K$.*

Proof. To prove (i), let $\mathfrak{a}$ be an $\mathcal{O}_K$-ideal prime to f. Since $\mathcal{O}/\mathfrak{a}\cap\mathcal{O}$ injects into $\mathcal{O}_K/\mathfrak{a}$ and $N(\mathfrak{a})$ is prime to f, so is $N(\mathfrak{a}\cap\mathcal{O})$, which proves that $\mathfrak{a}\cap\mathcal{O}$ is prime to f. As for norms, consider the natural map

$$\mathcal{O}/\mathfrak{a}\cap\mathcal{O}\longrightarrow\mathcal{O}_K/\mathfrak{a}.$$

It is injective, and since $\mathfrak{a}$ is prime to f, multiplication by f induces an isomorphism of $\mathcal{O}_K/\mathfrak{a}$. But $f\mathcal{O}_K\subset\mathcal{O}$, and surjectivity follows. This shows that the norms are equal, and (i) is proved.

To prove (ii), let $\mathfrak{a}$ be an $\mathcal{O}$-ideal prime to f. Since

$$\mathfrak{a}\mathcal{O}_K+f\mathcal{O}_K=(\mathfrak{a}+f\mathcal{O})\mathcal{O}_K=\mathcal{O}\mathcal{O}_K=\mathcal{O}_K$$

we see that $\mathfrak{a}\mathcal{O}_K$ is also prime to f. The statement about norms will be proved below.

Turning to (iii), we claim that

$$\begin{aligned}\mathfrak{a}\mathcal{O}_K\cap\mathcal{O}&=\mathfrak{a}\qquad\text{when }\mathfrak{a}\text{ is an }\mathcal{O}\text{-ideal prime to }f\\(\mathfrak{a}\cap\mathcal{O})\mathcal{O}_K&=\mathfrak{a}\qquad\text{when }\mathfrak{a}\text{ is an }\mathcal{O}_K\text{-ideal prime to }f.\end{aligned}\tag{7.21}$$

We start with the top line. If $\mathfrak{a}$ is an $\mathcal{O}$-ideal prime to f, then

$$\begin{aligned}\mathfrak{a}\mathcal{O}_K \cap \mathcal{O} &= (\mathfrak{a}\mathcal{O}_K \cap \mathcal{O})\mathcal{O}\\ &= (\mathfrak{a}\mathcal{O}_K \cap \mathcal{O})(\mathfrak{a} + f\mathcal{O})\\ &\subset \mathfrak{a} + f(\mathfrak{a}\mathcal{O}_K \cap \mathcal{O}) \subset \mathfrak{a} + \mathfrak{a}\cdot f\mathcal{O}_K.\end{aligned}$$

Since $f\mathcal{O}_K \subset \mathcal{O}$, this proves that $\mathfrak{a}\mathcal{O}_K \cap \mathcal{O} \subset \mathfrak{a}$. The other inclusion is obvious, so that equality follows. Turning to the second line of (7.21), let $\mathfrak{a}$ be an $\mathcal{O}_K$-ideal prime to f. Then

$$\mathfrak{a} = \mathfrak{a}\mathcal{O} = \mathfrak{a}(\mathfrak{a}\cap\mathcal{O} + f\mathcal{O}) \subset (\mathfrak{a}\cap\mathcal{O})\mathcal{O}_K + f\,\mathfrak{a}.$$

However, $f\,\mathfrak{a} \subset f\mathcal{O}_K \subset \mathcal{O}$, so that $f\,\mathfrak{a} \subset \mathfrak{a}\cap\mathcal{O} \subset (\mathfrak{a}\cap\mathcal{O})\mathcal{O}_K$, and $\mathfrak{a} \subset (\mathfrak{a}\cap \mathcal{O})\mathcal{O}_K$ follows. The other inclusion is obvious, which finishes the proof of (7.21). Notice that (7.21) and (i) imply the norm statement of (ii).

From (7.21) we get a bijection on the monoids of $\mathcal{O}_K$- and $\mathcal{O}$-ideals prime to f. If we can show that $\mathfrak{a} \mapsto \mathfrak{a}\cap\mathcal{O}$ preserves multiplication, then we get an isomorphism $I_K(f) \simeq I(\mathcal{O}, f)$ (see Exercise 7.25). But multiplicativity is easy, for the inverse map $\mathfrak{a} \mapsto \mathfrak{a}\mathcal{O}_K$ is obviously multiplicative:

$$(\mathfrak{a}\mathfrak{b})\mathcal{O}_K = \mathfrak{a}\mathcal{O}_K \cdot \mathfrak{b}\mathcal{O}_K.$$

This proves the proposition. Q.E.D.

Using this proposition, it follows that every $\mathcal{O}$-ideal prime to f has a unique decomposition as a product of prime $\mathcal{O}$-ideals which are prime to f (see Exercise 7.26).

We can now describe $C(\mathcal{O})$ in terms of the maximal order:

Proposition 7.22. *Let $\mathcal{O}$ be an order of conductor f in an imaginary quadratic field K. Then there are natural isomorphisms*

$$C(\mathcal{O}) \simeq I(\mathcal{O}, f)/P(\mathcal{O}, f) \simeq I_K(f)/P_{K,\mathbf{Z}}(f),$$

where $P_{K,\mathbf{Z}}(f)$ is the subgroup of $I_K(f)$ generated by principal ideals of the form $\alpha\mathcal{O}_K$, where $\alpha \in \mathcal{O}_K$ satisfies $\alpha \equiv a \bmod f\mathcal{O}_K$ for some integer a relatively prime to f.

Remark. To keep track of the various ideal groups, remember that the subscript K refers to the maximal order $\mathcal{O}_K$ (as in I_K, $I_K(f)$, etc.), while no subscript refers to the order $\mathcal{O}$ (as in $I(\mathcal{O})$, $I(\mathcal{O}, f)$, etc.).

Proof. The first isomorphism comes from Proposition 7.19. To prove the second, note that $\mathfrak{a} \mapsto \mathfrak{a}\mathcal{O}_K$ induces an isomorphism $I(\mathcal{O}, f) \simeq I_K(f)$ by Proposition 7.20. Under this isomorphism $P(\mathcal{O}, f) \subset I(\mathcal{O}, f)$ maps to a subgroup $\tilde{P} \subset I_K(f)$. It remains to prove $\tilde{P} = P_{K,\mathbf{Z}}(f)$.

We first show that for $\alpha \in \mathcal{O}_K$,

$$(7.23) \qquad \begin{aligned} &\alpha \equiv a \bmod f\mathcal{O}_K,\ a \in \mathbf{Z},\ \gcd(a,f) = 1 \\ &\iff \alpha \in \mathcal{O},\ \gcd(N(\alpha),f) = 1. \end{aligned}$$

Going one way, assume that $\alpha \equiv a \bmod f\mathcal{O}_K$, where $a \in \mathbf{Z}$ is relatively prime to f. Then $N(\alpha) \equiv a^2 \bmod f$ follows easily (see Exercise 7.27), so that $\gcd(N(\alpha),f) = \gcd(a^2,f) = 1$. Since $f\mathcal{O}_K \subset \mathcal{O}$, we also see that $\alpha \in \mathcal{O}$. Conversely, let $\alpha \in \mathcal{O} = [1, fw_K]$ have norm prime to f. Then $\alpha \equiv a \bmod f\mathcal{O}_K$ for some $a \in \mathbf{Z}$. Since $\gcd(N(\alpha),f) = 1$ and $N(\alpha) \equiv a^2 \bmod f$, we must have $\gcd(a,f) = 1$, and (7.23) is proved.

We know that $P(\mathcal{O},f)$ is generated by the ideals $\alpha\mathcal{O}$, where $\alpha \in \mathcal{O}$ and $N(\alpha)$ is relatively prime to f. Thus $\tilde{P}$ is generated by the corresponding ideals $\alpha\mathcal{O}_K$, and by (7.23), this implies that $\tilde{P} = P_K(\mathcal{O},f)$. Q.E.D.

In §9 we will use this proposition to link $C(\mathcal{O})$ to the class field theory of K. For other discussions of the relation between ideals of $\mathcal{O}$ and $\mathcal{O}_K$, see Deuring [24, §8] and Lang [73, §8.1].

D. The Class Number

One of the nicest applications of Proposition 7.22 is a formula for the class number $h(\mathcal{O})$ in terms of its conductor f and the class number $h(\mathcal{O}_K)$ of the maximal order. Before we can state the formula, we need to recall some terminology from §5. Given an odd prime p, we have the Legendre symbol (d_K/p), and for $p = 2$ we have the Kronecker symbol:

$$\left(\frac{d_K}{2}\right) = \begin{cases} 0 & \text{if } 2 \mid d_K \\ 1 & \text{if } d_K \equiv 1 \bmod 8 \\ -1 & \text{if } d_K \equiv 5 \bmod 8. \end{cases}$$

(Recall that $d_K \equiv 1 \bmod 4$ when d_K is odd.) We can now state our formula for $h(\mathcal{O})$:

Theorem 7.24. *Let $\mathcal{O}$ be the order of conductor f in an imaginary quadratic field K. Then*

$$h(\mathcal{O}) = \frac{h(\mathcal{O}_K)f}{[\mathcal{O}_K^* : \mathcal{O}^*]} \prod_{p \mid f} \left(1 - \left(\frac{d_K}{p}\right)\frac{1}{p}\right).$$

Furthermore, $h(\mathcal{O})$ is always an integer multiple of $h(\mathcal{O}_K)$.

Proof. By Theorem 7.7 and Proposition 7.22, we have

$$\begin{aligned} h(\mathcal{O}) &= |C(\mathcal{O})| = |I_K(f)/P_{K,\mathbf{Z}}(f)| \\ h(\mathcal{O}_K) &= |C(\mathcal{O}_K)| = |I_K/P_K|. \end{aligned}$$

Since $I_K(f) \subset I_K$ and $P_{K,\mathbf{Z}}(f) \subset I_K(f) \cap P_K$, we get an exact sequence

(7.25)

$$\begin{array}{ccccccc} 0 & \longrightarrow & I_K(f) \cap P_K/P_{K,\mathbf{Z}}(f) & \longrightarrow & I_K(f)/P_{K,\mathbf{Z}}(f) & \longrightarrow & I_K/P_K \\ & & & & \downarrow\wr & & \downarrow\wr \\ & & & & C(\mathcal{O}) & \longrightarrow & C(\mathcal{O}_K). \end{array}$$

We know from Corollary 7.17 that every class in $C(\mathcal{O}_K)$ contains an $\mathcal{O}_K$-ideal whose norm is relatively prime to f. This implies that $C(\mathcal{O}) \to C(\mathcal{O}_K)$ is surjective, which proves that $h(\mathcal{O}_K)$ divides $h(\mathcal{O})$. Furthermore, (7.25) then implies that

$$\frac{h(\mathcal{O})}{h(\mathcal{O}_K)} = |I_K(f) \cap P_K/P_{K,\mathbf{Z}}(f)|. \tag{7.26}$$

It remains to compute the order of $I_K(f) \cap P_K/P_{K,\mathbf{Z}}(f)$. The key idea is to relate this quotient to $(\mathcal{O}_K/f\mathcal{O}_K)^*$.

Given $[\alpha] \in (\mathcal{O}_K/f\mathcal{O}_K)^*$, the ideal $\alpha\mathcal{O}_K$ is prime to f and thus lies in $I_K(f) \cap P_K$. Furthermore, if $\alpha \equiv \beta \bmod f\mathcal{O}_K$, we can choose $u \in \mathcal{O}$ with $u\alpha \equiv u\beta \equiv 1 \bmod f\mathcal{O}_K$. Then the ideals $u\alpha\mathcal{O}_K$ and $u\beta\mathcal{O}_K$ lie in $P_{K,\mathbf{Z}}(f)$, and since

$$\alpha\mathcal{O}_K \cdot u\beta\mathcal{O}_K = \beta\mathcal{O}_K \cdot u\alpha\mathcal{O}_K,$$

$\alpha\mathcal{O}_K$ and $\beta\mathcal{O}_K$ lie in the same class in $I_K(f) \cap P_K/P_{K,\mathbf{Z}}(f)$. Consequently, the map

$$\phi : (\mathcal{O}_K/f\mathcal{O}_K)^* \longrightarrow I_K(f) \cap P_K/P_{K,\mathbf{Z}}(f)$$

sending $[\alpha]$ to $[\alpha\mathcal{O}_K]$ is a well-defined homomorphism.

We will first show that ϕ is surjective. An element of $I_K(f) \cap P_K$ can be written as $\alpha\mathcal{O}_K = \mathfrak{a}\mathfrak{b}^{-1}$, where $\alpha \in K$ and $\mathfrak{a}$ and $\mathfrak{b}$ are $\mathcal{O}_K$-ideals prime to f. Letting $m = N(\mathfrak{b})$, we've seen that $\overline{\mathfrak{b}} = m\mathfrak{b}^{-1}$, so that $m\alpha\mathcal{O}_K = \mathfrak{a}\overline{\mathfrak{b}}$, which implies that $m\alpha \in \mathcal{O}_K$. Note also that $m\alpha\mathcal{O}_K$ is prime to f. Since $m\mathcal{O}_K \in P_{K,\mathbf{Z}}(f)$, it follows $[\alpha\mathcal{O}_K] = [m\alpha\mathcal{O}_K] = \phi([m\alpha])$, proving that ϕ is surjective.

To determine the kernel of ϕ, we will assume that $\mathcal{O}_K^* = \{\pm 1\}$ (by Exercise 5.9, this means that $K \neq \mathbf{Q}(\sqrt{-3})$ or $\mathbf{Q}(i)$). In this case we will show that there is an exact sequence

$$1 \longrightarrow (\mathbf{Z}/f\mathbf{Z})^* \xrightarrow{\psi} (\mathcal{O}_K/f\mathcal{O}_K)^* \xrightarrow{\phi} I_K(f) \cap P_K/P_{K,\mathbf{Z}}(f) \longrightarrow 1 \tag{7.27}$$

where ψ is the obvious injection. The definition of $P_{K,\mathbf{Z}}(f)$ makes it clear that $\mathrm{im}(\psi) \subset \ker(\phi)$. Going the other way, let $[\alpha] \in \ker(\phi)$. Then $\alpha\mathcal{O}_K \in P_{K,\mathbf{Z}}(f)$, i.e., $\alpha\mathcal{O}_K = \beta\mathcal{O}_K \cdot \gamma^{-1}\mathcal{O}_K$, where β and γ satisfy $\beta \equiv b \bmod f\mathcal{O}_K$ and $\gamma \equiv c \bmod f\mathcal{O}_K$ for some $[b]$ and $[c]$ in $(\mathbf{Z}/f\mathbf{Z})^*$. Since $\mathcal{O}_K^* = \{\pm 1\}$, it follows that $\alpha = \pm\beta\gamma^{-1}$, and one then easily sees that $[b][c]^{-1} \in (\mathbf{Z}/f\mathbf{Z})^*$ maps to $[\alpha] \in (\mathcal{O}_K/f\mathcal{O}_K)^*$. This proves exactness.

It is well-known that

$$|(\mathbf{Z}/f\mathbf{Z})^*| = f\prod_{p|f}\left(1-\frac{1}{p}\right),$$

and in Exercises 7.28 and 7.29 we will show that

$$|(\mathcal{O}_K/f\mathcal{O}_K)^*| = f^2\prod_{p|f}\left(1-\frac{1}{p}\right)\left(1-\left(\frac{d_K}{p}\right)\frac{1}{p}\right).$$

Using these formulas and (7.26), we obtain

$$\frac{h(f^2 d_K)}{h(d_K)} = |I_K(f)\cap P_K/P_{K,\mathbf{Z}}(f)| = f\prod_{p|f}\left(1-\left(\frac{d_K}{p}\right)\frac{1}{p}\right),$$

which proves the desired formula since $|\mathcal{O}_K^*| = |\mathcal{O}^*| = 2$. In Exercise 7.30 we will indicate how to modify this argument when $\mathcal{O}_K^* \neq \{\pm 1\}$. Q.E.D.

This theorem may also be proved by analytic methods—see, for example, Zagier [111, §8, Exercise 8].

Using Theorem 7.24, we can relate the class numbers $h(m^2D)$ and $h(D)$ as follows:

Corollary 7.28. *Let $D \equiv 0, 1 \bmod 4$ be negative, and let m be a positive integer. Then*

$$h(m^2D) = \frac{h(D)m}{[\mathcal{O}^*:\mathcal{O}'^*]}\prod_{p|m}\left(1-\left(\frac{D}{p}\right)\frac{1}{p}\right),$$

where $\mathcal{O}$ and $\mathcal{O}'$ are the orders of discriminant D and m^2D, respectively (and $\mathcal{O}'$ has index m in $\mathcal{O}$).

Proof. Suppose that the order $\mathcal{O}$ has discriminant D and conductor f. Then the order $\mathcal{O}' \subset \mathcal{O}$ of index m has discriminant m^2D and conductor mf, and the corollary follows from Theorem 7.24 (see Exercise 7.31). This corollary is due to Gauss, and his proof may be found in *Disquisitiones* [41, §§254–256]. Q.E.D.

The only method we learned in §2 for computing class numbers $h(D)$ for $D < 0$ was to count reduced forms. This becomes awkward as $|D|$ gets large, but other methods are available. By Theorem 7.24, we are reduced to computing $h(d_K)$, and here one has the classic formula

$$h(d_K) = \sum_{n=1}^{|d_K|-1}\left(\frac{d_K}{n}\right)n, \tag{7.29}$$

where (d_K/n) is defined for $n = p_1 \cdots p_r$, p_i prime, by $(d_K/n) = \prod_{i=1}^r (d_K/p_i)$. This formula is usually proved by analytic methods (see Borevich and Shafarevich [8, Chapter 5, Section 4], or Zagier [111, §9]), but there is also a purely algebraic proof (see Orde [83]).

While (7.29) enables us to compute $h(d_K)$ for a given imaginary quadratic field, it doesn't reveal the way $h(d_K)$ grows as $|d_K|$ gets large. Gauss noticed this empirically in *Disquisitiones* [41, §302], but there were no complete proofs until the 1930s. The best result is due to Siegel [92], who proved in 1935 that

$$\lim_{d_K \to -\infty} \frac{\log h(d_K)}{\log |d_K|} = \frac{1}{2}.$$

This implies that given any $\epsilon > 0$, there is a constant $C(\epsilon)$ such that

$$h(d_K) > C(\epsilon) |d_K|^{(1/2)-\epsilon}$$

for all field discriminants $d_K < 0$. Unfortunately, the constant $C(\epsilon)$ in Siegel's proof is not effectively computable given what we currently know about L-series (these difficulties are related to the Riemann Hypothesis). However, recent work by Goldfeld, Gross, Zagier and Oesterlé has led to the weaker formula

$$h(d_K) > \frac{1}{7000} \prod_{p|d_K} \left(1 - \frac{[2\sqrt{p}]}{p+1}\right) \log |d_K|,$$

where [ ] is the greatest integer function. For a fuller discussion of this result and its implications, see Oesterlé [81] or Zagier [112].

These results on the growth of $h(d_K)$ imply that there are only finitely many orders with given class number h (see Exercise 7.32). Nevertheless, even when h is small, determining exactly which orders have class number h remains a difficult problem. For the case of class number 1, the answer is given by the following theorem due independently to Baker [3], Heegner [52] and Stark [96]:

Theorem 7.30.

(i) *If K is an imaginary quadratic field of discriminant d_K, then*

$$h(d_K) = 1 \iff d_K = -3, -4, -7, -8, -11, -19, -43, -67, -163.$$

(ii) *If $D \equiv 0, 1 \bmod 4$ is negative, then*

$$h(D) = 1 \iff D = -3, -4, -7, -8, -11, -12, -16, -19, -27, -28, -43, -67, -163.$$

Proof. First note that (i) $\Rightarrow$ (ii). To see this, assume $h(D) = 1$. If we write $D = f^2 d_K$, then Theorem 7.24 tells us that $h(d_K)|h(D)$, and thus $h(d_K) = 1$. By (i), this determines the possibilities for d_K, but we still need to see which conductors $f > 1$ can occur. First, suppose that $\mathcal{O}_K^* = \{\pm 1\}$. If $f > 2$, then

$$f \prod_{p|f} \left(1 - \left(\frac{d_K}{p}\right)\frac{1}{p}\right) > 1,$$

so that by Theorem 7.24, this case can be excluded. One then calculates directly (using (i) and Theorem 7.24) that $f = 2$ happens only when $d_K = -7$, i.e., $D = -28$. The argument when $\mathcal{O}_K^* \neq \{\pm 1\}$ is similar and is left to the reader (see Exercise 7.33).

The proof of (i) is a different matter. When the discriminant is even, the theorem was proved in §2 by an elementary argument due to Landau (see Theorem 2.18). But when the discriminant is odd, the proof is *much* more difficult. In §12 we will use modular functions and complex multiplication to give a complete proof of (i). Q.E.D.

E. Exercises

7.1. Let K be a finite extension of $\mathbb{Q}$ of degree n, and let $M \subset K$ be a finitely generated $\mathbb{Z}$-module.

(a) Prove that M is a free $\mathbb{Z}$-module.

(b) Prove that M has rank n if and only if M contains a $\mathbb{Q}$-basis of K.

7.2. Let $\mathcal{O}$ be an order in a quadratic field K. Prove that $\mathcal{O} \subset \mathcal{O}_K$.

7.3. This exercise is concerned with the conductor and discriminant of an order $\mathcal{O}$ in a quadratic field K. Let $\alpha \mapsto \alpha'$ be the nontrivial automorphism of K.

(a) If $\mathcal{O} = [\alpha, \beta]$, then the discriminant is defined to be

$$D = \left(\det\begin{pmatrix} \alpha & \beta \\ \alpha' & \beta' \end{pmatrix}\right)^2.$$

Prove that the D is independent of the basis used and hence depends only on $\mathcal{O}$.

(b) Use the basis $\mathcal{O} = [1, f w_K]$ from Lemma 7.2 to prove that $D = f^2 d_K$.

(c) Use (c) and Lemma 7.2 to prove that an order in a quadratic field is uniquely determined by its discriminant.

(d) If $D \equiv 0, 1 \bmod 4$ is nonsquare, then show that there is an order in a quadratic field whose discriminant is D.

7.4. Let $\mathcal{O}$ be an order in a quadratic field K.

(a) If $\mathfrak{a}$ is a nonzero ideal of $\mathcal{O}$, prove that $\mathfrak{a}$ contains a nonzero integer m. Hint: take $\alpha \in \mathfrak{a}$, and use Lemma 7.2 to show that $\alpha' \in \mathcal{O}$, where $\alpha \mapsto \alpha'$ is the nontrivial automorphism of K.

(b) If $\mathfrak{a}$ is a nonzero ideal of $\mathcal{O}$, show that $\mathcal{O}/\mathfrak{a}$ is finite. Hint: take the integer m from (a) and show that $\mathcal{O}/m\mathcal{O}$ is finite.

(c) Use (b) to show that every nonzero prime ideal of $\mathcal{O}$ is maximal.

(d) Use (b) to show that $\mathcal{O}$ is Noetherian.

7.5. Let $K = \mathbb{Q}(\sqrt{-3})$, and let $\mathfrak{a}$ be the ideal of $\mathcal{O} = \mathbb{Z}[\sqrt{-3}]$ generated by 2 and $1 + \sqrt{-3}$. Show that

$$\{\beta \in K : \beta\mathfrak{a} \subset \mathfrak{a}\} = \mathcal{O}_K \neq \mathcal{O}.$$

7.6. Let K be a quadratic field.

(a) Show that for any order of K, principal ideals are always proper.

(b) Show that for the maximal order $\mathcal{O}_K$, all ideals are proper.

7.7. Let $\mathcal{O}$ be an order of K, and let $\mathfrak{b} \subset K$ be an $\mathcal{O}$-module (note that $\mathfrak{b}$ need not be contained in $\mathcal{O}$). Show that $\mathfrak{b}$ is finitely generated as an $\mathcal{O}$-module if and only if $\mathfrak{b}$ is of the form $\alpha\mathfrak{a}$, where $\alpha \in K$ and $\mathfrak{a}$ is an $\mathcal{O}$-ideal.

7.8. Show that a nonzero fractional $\mathcal{O}$-ideal $\mathfrak{a}$ is a free $\mathbb{Z}$-module of rank 2. Hint: use the previous exercise and part (b) of Exercise 7.4.

7.9. Let $\mathcal{O} = \mathbb{Z}[\sqrt{-3}]$, which is an order of conductor 2 in the imaginary quadratic field $K = \mathbb{Q}(\sqrt{-3})$.

(a) Show that $C(\mathcal{O}) \simeq \{1\}$, so that the proper ideals of $\mathcal{O}$ are exactly the principal ideals. Hint: use Theorem 7.7 and what we know from §2.

(b) Show that if unique factorization holds for proper ideals of $\mathcal{O}$, then $\mathcal{O}$ is a UFD.

(c) Show that 2, $1 + \sqrt{-3}$ and $1 - \sqrt{-3}$ are irreducible (in the sense of §4) in $\mathcal{O}$. Since $4 = 2 \cdot 2 = (1 + \sqrt{-3})(1 - \sqrt{-3})$, this shows that $\mathcal{O}$ is not a UFD.

This example shows that unique factorization can fail for proper ideals.

7.10. If $\mathfrak{a}$ and $\mathfrak{b}$ are invertible fractional ideals for an order $\mathcal{O}$, then prove that $\mathfrak{a}\mathfrak{b}$ and $\mathfrak{a}^{-1}$ (where $\mathfrak{a}^{-1}$ is the fractional $\mathcal{O}$-ideal such that $\mathfrak{a}\mathfrak{a}^{-1} = \mathcal{O}$) are also invertible fractional $\mathcal{O}$-ideals.

7.11. Prove (7.10).

7.12. Let $f(x,y) = ax^2 + bxy + cy^2$ be a quadratic form with integer coefficients, and let τ be a root of $ax^2 + bx + c = 0$.

(a) Prove that $f(x,y)$ is positive definite if and only if $a > 0$ and $\tau \notin \mathbf{R}$.

(b) When $f(x,y)$ is positive definite and $\gcd(a,b,c) = 1$, prove that the discriminant of $f(x,y)$ is D, where D is the discriminant of the order $\mathcal{O} = [1, a\tau]$.

(c) Prove that two primitive positive definite forms which have the same root τ must be equal.

7.13. Let $ax^2 + bxy + cy^2$ and $a'x^2 + b'xy + c'y^2$ be two primitive positive definite forms of the same discriminant. Assume that $\gcd(a, a', (b+b')/2) = 1$, and let B be the unique integer modulo $2aa'$ which satisfies the three conditions of (7.12). Prove that $\gcd(a, a', B) = 1$.

7.14. Let $\mathcal{O} = [1, u]$ be an order in a quadratic field, and pick $\alpha = a + bu \in \mathcal{O}$, $\alpha \neq 0$. Since $\mathcal{O}$ is a ring, αu can be written $\alpha u = c + du$.

(a) Show that $N(\alpha) = ad - bc \neq 0$.

(b) Since $\alpha\mathcal{O} = [\alpha, \alpha u] = [a + bu, c + du] \subset \mathcal{O} = [1, u]$ and $ad - bc \neq 0$, it is a standard fact (proved in Exercise 7.15) that $|\mathcal{O}/\alpha\mathcal{O}| = |ad - bc|$. Thus (a) proves the general relation that $N(\alpha\mathcal{O}) = |N(\alpha)|$.

7.15. Let $M = \mathbf{Z}^2$, and let $A = \begin{pmatrix} a & b \\ c & d \end{pmatrix}$ be an integer matrix with $\det(A) = ad - bc \neq 0$. Writing $M = [e_1, e_2]$, note that $AM = [ae_1 + be_2, ce_1 + de_2]$. Our goal is to prove that $|M/AM| = |\det(A)|$. Let $A' = \begin{pmatrix} d & -b \\ -c & a \end{pmatrix}$, and note that $AA' = A'A = \det(A)I$.

(a) Show that $\det(A)M \subset AM$ and that $AM/\det(A)M \simeq M/A'M$.

(b) Use (a) and the exact sequence

$$0 \longrightarrow AM/\det(A)M \longrightarrow M/\det(A)M \longrightarrow M/AM \longrightarrow 0$$

to show that $|M/AM||M/A'M| = (\det(A))^2$.

(c) Let $\Theta = \begin{pmatrix} 0 & -1 \\ 1 & 0 \end{pmatrix}$. Using $\Theta A \Theta^{-1} = A'$, show that $\Theta : M \to M$ induces an isomorphism $M/AM \xrightarrow{\sim} M/A'M$.

(d) Conclude that $|M/AM| = |\det(A)|$.

7.16. Let $\mathcal{O}$ be the order of discriminant D in an imaginary quadratic field K, and let $\mathfrak{a}$ be a proper $\mathcal{O}$-ideal. In this exercise we will give two proofs that the class of $\overline{\mathfrak{a}}$ is the inverse of the class of $\mathfrak{a}$ in $C(\mathcal{O})$.

(a) Prove this assertion using part (iii) of Lemma 7.14.

(b) In §3, we proved that the class of the opposite $f'(x,y) = ax^2 - bxy + cy^2$ is the inverse of the class of $f(x,y) = ax^2 + bxy + cy^2$. Using the isomorphism $C(D) \simeq C(\mathcal{O})$ from Theorem 7.7, show that the class of $\overline{\mathfrak{a}}$ is the inverse of the class of $\mathfrak{a}$ in $C(\mathcal{O})$.

7.17. Let $\mathcal{O}$ be the order of discriminant D in the imaginary quadratic field K.

(a) Show that the map sending the proper $\mathcal{O}$-ideal $\mathfrak{a} = [\alpha,\beta]$ to the quadratic form

$$f(x,y) = \frac{N(\alpha x - \beta y)}{N(\mathfrak{a})}$$

induces a well-defined map $C(\mathcal{O}) \to C(D)$ which is the inverse of the map $ax^2 + bxy + cy^2 \mapsto [a,(-b+\sqrt{D})/2]$ of Theorem 7.7. Hint: use (7.16) and Exercise 7.12.

(b) Give examples to show that the map $ax^2 + bxy + cy^2 \mapsto [a,(-b+\sqrt{D})/2]$ of Theorem 7.7 is neither injective nor surjective on the level of forms and ideals.

7.18. Let $K = \mathbf{Q}(\sqrt{3})$, a field of discriminant $d_K = 12$. By (5.13), we know that $\mathcal{O}_K = \mathbf{Z}[\sqrt{3}]$.

(a) Use the absolute value of the norm function to show that $\mathcal{O}_K$ is Euclidean, and conclude that $C(\mathcal{O}_K) \simeq \{1\}$.

(b) Show that the form class group $C(d_K) = C(12)$ is nontrivial. Hint: show that the forms $\pm(x^2 - 3y^2)$ are not properly equivalent. You will need to show that the equation $a^2 - 3c^2 = -1$ has no solutions.

This shows that $C(d_K) \not\simeq C(\mathcal{O}_K)$ for $K = \mathbf{Q}(\sqrt{3})$.

7.19. In Exercises 7.19–7.24 we will explore two versions of Theorem 7.7 that hold for real quadratic fields K. To begin, we will study the orientation of a basis α, β of a proper ideal $\mathfrak{a} = [\alpha,\beta]$ of an order $\mathcal{O}$ in K. Let $\alpha \mapsto \alpha'$ denote the nontrivial automorphism of K.

(a) Prove that $\alpha'\beta - \alpha\beta' \in \mathbf{R}^*$. We then define $\mathrm{sgn}(\alpha,\beta)$ to be the sign of the nonzero real number $\alpha'\beta - \alpha\beta'$.

(b) Let $\left(\begin{smallmatrix} p & q \\ r & s \end{smallmatrix}\right) \in \mathrm{GL}(2,\mathbf{Z})$, and set $\tilde{\alpha} = p\alpha + q\beta$, $\tilde{\beta} = r\alpha + s\beta$. Note that $\mathfrak{a} = [\alpha,\beta] = [\tilde{\alpha},\tilde{\beta}]$. Prove that

$$\mathrm{sgn}(\tilde{\alpha},\tilde{\beta}) = \det\begin{pmatrix} p & q \\ r & s \end{pmatrix} \mathrm{sgn}(\alpha,\beta).$$

We say that α, β are positively oriented if $\mathrm{sgn}(\alpha,\beta) > 0$ and negatively oriented otherwise. By (b), two bases of $\mathfrak{a}$ have the same orientation if and only if their transition matrix is in $\mathrm{SL}(2,\mathbf{Z})$.

7.20. Theorem 7.7 was proved using a map from quadratic forms to ideals. In the real quadratic case, such a map is harder to describe (see Exercise 7.24), but it is relatively easy to go from ideals to forms. The goal of this exercise is to show how this is done. Let $\mathcal{O}$ be an order in a real quadratic field K, and let $\mathfrak{a} = [\alpha, \beta]$ be a proper $\mathcal{O}$-ideal. Then define the quadratic form $f(x,y)$ by the formula

$$f(x,y) = \frac{N(\alpha x - \beta y)}{N(\mathfrak{a})}.$$

At this point, all we know is that $f(x,y)$ has rational coefficients. Let $\tau = \beta/\alpha$, and let $ax^2 + bx + c$ be the minimal polynomial of τ. We can assume that $a, b, c \in \mathbf{Z}$, $a > 0$ and $\gcd(a,b,c) = 1$.

(a) Prove that $N(\mathfrak{a}) = |N(\alpha)|/a$. Hint: adapt the proof of (7.16) to the real quadratic case. Exercise 7.14 will be useful.

(b) Use (a) to prove that $f(x,y) = \operatorname{sgn}(N(\alpha))(ax^2 + bxy + cy^2)$. Thus $f(x,y)$ has relatively prime integer coefficients.

(c) Prove that the discriminant of $f(x,y)$ is D, where D is the discriminant of $\mathcal{O}$. Hint: see Exercise 7.12.

7.21. In this exercise we will construct a bijection $C^+(\mathcal{O}) \simeq C(D)$, where $C^+(\mathcal{O})$ is defined in the text.

(a) Let $\mathfrak{a}$ be a proper $\mathcal{O}$-ideal, and write $\mathfrak{a} = [\alpha, \beta]$ where $\operatorname{sgn}(\alpha, \beta) > 0$ (see Exercise 7.19). Then let $f(x,y)$ be the corresponding quadratic form defined in Exercise 7.20. If $\tilde{\alpha}$, $\tilde{\beta}$ is another positively oriented basis of $\mathfrak{a}$, then show that the corresponding form $g(x,y)$ from Exercise 7.20 is properly equivalent to $f(x,y)$. Furthermore, show that *all* forms properly equivalent to $f(x,y)$ arise in this way.

(b) If $\lambda \in \mathcal{O}$ and $N(\lambda) > 0$, then show that $\lambda\mathfrak{a}$ gives the same class of forms as $\mathfrak{a}$. Hint: show that $\operatorname{sgn}(\lambda\alpha, \lambda\beta) = \operatorname{sgn}(N(\lambda))\operatorname{sgn}(\alpha, \beta)$.

(c) From (a), (b) and Exercise 7.20 we get a well-defined map $C^+(\mathcal{O}) \to C(D)$. To show that the map is injective, suppose that $\mathfrak{a}$ and $\tilde{\mathfrak{a}}$ give the same class in $C(D)$. By (a), we can choose positively oriented bases $\mathfrak{a} = [\alpha, \beta]$ and $\tilde{\mathfrak{a}} = [\tilde{\alpha}, \tilde{\beta}]$ which give the same form $f(x,y)$.

  (i) Using Exercise 7.19, show that $\operatorname{sgn}(N(\alpha)) = \operatorname{sgn}(N(\tilde{\alpha}))$, i.e., $N(\alpha\tilde{\alpha}) > 0$. Then replacing $\mathfrak{a}$ and $\tilde{\mathfrak{a}}$ by $\alpha\tilde{\alpha}\mathfrak{a}$ and $\alpha^2\tilde{\mathfrak{a}}$ respectively allows us to assume that $\alpha = \tilde{\alpha}$, i.e., $\mathfrak{a} = [\alpha, \beta]$ and $\tilde{\mathfrak{a}} = [\alpha, \tilde{\beta}]$.

  (ii) Let $\tau = \beta/\alpha$ and $\tilde{\tau} = \tilde{\beta}/\alpha$. Show that $f(\tau, 1) = f(\tilde{\tau}, 1) = 0$, so that $\tilde{\tau} = \tau$ or τ'. Then show that $\tilde{\tau} = \tau'$ contradicts $\operatorname{sgn}(\alpha, \tilde{\beta}) > 0$, which proves that $\beta = \tilde{\beta}$.

(d) To prove surjectivity, let $f(x,y) = ax^2 + bxy + cy^2$ be a form of discriminant D, and let τ be either of the roots of $ax^2 + bx + c = 0$. First show that $a\tau \in \mathcal{O}$. Then define an $\mathcal{O}$-ideal $\mathfrak{a}$ as follows: if $a > 0$, then

$$\mathfrak{a} = [a, a\tau] \qquad \text{where } f(\tau,1) = 0,\ \operatorname{sgn}(1,\tau) > 0,$$

and if $a < 0$, then

$$\mathfrak{a} = \sqrt{d_K}[a, a\tau] \qquad \text{where } f(\tau,1) = 0,\ \operatorname{sgn}(1,\tau) < 0.$$

Show that $\mathfrak{a}$ is a proper $\mathcal{O}$-ideal and that the form corresponding to $\mathfrak{a}$ from Exercise 7.20 is exactly $f(x,y)$.

This completes the proof that $C^+(\mathcal{O}) \to C(D)$ is a bijection.

7.22. In this exercise we will construct a bijection $C(\mathcal{O}) \simeq C_s(D)$, where $C_s(D)$ is defined in the text. Our treatment of $C_s(D)$ is based on Zagier [111, §8].

(a) Let $\mathfrak{a} = [\alpha, \beta]$ be a proper $\mathcal{O}$-ideal, where this time we make no assumptions about $\operatorname{sgn}(\alpha,\beta)$. Define $f(x,y)$ to be the quadratic form

$$f(x,y) = \operatorname{sgn}(\alpha,\beta)\frac{N(\alpha x - \beta y)}{N(\mathfrak{a})},$$

which by Exercise 7.20 has relatively prime integer coefficients and discriminant D. Show that as we vary over all bases of $\mathfrak{a}$, the corresponding forms vary over all forms signed equivalent to $f(x,y)$.

(b) Show that the map $\mathfrak{a} \mapsto f(x,y)$ of (a) induces a well-defined bijection $C(\mathcal{O}) \simeq C_s(D)$. Hint: adapt the arguments of parts (b)–(d) of Exercise 7.21.

7.23. This exercise will explore the relations between $C(\mathcal{O})$, $C^+(\mathcal{O})$, $C(D)$ and $C_s(D)$.

(a) Let K be an imaginary quadratic field.

(i) Show that $P^+(\mathcal{O}) = P(\mathcal{O})$, so that $C^+(\mathcal{O})$ always equals $C(\mathcal{O})$.

(ii) The relation between $C(D)$ and $C_s(D)$ is more interesting. Namely, in $C(D)$, we had to explicitly assume that we were only dealing with positive definite forms. However, in $C_s(D)$, one uses both positive definite and negative definite forms. Show that any negative definite form is signed equivalent to a positive definite one, and conclude that $C(D) \simeq C_s(D)$.

(b) Now assume that K is a real quadratic field.

(i) Show that there are natural surjections

$$C^+(\mathcal{O}) \longrightarrow C(\mathcal{O})$$
$$C(D) \longrightarrow C_s(D)$$

which fit together with the bijections of Exercises 7.21 and 7.22 to give a commutative diagram

$$\begin{array}{ccc} C^+(\mathcal{O}) & \xrightarrow{\sim} & C(D) \\ \downarrow & & \downarrow \\ C(\mathcal{O}) & \xrightarrow{\sim} & C_s(D) \end{array} .$$

(ii) Show that the kernel of $C^+(\mathcal{O}) \to C(\mathcal{O})$ is $P(\mathcal{O})/P^+(\mathcal{O})$ and that $P(\mathcal{O}) = P^+(\mathcal{O}) \cup \sqrt{d_K}P^+(\mathcal{O})$. Then conclude that

$$\frac{|C^+(\mathcal{O})|}{|C(\mathcal{O})|} = \begin{cases} 1 & \text{if } \mathcal{O} \text{ has a unit of norm } -1 \\ 2 & \text{otherwise.} \end{cases}$$

(iii) From (i) and (ii), conclude that

$$\frac{|C(D)|}{|C_s(D)|} = \begin{cases} 1 & \text{if } \mathcal{O} \text{ has a unit of norm } -1 \\ 2 & \text{otherwise.} \end{cases}$$

7.24. Write down inverses to the bijections $C^+(\mathcal{O}) \xrightarrow{\sim} C(D)$ and $C(\mathcal{O}) \xrightarrow{\sim} C_s(D)$ of Exercises 7.21 and 7.22. Hint: see part (d) of Exercise 7.21. Note that the answer is more complicated than the map $ax^2 + bxy + cy^2 \to [a, (-b+\sqrt{D})/2]$ of Theorem 7.7.

7.25. Let $\phi : \{\mathcal{O}_K\text{-ideals prime to } f\} \to \{\mathcal{O}\text{-ideals prime to } f\}$ be a bijection which preserves multiplication. Show that ϕ extends to an isomorphism $\overline{\phi} : I_K(f) \xrightarrow{\sim} I(\mathcal{O}, f)$.

7.26. Let $\mathcal{O}$ be an order of conductor f.

(a) Let $\mathfrak{a}$ be an ideal of $\mathcal{O}$ which is relatively prime to f. Prove that $\mathfrak{a}$ is a prime $\mathcal{O}$-ideal if and only if $\mathfrak{a}\mathcal{O}_K$ is a prime $\mathcal{O}_K$-ideal. Hint: use Proposition 7.20 to show that $\mathcal{O}/\mathfrak{a} \simeq \mathcal{O}_K/\mathfrak{a}\mathcal{O}_K$.

(b) Use (a) and the unique factorization of ideals in $\mathcal{O}_K$ to show that $\mathcal{O}$-ideals relatively prime to the conductor can be factored uniquely into prime $\mathcal{O}$-ideals (which are also relatively prime to f).

7.27. If $\alpha, \beta \in \mathcal{O}_K$ and $\alpha \equiv \beta \bmod m\mathcal{O}_K$ for some integer m, then prove that $N(\alpha) \equiv N(\beta) \bmod m$.

7.28. Let K be a quadratic field, and let $\mathfrak{p}$ be prime in $\mathcal{O}_K$. The goal of this exercise is to prove that

$$|(\mathcal{O}_K/\mathfrak{p}^n)^*| = N(\mathfrak{p})^{n-1}(N(\mathfrak{p}) - 1).$$

The formula is true if $n = 1$, and the general case follows easily by induction once we prove that there is an exact sequence

$$1 \longrightarrow \mathcal{O}_K/\mathfrak{p} \xrightarrow{\phi} (\mathcal{O}_K/\mathfrak{p}^n)^* \longrightarrow (\mathcal{O}_K/\mathfrak{p}^{n-1})^* \longrightarrow 1$$

for $n \geq 2$. For the rest of the exercise fix an integer $n \geq 2$.

(a) Show that $(\mathcal{O}_K/\mathfrak{p}^n)^* \to (\mathcal{O}_K/\mathfrak{p}^{n-1})^*$ is onto. Hint: take $[\alpha] \in (\mathcal{O}_K/\mathfrak{p}^{n-1})^*$, which means that $\alpha\beta = 1+\gamma$, where $\beta \in \mathcal{O}_K$ and $\gamma \in \mathfrak{p}^{n-1}$. Then show that $\alpha(\beta+\gamma\delta)-1 \in \mathfrak{p}^n$ for an appropriately chosen δ.

(b) By unique factorization, we know that $\mathfrak{p}^n$ is a proper subset of $\mathfrak{p}^{n-1}$. Pick $u \in \mathfrak{p}^{n-1}$ such that $u \notin \mathfrak{p}^n$.

  (i) Given $\alpha \in \mathcal{O}_K$, show that $[1+\alpha u] \in (\mathcal{O}_K/\mathfrak{p}^n)^*$.

  (ii) From (i), it is easy to define a map $\phi : \mathcal{O}_K/\mathfrak{p} \to (\mathcal{O}_K/\mathfrak{p}^n)^*$. With this definition of ϕ, show that the above sequence is exact.

7.29. Let K be an imaginary quadratic field.

(a) Let $\mathfrak{a} = \prod_{i=1}^r \mathfrak{p}_i^{n_i}$ be the factorization of $\mathfrak{a}$ into primes. Show that there is a natural isomorphism

$$\mathcal{O}_K/\mathfrak{a} \simeq \prod_{i=1}^r (\mathcal{O}_K/\mathfrak{p}_i)^{n_i}.$$

This is the Chinese Remainder Theorem for $\mathcal{O}_K$. Hint: it is easy to construct a map and show it is injective. Then use part (ii) of Lemma 7.14.

(b) Use (a) and the previous exercise to show that if $\mathfrak{a}$ is a nonzero ideal of $\mathcal{O}_K$, then

$$|(\mathcal{O}_K/\mathfrak{a})^*| = N(\mathfrak{a})\prod_{\mathfrak{p}|\mathfrak{a}}\left(1-\frac{1}{N(\mathfrak{p})}\right).$$

Notice the similarity to the usual formula for $\phi(n) = |(\mathbf{Z}/n\mathbf{Z})^*|$.

(c) If m is a positive integer, conclude that

$$|(\mathcal{O}_K/m\mathcal{O}_K)^*| = m^2\prod_{p|m}\left(1-\frac{1}{p}\right)\left(1-\left(\frac{d_K}{p}\right)\frac{1}{p}\right),$$

where (d_K/p) is the Kronecker symbol when $p = 2$.

7.30. Let K be any quadratic field, and let f be a positive integer.

(a) Use the obvious maps

$$\{\pm 1\} \longrightarrow (\mathbf{Z}/f\mathbf{Z})^* \times \mathcal{O}_K^*$$
$$(\mathbf{Z}/f\mathbf{Z})^* \times \mathcal{O}_K^* \longrightarrow (\mathcal{O}_K/f\mathcal{O}_K)^*$$

and the maps from (7.27) to prove that there is an exact sequence

$$1 \longrightarrow \{\pm 1\} \longrightarrow (\mathbf{Z}/f\mathbf{Z})^* \times \mathcal{O}_K^* \longrightarrow (\mathcal{O}_K/f\mathcal{O}_K)^*$$
$$\longrightarrow I_K(f) \cap P_K/P_{K,\mathbf{Z}}(f) \longrightarrow 1.$$

Notice that when $\mathcal{O}_K^* = \{\pm 1\}$, this sequence is equivalent to (7.27).

(b) Use the exact sequence of (a) to prove Theorem 7.24 for all imaginary quadratic fields.

7.31. Prove Corollary 7.28.

7.32. In this exercise we will use the inequality

$$(*) \qquad h(d_K) > \frac{1}{7000} \prod_{p \mid d_K} \left(1 - \frac{[2\sqrt{p}]}{p+1}\right) \log |d_K|$$

to study the equation $h(d_K) = h$, where $h > 0$ is a fixed integer and d_K varies over all negative discriminants.

(a) Show that $1 - [2\sqrt{p}]/(p+1) \geq 1/2$ when $p \geq 11$.

(b) If $h(d_K) = h$, then use (a) and genus theory to conclude that

$$\prod_{p \mid d_K} \left(1 - \frac{[2\sqrt{p}]}{p+1}\right) \geq \frac{1}{3 \cdot 2^{\nu_2(h)+2}},$$

where $\nu_2(h)$ is the highest power of 2 dividing h. Hint: use Theorem 3.15 or 6.1 to show that d_K is divisible by at most $\nu_2(h) + 1$ distinct primes.

(c) If $h(d_K) = h$, then show that $(*)$ gives us the following estimate for $|d_K|$:

$$|d_K| \leq e^{21000 \cdot 2^{\nu_2(h)+2} h}.$$

This proves that there are only finitely many negative discriminants with class number at most h. Better bounds for $|d_K|$ can be derived from $(*)$ (see Oesterlé [81]), but the constant $1/7000$ in $(*)$ limits their usefulness. For discriminants prime to 5077, Oesterlé hopes to improve this constant from $1/7000$ to $1/55$, which would give an estimate strong enough to solve the class number 3 problem (see [81]).

(d) If h is fixed and $D \equiv 0, 1 \bmod 4$ varies over all negative integers, show that the equation $h(D) = h$ has only finitely many solutions. Hint: use genus theory to bound the number of primes dividing D, and then use Theorem 7.24.

7.33. In Theorem 7.30, complete the proof of (i) $\Rightarrow$ (ii) sketched in the text.

§8. CLASS FIELD THEORY AND THE ČEBOTAREV DENSITY THEOREM

In this section we will present a classical formulation of class field theory, where Abelian extensions of a number field are described in terms of certain *generalized ideal class groups*. After stating the main theorems (without proof), we will illustrate their use by proving the Kronecker–Weber Theorem and the existence of the Hilbert class field. We will then discuss generalized reciprocity theorems for the nth power Legendre symbol $(\alpha/\mathfrak{p})_n$ and show how quadratic reciprocity follows from class field theory.

The Čebotarev Density Theorem hasn't been mentioned before, but it provides some important information about the behavior of the Artin map. One of its classic applications is Dirichlet's theorem on primes in arithmetic progressions, and in §9 we will use the same methods to study primes represented by a given quadratic form. Another consequence of the Density Theorem is that a Galois extension of a number field is determined uniquely by the primes in the base field that split completely in the extension. As we will see, this is closely related to our basic problem of characterizing the primes represented by $x^2 + ny^2$.

Our account of class field theory will be incomplete in several ways, and at the end of the section we will discuss two of the most obvious omissions, norms and ideles.

A. The Theorems of Class Field Theory

We begin our treatment of class field theory with the notion of a *modulus*. Given a number field K, a *modulus* in K is a formal product

$$\mathfrak{m} = \prod_{\mathfrak{p}} \mathfrak{p}^{n_{\mathfrak{p}}}$$

over all primes $\mathfrak{p}$, finite or infinite, of K, where the exponents must satisfy:

(i) $n_{\mathfrak{p}} \geq 0$, and at most finitely many are nonzero.
(ii) $n_{\mathfrak{p}} = 0$ wherever $\mathfrak{p}$ is a complex infinite prime.
(iii) $n_{\mathfrak{p}} \leq 1$ whenever $\mathfrak{p}$ is a real infinite prime.

A modulus $\mathfrak{m}$ may thus be written $\mathfrak{m}_0\mathfrak{m}_\infty$, where $\mathfrak{m}_0$ is an $\mathcal{O}_K$-ideal and $\mathfrak{m}_\infty$ is a product of distinct real infinite primes of K. When all of the exponents $n_{\mathfrak{p}} = 0$, we set $\mathfrak{m} = 1$. Note that for a purely imaginary field K

(the case we're most interested in), a modulus may be regarded simply as an ideal of $\mathcal{O}_K$.

Given a modulus $\mathfrak{m}$, let $I_K(\mathfrak{m})$ be the group of all fractional $\mathcal{O}_K$-ideals relatively prime to $\mathfrak{m}$ (which means relatively prime to $\mathfrak{m}_0$), and let $P_{K,1}(\mathfrak{m})$ be the subgroup of $I_K(\mathfrak{m})$ generated by the principal ideals $\alpha\mathcal{O}_K$, where $\alpha \in \mathcal{O}_K$ satisfies

$$\alpha \equiv 1 \bmod \mathfrak{m}_0 \text{ and } \sigma(\alpha) > 0 \text{ for every real infinite prime } \sigma \text{ dividing } \mathfrak{m}_\infty.$$

A basic result is that $P_{K,1}(\mathfrak{m})$ has finite index in $I_K(\mathfrak{m})$. When K is imaginary quadratic, this is proved in Exercise 8.1, while the general case may be found in Janusz [62, Chapter IV.1]. A subgroup $H \subset I_K(\mathfrak{m})$ is called a *congruence subgroup* for $\mathfrak{m}$ if it satisfies

$$P_{K,1}(\mathfrak{m}) \subset H \subset I_K(\mathfrak{m}),$$

and the quotient

$$I_K(\mathfrak{m})/H$$

is called a *generalized ideal class group* for $\mathfrak{m}$.

For an example of these concepts, consider the modulus $\mathfrak{m} = 1$. Then $P_K = P_{K,1}(1)$ is a congruence subgroup, so that the ideal class group $C(\mathcal{O}_K) = I_K/P_K$ is a generalized ideal class group. We also get some interesting examples from §7. Let $\mathcal{O}$ be an order of conductor f in an imaginary quadratic field K. In Proposition 7.22 we proved that the ideal class group $C(\mathcal{O})$ can be written

$$C(\mathcal{O}) \simeq I_K(f)/P_{K,\mathbf{Z}}(f),$$

where $P_{K,\mathbf{Z}}(f)$ is generated by the principal ideals $\alpha\mathcal{O}_K$ for $\alpha \equiv a \bmod f\mathcal{O}_K$, $a \in \mathbf{Z}$ and $\gcd(a,f) = 1$. If we use the modulus $f\mathcal{O}_K$, then the definition of $P_{K,1}(f\mathcal{O}_K)$ shows that

$$P_{K,1}(f\mathcal{O}_K) \subset P_{K,\mathbf{Z}}(f) \subset I_K(f) = I_K(f\mathcal{O}_K), \tag{8.1}$$

and thus $P_{K,\mathbf{Z}}(f)$ is a congruence subgroup for $f\mathcal{O}_K$. This proves that $C(\mathcal{O})$ is a generalized ideal class group of K for the modulus $f\mathcal{O}_K$. In §7, the group $P_{K,\mathbf{Z}}(f)$ seemed awkward, but it's a very natural object from the point of view of class field theory.

The basic idea of class field theory is that the generalized ideal class groups are the Galois groups of all Abelian extensions of K, and the link between these two is provided by the Artin map. To make this precise, we need to define the *Artin map* of an Abelian extension of K.

Let $\mathfrak{m}$ be a modulus divisible by all ramified primes of an Abelian extension $K \subset L$. Given a prime $\mathfrak{p}$ not dividing $\mathfrak{m}$, we have the Artin symbol

$$\left(\frac{L/K}{\mathfrak{p}}\right) \in \mathrm{Gal}(L/K)$$

from §5. As in the discussion preceding Theorem 5.23, the Artin symbol extends by multiplicativity to give us a homomorphism

$$\Phi_{\mathfrak{m}} : I_K(\mathfrak{m}) \longrightarrow \mathrm{Gal}(L/K)$$

which is called the *Artin map* for $K \subset L$ and $\mathfrak{m}$. When we want to refer explicitly to the extension involved, we will write $\Phi_{L/K,\mathfrak{m}}$ instead of $\Phi_{\mathfrak{m}}$.

The first theorem of class field theory tells us that $\mathrm{Gal}(L/K)$ is a generalized ideal class group for some modulus:

Theorem 8.2. *Let $K \subset L$ be an Abelian extension, and let $\mathfrak{m}$ be a modulus divisible by all primes of K, finite or infinite, that ramify in L. Then:*

(i) *The Artin map $\Phi_{\mathfrak{m}}$ is surjective.*

(ii) *If the exponents of the finite primes dividing $\mathfrak{m}$ are sufficiently large, then $\ker(\Phi_{\mathfrak{m}})$ is a congruence subgroup for $\mathfrak{m}$, i.e.,*

$$P_{K,1}(\mathfrak{m}) \subset \ker(\Phi_{\mathfrak{m}}) \subset I_K(\mathfrak{m}),$$

and consequently the isomorphism

$$I_K(\mathfrak{m})/\ker(\Phi_{\mathfrak{m}}) \xrightarrow{\sim} \mathrm{Gal}(L/K)$$

shows that $\mathrm{Gal}(L/K)$ *is a generalized ideal class group for the modulus* $\mathfrak{m}$.

Proof. See Janusz [62, Chapter V, Theorem 5.7]. Q.E.D.

This theorem is sometimes called the *Artin Reciprocity Theorem*. The key ingredient is the condition $P_{K,1}(\mathfrak{m}) \subset \ker(\Phi_{\mathfrak{m}})$, for it says (roughly) that the Artin symbol $((L/K)/\mathfrak{p})$ depends only on $\mathfrak{p}$ up to multiplication by α, $\alpha \equiv 1 \bmod \mathfrak{m}$. Later in this section we will see how Artin Reciprocity relates to quadratic, cubic and biquadratic reciprocity.

Let's work out an example of Theorem 8.2. Consider the extension $\mathbf{Q} \subset \mathbf{Q}(\zeta_m)$, where $\zeta_m = e^{2\pi i/m}$ is a primitive mth of unity, and let $\mathfrak{m}$ be the modulus $m\infty$, where ∞ is the real infinite prime of $\mathbf{Q}$. Using Proposition 5.11, one sees that any prime not dividing m is unramified in $\mathbf{Q}(\zeta_m)$ (see Exercise 8.2), and it follows that the Artin map

$$\Phi_{\mathfrak{m}} : I_{\mathbf{Q}}(\mathfrak{m}) \longrightarrow \mathrm{Gal}(\mathbf{Q}(\zeta_m)/\mathbf{Q}) \simeq (\mathbf{Z}/m\mathbf{Z})^*$$

is defined. $\Phi_{\mathfrak{m}}$ can be described as follows: given $(a/b)\mathbf{Z} \in I_{\mathbf{Q}}(\mathfrak{m})$, where $(a/b) > 0$ and $\gcd(a,m) = \gcd(b,m) = 1$, then

$$\Phi_{\mathfrak{m}}\left(\frac{a}{b}\mathbf{Z}\right) = [a][b]^{-1} \in (\mathbf{Z}/m\mathbf{Z})^*. \tag{8.3}$$

It follows easily that

$$\text{(8.4)} \qquad \ker(\Phi_{\mathfrak{m}}) = P_{\mathbf{Q},1}(\mathfrak{m})$$

(see Exercise 8.2). The importance of this computation will soon become clear.

One difficulty with Theorem 8.2 is that the $\mathfrak{m}$ for which $\ker(\Phi_{\mathfrak{m}})$ is a congruence subgroup is not unique. In fact, if $P_{K,1}(\mathfrak{m}) \subset \ker(\Phi_{\mathfrak{m}})$ and $\mathfrak{n}$ is any modulus divisible by $\mathfrak{m}$ (it's clear what this means), then

$$P_{K,1}(\mathfrak{m}) \subset \ker(\Phi_{\mathfrak{m}}) \Rightarrow P_{K,1}(\mathfrak{n}) \subset \ker(\Phi_{\mathfrak{n}})$$

(see Exercise 8.4), so that $\text{Gal}(L/K)$ is a generalized ideal class group for infinitely many moduli. However, there is one modulus which is better than the others:

Theorem 8.5. *Let $K \subset L$ be an Abelian extension. Then there is a modulus $\mathfrak{f} = \mathfrak{f}(L/K)$ such that*

(i) *A prime of K, finite or infinite, ramifies in L if and only if it divides $\mathfrak{f}$.*

(ii) *Let $\mathfrak{m}$ be a modulus divisible by all primes of K which ramify in L. Then $\ker(\Phi_{\mathfrak{m}})$ is a congruence subgroup for $\mathfrak{m}$ if and only if $\mathfrak{f} \mid \mathfrak{m}$.*

Proof. See Janusz [62, Chapter V, §6 and Theorem 12.7]. Q.E.D.

The modulus $\mathfrak{f}(L/K)$ is uniquely determined by $K \subset L$ and is called the *conductor* of the extension, and for this reason Theorem 8.5 is often called the *Conductor Theorem*. In Exercise 8.5 we will compute the conductor of $\mathbf{Q} \subset \mathbf{Q}(\zeta_m)$ (it need not be m), and in §9 we will compute the conductor of a ring class field.

The final theorem of class field theory is the *Existence Theorem*, which asserts that *every* generalized ideal class group is the Galois group of some Abelian extension $K \subset L$. More precisely:

Theorem 8.6. *Let $\mathfrak{m}$ be a modulus of K, and let H be a congruence subgroup for $\mathfrak{m}$, i.e.,*

$$P_{K,1}(\mathfrak{m}) \subset H \subset I_K(\mathfrak{m}).$$

Then there is a unique Abelian extension L of K, all of whose ramified primes, finite or infinite, divide $\mathfrak{m}$, such that if

$$\Phi_{\mathfrak{m}} : I_K(\mathfrak{m}) \longrightarrow \text{Gal}(L/K)$$

is the Artin map of $K \subset L$, then

$$H = \ker(\Phi_{\mathfrak{m}}).$$

Proof. See Janusz [62, Chapter V, Theorem 9.16]. Q.E.D.

The importance of this theorem is that it allows us to construct Abelian extensions of K with specified Galois group and restricted ramification. This will be very useful in the applications that follow.

Now that we've stated the basic theorems of class field theory, the next step is to indicate how they are used. We will start with two of the nicest applications: proofs of the Kronecker–Weber Theorem and the existence of the Hilbert class field. A key tool in both proofs is the following corollary of the uniqueness part of Theorem 8.6:

Corollary 8.7. *Let L and M be Abelian extensions of K. Then $L \subset M$ if and only if there is a modulus $\mathfrak{m}$, divisible by all primes of K ramified in either L or M, such that*

$$P_{K,1}(\mathfrak{m}) \subset \ker(\Phi_{M/K,\mathfrak{m}}) \subset \ker(\Phi_{L/K,\mathfrak{m}}).$$

Proof. First, assume that $L \subset M$, and let $r : \mathrm{Gal}(M/K) \to \mathrm{Gal}(L/K)$ be the restriction map. By Theorem 8.2 and Exercise 8.4, there is a modulus $\mathfrak{m}$ for which $\Phi_{L/K,\mathfrak{m}}$ and $\Phi_{M/K,\mathfrak{m}}$ are both congruence subgroups for $\mathfrak{m}$. The proof of Exercise 5.13 shows that $r \circ \Phi_{M/K,\mathfrak{m}} = \Phi_{L/K,\mathfrak{m}}$, and then $\ker(\Phi_{M/K,\mathfrak{m}}) \subset \ker(\Phi_{L/K,\mathfrak{m}})$ follows immediately.

Going the other way, assume that $P_{K,1}(\mathfrak{m}) \subset \ker(\Phi_{M/K,\mathfrak{m}}) \subset \ker(\Phi_{L/K,\mathfrak{m}})$. Then, under the map $\Phi_{M/K,\mathfrak{m}} : I_K(\mathfrak{m}) \to \mathrm{Gal}(M/K)$, the subgroup $\ker(\Phi_{L/K,\mathfrak{m}}) \subset I_K(\mathfrak{m})$ maps to a subgroup $H \subset \mathrm{Gal}(M/K)$. By Galois theory, H corresponds to an intermediate field $K \subset \tilde{L} \subset M$. The first part of the proof, applied to $\tilde{L} \subset M$, shows that $\ker(\Phi_{\tilde{L}/K,\mathfrak{m}}) = \ker(\Phi_{L/K,\mathfrak{m}})$. Then the uniqueness part of Theorem 8.6 shows that $L = \tilde{L} \subset M$, and we are done. Q.E.D.

We can now prove the Kronecker–Weber Theorem, which classifies all Abelian extensions of $\mathbb{Q}$:

Theorem 8.8. *Let L be an Abelian extension of $\mathbb{Q}$. Then there is a positive integer m such that $L \subset \mathbb{Q}(\zeta_m)$, $\zeta_m = e^{2\pi i/m}$.*

Proof. By the Artin Reciprocity Theorem (Theorem 8.2), there is a modulus $\mathfrak{m}$ such that $P_{\mathbb{Q},1}(\mathfrak{m}) \subset \ker(\Phi_{L/\mathbb{Q},\mathfrak{m}})$, and by Exercise 8.4, we may assume that $\mathfrak{m} = m\infty$. By (8.4) we know that $P_{\mathbb{Q},1}(\mathfrak{m}) = \ker(\Phi_{\mathbb{Q}(\zeta_m)/\mathbb{Q},\mathfrak{m}})$, so that

$$P_{\mathbb{Q},1}(\mathfrak{m}) = \ker(\Phi_{\mathbb{Q}(\zeta_m)/\mathbb{Q},\mathfrak{m}}) \subset \ker(\Phi_{L/K,\mathfrak{m}}).$$

Then $L \subset \mathbb{Q}(\zeta_m)$ follows from Corollary 8.7. Q.E.D.

We should mention that the Kronecker–Weber Theorem can be proved without using class field theory (see Marcus [77, Chapter 4, Exercises 29–36]).

Next, let's discuss the Hilbert class field. To define it, apply the Existence Theorem (Theorem 8.6) to the modulus $\mathfrak{m} = 1$ and the subgroup $P_K \subset I_K$ (note that $P_K = P_{K,1}(\mathfrak{m})$ in this case). Thus there is a unique Abelian extension L of K, unramified since $\mathfrak{m} = 1$, such that the Artin map induces an isomorphism

$$C(\mathcal{O}_K) = I_K/P_K \xrightarrow{\sim} \mathrm{Gal}(L/K). \tag{8.9}$$

L is the *Hilbert class field* of K, and its main property is the following:

Theorem 8.10. *The Hilbert class field L is the maximal unramified Abelian extension of K.*

Proof. We already know that L is an unramified extension. Let M be another unramified extension. The first part of the Conductor Theorem (Theorem 8.5) implies that $\mathfrak{f}(M/K) = 1$ since a prime ramifies if and only if it divides the conductor, and then the second part tells us that $\ker(\Phi_{M/K,1})$ is a congruence subgroup for the modulus 1, so that

$$P_K \subset \ker(\Phi_{M/K,1}).$$

By the definition of the Hilbert class field, this becomes

$$P_K = \ker(\Phi_{L/K,1}) \subset \ker(\Phi_{M/K,1}),$$

and then $M \subset L$ follows from Corollary 8.7. Q.E.D.

Notice that Theorems 5.18 and 5.23 from §5 are immediate consequences of (8.9) and Theorem 8.10.

There is a generalization of the Hilbert class field called the *ray class field*. Namely, given any modulus $\mathfrak{m}$, the Existence Theorem shows that there is a unique Abelian extension $K_{\mathfrak{m}}$ of K such that

$$P_{K,1}(\mathfrak{m}) = \ker(\Phi_{K_{\mathfrak{m}}/K,\mathfrak{m}}).$$

$K_{\mathfrak{m}}$ is called the *ray class field* for the modulus $\mathfrak{m}$, and when $\mathfrak{m} = 1$, this reduces to the Hilbert class field. Another example is given by the cyclotomic field $\mathbb{Q}(\zeta_m)$: here, (8.4) shows that $\mathbb{Q}(\zeta_m)$ is the ray class field of $\mathbb{Q}$ for the modulus $m\infty$. We also get a nice interpretation of the conductor $\mathfrak{f}(L/K)$ of an arbitrary Abelian extension L of K: it's the smallest modulus $\mathfrak{m}$ for which L is contained in the ray class field $K_{\mathfrak{m}}$ (see Exercise 8.6).

Besides proving these classical results, class field theory is also the source of most reciprocity theorems. In particular, we will discuss some reciprocity theorems for the nth power Legendre symbol $(\alpha/\mathfrak{p})_n$ mentioned in §5. To define this symbol, let K be a number field containing a primitive nth root of unity ζ, and let $\mathfrak{p}$ be a prime ideal of $\mathcal{O}_K$. Then, for $\alpha \in \mathcal{O}_K$ prime to $\mathfrak{p}$, we have Fermat's Little Theorem

$$\alpha^{N(\mathfrak{p})-1} \equiv 1 \bmod \mathfrak{p}.$$

Suppose that in addition $\mathfrak{p}$ is prime to n. It can be shown that $n \mid N(\mathfrak{p})-1$ (see Exercise 5.13), and it follows that $x = \alpha^{(N(\mathfrak{p})-1)/n}$ is a solution of the congruence $x^n \equiv 1 \bmod \mathfrak{p}$. Consequently

$$\alpha^{(N(\mathfrak{p})-1)/n} \equiv 1, \zeta, \ldots, \zeta^{n-1} \bmod \mathfrak{p}.$$

Since the nth roots of unity are distinct modulo $\mathfrak{p}$ (see Exercise 5.13), $\alpha^{(N(\mathfrak{p})-1)/n}$ is congruent modulo $\mathfrak{p}$ to a *unique* nth root of unity. This root of unity is defined to be the nth *power Legendre symbol* $(\alpha/\mathfrak{p})_n$, so that $(\alpha/\mathfrak{p})_n$ satisfies the congruence

$$\alpha^{(N(\mathfrak{p})-1)/n} \equiv \left(\frac{\alpha}{\mathfrak{p}}\right)_n \bmod \mathfrak{p}.$$

This symbol is a natural generalization of the Legendre symbols $(\alpha/\pi)_3$ and $(\alpha/\pi)_4$ from cubic and biquadratic reciprocity.

The nth power Legendre symbol can be defined for more general ideals as follows: given an ideal $\mathfrak{a}$ of $\mathcal{O}_K$ which is prime to n and α, we set $(\alpha/\mathfrak{a})_n$ to be the product

$$\left(\frac{\alpha}{\mathfrak{a}}\right)_n = \prod_{i=1}^{r} \left(\frac{\alpha}{\mathfrak{p}_i}\right)_n,$$

where $\mathfrak{a} = \mathfrak{p}_1 \cdots \mathfrak{p}_r$ is the prime factorization of $\mathfrak{a}$. Thus, if $\mathfrak{m}$ is a modulus of K such that every prime containing $n\alpha$ divides $\mathfrak{m}$, then the nth power Legendre symbol gives a homomorphism

$$\left(\frac{\alpha}{\cdot}\right)_n : I_K(\mathfrak{m}) \longrightarrow \mu_n,$$

where $\mu_n \subset \mathbf{C}^*$ is the group of nth roots of unity.

We will prove two reciprocity theorems for the nth power Legendre symbol, but first we need to recall a fact from Galois theory. If K has a primitive nth root of unity, then for $\alpha \in K$, the extension $K \subset L = K(\sqrt[n]{\alpha})$ is Galois, and if $\sigma \in \mathrm{Gal}(L/K)$, then $\sigma(\sqrt[n]{\alpha}) = \zeta\sqrt[n]{\alpha}$ for some nth root of unity ζ. This gives us a map $\sigma \mapsto \zeta$, which defines an injective homomorphism

$$\mathrm{Gal}(L/K) \hookrightarrow \mu_n.$$

We can now state our first reciprocity theorem for $(\alpha/\mathfrak{a})_n$:

Theorem 8.11 (Weak Reciprocity). *Let K be a number field containing a primitive nth root of unity, and let $L = K(\sqrt[n]{\alpha})$, where $\alpha \in \mathcal{O}_K$ is nonzero.*

Assume that $\mathfrak{m}$ *is a modulus divisible by all primes of* K *containing* $n\alpha$, *and assume in addition that* $\ker(\Phi_{L/K,\mathfrak{m}})$ *is a congruence subgroup for* $\mathfrak{m}$. *Then there is a commutative diagram*

$$\begin{array}{ccc} I_K(\mathfrak{m}) & \xrightarrow{\Phi_{L/K,\mathfrak{m}}} & \mathrm{Gal}(L/K) \\ & \searrow & \downarrow \\ & (\alpha/\cdot)_n & \mu_n \end{array},$$

where $\mathrm{Gal}(L/K) \to \mu_n$ *is the natural injection. Thus, if* G *is the image of* $\mathrm{Gal}(L/K)$ *in* μ_n, *then the nth power Legendre symbol* $(\alpha/\mathfrak{a})_n$ *induces a surjective homomorphism*

$$\left(\frac{\alpha}{\cdot}\right)_n : I_K(\mathfrak{m})/P_{K,1}(\mathfrak{m}) \longrightarrow G \subset \mu_n.$$

Proof. To prove that the diagram commutes, it suffices to show

$$\left(\frac{L/K}{\mathfrak{p}}\right)(\sqrt[n]{\alpha}) = \left(\frac{\alpha}{\mathfrak{p}}\right)_n \sqrt[n]{\alpha}.$$

This is an easy consequence of the definition of the Artin symbol (from Lemma 5.19). The case $n = 3$ was proved in (5.22), and for general n, see Exercise 5.14.

Turning to the final statement of the theorem, recall that $\ker(\Phi_{L/K,\mathfrak{m}})$ is a congruence subgroup for $\mathfrak{m}$. Thus $P_{K,1}(\mathfrak{m}) \subset \ker(\Phi_{L/K,\mathfrak{m}}) \subset I_K(\mathfrak{m})$, so that the Artin map $\Phi_{L/K,\mathfrak{m}}$ induces a surjective homomorphism

$$I_K(\mathfrak{m})/P_{K,1}(\mathfrak{m}) \longrightarrow I_K(\mathfrak{m})/\ker(\Phi_{L/K,\mathfrak{m}}) \xrightarrow{\sim} \mathrm{Gal}(L/K).$$

Using the above commutative diagram, the theorem follows immediately. Q.E.D.

This result is called "Weak Reciprocity" because rather than giving formulas for computing $(\alpha/\mathfrak{a})_n$, the theorem simply asserts that the symbol is a homomorphism on an appropriate group. Nevertheless, Weak Reciprocity is a powerful result. For example, let's use it to prove quadratic reciprocity:

Theorem 8.12. *Let* p *and* q *be distinct odd primes. Then*

$$\left(\frac{p}{q}\right)\left(\frac{q}{p}\right) = (-1)^{(p-1)(q-1)/4}.$$

Proof. Recall from §1 that quadratic reciprocity can be written in the form

$$\left(\frac{p^*}{q}\right) = \left(\frac{q}{p}\right)$$

where $p^* = (-1)^{(p-1)/2}$.

The first step is to study $\mathbf{Q} \subset \mathbf{Q}(\sqrt{p^*})$. By (8.3) and (8.4), $\mathrm{Gal}(\mathbf{Q}(\zeta_p)/\mathbf{Q})$ is a generalized ideal class group for the modulus $p\infty$, which implies that the same is true for any subfield of $\mathbf{Q}(\zeta_p)$ (see Exercise 8.7). Since $\mathrm{Gal}(\mathbf{Q}(\zeta_p)/\mathbf{Q})$ is cyclic of order $p-1$, there is a unique subfield $\mathbf{Q} \subset K \subset \mathbf{Q}(\zeta_p)$ which is quadratic over $\mathbf{Q}$. Then $\mathrm{Gal}(K/\mathbf{Q})$ is a generalized ideal class group for $p\infty$, which implies that p is the only finite prime of $\mathbf{Q}$ that ramifies in K. If we write $K = \mathbf{Q}(\sqrt{m})$, m squarefree, then Corollary 5.17 implies that $m = p^*$, and hence $K = \mathbf{Q}(\sqrt{p^*})$ (see Exercise 8.7).

It follows that $\Phi_{\mathbf{Q}(\sqrt{p^*})/\mathbf{Q},p\infty}$ is a congruence subgroup for $p\infty$, and thus by Weak Reciprocity, the Legendre symbol $(p^*/\cdot)$ gives a surjective homomorphism

$$I_{\mathbf{Q}}(p\infty)/P_{\mathbf{Q},1}(p\infty) \longrightarrow \{\pm 1\}. \tag{8.13}$$

However, the map sending $[a] \in (\mathbf{Z}/p\mathbf{Z})^*$ to $[a\mathbf{Z}] \in I_{\mathbf{Q}}(p\infty)/P_{\mathbf{Q},1}(p\infty)$ induces an isomorphism $(\mathbf{Z}/p\mathbf{Z})^* \xrightarrow{\sim} I_{\mathbf{Q}}(p\infty)/P_{\mathbf{Q},1}(p\infty)$ (see Exercise 8.7). Composing this map with (8.13) shows that $(p^*/\cdot)$ induces a surjective homomorphism from $(\mathbf{Z}/p\mathbf{Z})^*$ to $\{\pm 1\}$. But the Legendre symbol $(\cdot/p)$ is also a surjective homomorphism between the same two groups, and since $(\mathbf{Z}/p\mathbf{Z})^*$ is cyclic, there is only one such homomorphism. This proves that

$$\left(\frac{p^*}{q}\right) = \left(\frac{q}{p}\right),$$

and we are done. Q.E.D.

The proof just given is closely related to the discussion of quadratic reciprocity from §1. Recall that a key result implicit in Euler's work was Lemma 1.14, which showed that $(D/\cdot)$ gives a well defined homomorphism defined on $(\mathbf{Z}/D\mathbf{Z})^*$ when $D \equiv 0, 1 \bmod 4$. The above argument uses Weak Reciprocity to prove this when $D = p^*$. In this way Weak Reciprocity (or more generally, Artin Reciprocity) may be regarded as a far-reaching generalization of Lemma 1.14.

Before we can state our second reciprocity theorem for the nth power Legendre symbol, we need some notation: if α and β are in $\mathcal{O}_K$, then $(\alpha/\beta\mathcal{O}_K)_n$ is written simply $(\alpha/\beta)_n$ when defined. Then we have the following reciprocity theorem for $(\alpha/\beta)_n$:

Theorem 8.14 (Strong Reciprocity). *Let K be a number field containing a primitive nth root of unity, and suppose that $\alpha, \beta \in \mathcal{O}_K$ are relatively to each other and to n. Then*

$$\left(\frac{\alpha}{\beta}\right)_n \left(\frac{\beta}{\alpha}\right)_n^{-1} = \prod_{\mathfrak{p} \mid n\infty} \left(\frac{\alpha, \beta}{\mathfrak{p}}\right)_n,$$

where $(\alpha,\beta/\mathfrak{p})_n$ is the nth power Hilbert symbol (to be discussed below) and ∞ is the product of the real infinite primes of K (which can occur only when $n = 2$).

Proof. While Weak Reciprocity was an immediate consequence of Artin reciprocity, Strong Reciprocity is a different matter, for here one must first study the nth power Hilbert symbol

$$\left(\frac{\alpha,\beta}{\mathfrak{p}}\right)_n.$$

This symbol is an nth root of unity defined using the local class field theory of the completion $K_{\mathfrak{p}}$ of K at the prime $\mathfrak{p}$. Since we haven't discussed local methods, we can't even give a precise definition. A full discussion of the Hilbert symbol is given in Hasse [49, Part II, §§11–12, pp. 53–64] and Neukirch [80, §§III.5 and IV.9, pp. 50–55 and 110–112], and both references present a complete proof of the Strong Reciprocity theorem. In Exercise 8.9 we will list the main properties of the Hilbert symbol. Q.E.D.

To get a better idea of how Strong Reciprocity works, let's apply it to cubic reciprocity. Here, $n = 3$ and $K = \mathbb{Q}(\omega)$, $\omega = e^{2\pi i/3}$, and the only prime of $\mathcal{O}_K$ dividing 3 is $\lambda = 1-\omega$. Thus, given nonassociate primes π and θ in $\mathcal{O}_K$, Strong Reciprocity tells us that

$$\left(\frac{\pi}{\theta}\right)_3\left(\frac{\theta}{\pi}\right)_3^{-1} = \left(\frac{\pi,\theta}{\lambda}\right)_3.$$

Hence, to prove cubic reciprocity, it suffices to show that

$$\pi,\ \theta \text{ primary} \Rightarrow \left(\frac{\pi,\theta}{\lambda}\right)_3 = 1. \tag{8.15}$$

The proof of cubic reciprocity is thus reduced to a purely local computation in the completion K_λ of K at λ. Given the properties of the Hilbert symbol, (8.15) is not difficult to prove (see Exercise 8.9). Biquadratic reciprocity can be proved similarly, though the proof is a bit more complicated (see Hasse [49, Part II, §20, pp. 105–106]). This shows that class field theory encompasses all of the reciprocity theorems we've seen so far.

B. The Čebotarev Density Theorem

The Čebotarev Density Theorem will provide some very useful information about the Artin map. But first, we need to define the notion of *Dirichlet density*.

Let K be a number field, and let $\mathcal{P}_K$ be the set of all finite primes of K. Given a subset $\mathcal{S} \subset \mathcal{P}_K$, the *Dirichlet density* of $\mathcal{S}$ is defined to be

$$\delta(\mathcal{S}) = \lim_{s \to 1^+} \frac{\sum_{\mathfrak{p} \in \mathcal{S}} N(\mathfrak{p})^{-s}}{-\log(s-1)},$$

provided the limit exists. The basic properties of the Dirichlet density are:

(i) $\delta(\mathcal{P}_K) = 1$.
(ii) If $\mathcal{S} \subset \mathcal{T}$ and $\delta(\mathcal{S})$ and $\delta(\mathcal{T})$ exist, then $\delta(\mathcal{S}) \leq \delta(\mathcal{T})$.
(iii) If $\delta(\mathcal{S})$ exists, then $0 \leq \delta(\mathcal{S}) \leq 1$.
(iv) If $\mathcal{S}$ and $\mathcal{T}$ are disjoint and $\delta(\mathcal{S})$ and $\delta(\mathcal{T})$ exist, then $\delta(\mathcal{S} \cup \mathcal{T}) = \delta(\mathcal{S}) + \delta(\mathcal{T})$.
(v) If $\mathcal{S}$ is finite, then $\delta(\mathcal{S}) = 0$.
(vi) If $\delta(\mathcal{S})$ exists and $\mathcal{T}$ differs from $\mathcal{S}$ by finitely many elements, then $\delta(\mathcal{T}) = \delta(\mathcal{S})$.

To prove these properties, one first must study the Dirichlet zeta function $\zeta_K(s)$ of K. This function is defined by

$$\zeta_K(s) = \sum_{\mathfrak{a} \subset \mathcal{O}_K} N(\mathfrak{a})^{-s} = \prod_{\mathfrak{p} \in \mathcal{P}_K} \left(1 - N(\mathfrak{p})^{-s}\right)^{-1}.$$

One can prove without difficulty that $\zeta_K(s)$ converges absolutely for $\mathrm{Re}(s) > 1$ (see Janusz [62, §IV.4] or Neukirch [80, §V.6]). This implies that for any $\mathcal{S} \subset \mathcal{P}_K$, the sum $\sum_{\mathfrak{p} \in \mathcal{S}} N(\mathfrak{p})^{-s}$ converges absolutely for $\mathrm{Re}(s) > 1$ (see Exercise 8.10). A much deeper property of $\zeta_K(s)$ is that it has a simple pole at $s = 1$, which enables one to prove

$$1 = \lim_{s \to 1^+} \frac{\log(\zeta_K(s))}{-\log(s-1)} = \lim_{s \to 1^+} \frac{\sum_{\mathfrak{p} \in \mathcal{P}_K} N(\mathfrak{p})^{-s}}{-\log(s-1)}$$

(see Janusz [62, §IV.4] or Neukirch [80, §V.6]). This proves (i), and it is now straightforward to prove (ii)–(vi) (see Exercise 8.10).

There is one more property of the Dirichlet density which is sometimes useful. Let $\mathcal{P}_{K,1} = \{\mathfrak{p} \in \mathcal{P}_K : N(\mathfrak{p}) \text{ is prime}\}$. $\mathcal{P}_{K,1}$ is sometimes called the degree 1 primes in K (recall that in general, $N(\mathfrak{p}) = p^f$, where f is the inertial degree of $p \in \mathfrak{p}$ in the extension $\mathbf{Q} \subset K$). Then one can prove that

$$\delta(\mathcal{S}) = \delta(\mathcal{S} \cap \mathcal{P}_{K,1}) \tag{8.16}$$

whenever $\delta(\mathcal{S})$ exists (see Janusz [62, §IV.4] or Neukirch [80, §V.6]).

Now let L be a Galois extension of K, possibly non-Abelian. If $\mathfrak{p}$ is a prime of K unramified in L, then different primes $\mathfrak{P}$ of L containing $\mathfrak{p}$ may give us different Artin symbols $((L/K)/\mathfrak{P})$. But all of the $((L/K)/\mathfrak{P})$ are conjugate by Corollary 5.21, and in fact they form a complete conjugacy class in $\mathrm{Gal}(L/K)$ (see Exercise 5.12). Thus we can define the Artin symbol $((L/K)/\mathfrak{p})$ of $\mathfrak{p}$ to be this conjugacy class in $\mathrm{Gal}(L/K)$. We can now state the *Čebotarev Density Theorem*:

Theorem 8.17. *Let L be a Galois extension of K, and let $\langle\sigma\rangle$ be the conjugacy class of an element $\sigma \in \mathrm{Gal}(L/K)$. Then the set*

$$\mathcal{S} = \{\mathfrak{p} \in \mathcal{P}_K : \mathfrak{p} \text{ is unramified in } L \text{ and } ((L/K)/\mathfrak{p}) = \langle\sigma\rangle\}$$

has Dirichlet density

$$\delta(\mathcal{S}) = \frac{|\langle\sigma\rangle|}{|\mathrm{Gal}(L/K)|} = \frac{|\langle\sigma\rangle|}{[L:K]}.$$

Proof. See Janusz [62, Chapter V, Theorem 10.4] or Neukirch [80, Chapter V, Theorem 6.4]. Q.E.D.

Notice that the set $\mathcal{S}$ of the theorem must be infinite since it has positive density (this follows from property (v) above). In particular, we get the following corollary for Abelian extensions:

Corollary 8.18. *Let L be an Abelian extension of K, and let $\mathfrak{m}$ be a modulus divisible by all primes that ramify in L. Then, given any element $\sigma \in \mathrm{Gal}(L/K)$, the set of primes $\mathfrak{p}$ not dividing $\mathfrak{m}$ such that $((L/K)/\mathfrak{p}) = \sigma$ has density $1/[L:K]$ and hence is infinite.*

Proof. When $\mathrm{Gal}(L/K)$ is Abelian, the conjugacy class $\langle\sigma\rangle$ is just the set $\{\sigma\}$. Q.E.D.

This corollary shows that the Artin map $\Phi_{L/K,\mathfrak{m}} : I_K(\mathfrak{m}) \to \mathrm{Gal}(L/K)$ is surjective in a very strong sense.

An especially nice case is when $K = \mathbb{Q}$ and $L = \mathbb{Q}(\zeta_m)$, for here Corollary 8.18 gives a quick proof of Dirichlet's theorem on primes in arithmetic progressions (the details are left to the reader—see Exercise 8.11). In §9 we will apply these same ideas to study the primes represented by a fixed quadratic form $ax^2 + bxy + cy^2$.

Another application of Čebotarev Density concerns primes that split completely in a Galois extension $K \subset L$. Namely, if we apply Theorem 8.17 to the conjugacy class of the identity element, we see that the primes in K for which $((L/K)/\mathfrak{p}) = 1$ have density $1/[L:K]$. However, from Corollary 5.21, we know that

$$\left(\frac{L/K}{\mathfrak{p}}\right) = 1 \iff \mathfrak{p} \text{ splits completely in } L.$$

Thus the primes that split completely in L have density $1/[L:K]$, and in particular there are infinitely many of them. The unexpected fact is that these primes characterize the extension $K \subset L$ uniquely. Before we can prove this, we need to introduce some terminology.

Given two sets $\mathcal{S}$ and $\mathcal{T}$, we say that $\mathcal{S} \dot{\subset} \mathcal{T}$ if $\mathcal{S} \subset \mathcal{T} \cup \Sigma$ for some finite set Σ, and $\mathcal{S} \doteq \mathcal{T}$ means that $\mathcal{S} \dot{\subset} \mathcal{T}$ and $\mathcal{T} \dot{\subset} \mathcal{S}$. Also, given a finite extension $K \subset L$, we set

$$\mathcal{S}_{L/K} = \{\mathfrak{p} \in \mathcal{P}_K : \mathfrak{p} \text{ splits completely in } L\}.$$

We can now state our result:

Theorem 8.19. *Let L and M be Galois extensions of K. Then*
(i) $L \subset M \iff \mathcal{S}_{M/K} \dot{\subset} \mathcal{S}_{L/K}$.
(ii) $L = M \iff \mathcal{S}_{M/K} \doteq \mathcal{S}_{L/K}$.

Proof. Notice that (ii) is an immediate consequence of (i). As for (i), we will prove the following more general result which applies when only one of L or M is Galois over K. This will be useful in §§9 and 11.

Proposition 8.20. *Let L and M be finite extensions of K.*
(i) *If M is Galois over K, then $L \subset M \iff \mathcal{S}_{M/K} \dot{\subset} \mathcal{S}_{L/K}$.*
(ii) *If L is Galois over K, then $L \subset M \iff \tilde{\mathcal{S}}_{M/K} \dot{\subset} \mathcal{S}_{L/K}$, where $\tilde{\mathcal{S}}_{M/K}$ is defined by*

$$\tilde{\mathcal{S}}_{M/K} = \{\mathfrak{p} \in \mathcal{P}_K : \mathfrak{p} \text{ unramified in } M,\ f_{\mathfrak{P}|\mathfrak{p}} = 1 \text{ for some prime } \mathfrak{P} \text{ of } M\}.$$

Remark. If M is Galois over K, then $\tilde{\mathcal{S}}_{M/K}$ reduces to $\mathcal{S}_{M/K}$ (see Exercise 8.12), and thus either part of Proposition 8.20 implies Theorem 8.19.

Proof. We start with the proof of (ii). When $L \subset M$, it is easy to see that $\tilde{\mathcal{S}}_{M/K} \dot{\subset} \mathcal{S}_{L/K}$ (see Exercise 8.12). Conversely, assume that $\tilde{\mathcal{S}}_{M/K} \dot{\subset} \mathcal{S}_{L/K}$, and let N be a Galois extension of K containing both L and M. By Galois theory, it suffices to show that $\mathrm{Gal}(N/M) \subset \mathrm{Gal}(N/L)$. Thus, given $\sigma \in \mathrm{Gal}(N/M)$, we need to prove that $\sigma_{|L}$ is the identity.

By the Čebotarev Density Theorem, there is a prime $\mathfrak{p}$ in K, unramified in N, such that $((N/K)/\mathfrak{p})$ is the conjugacy class of σ. Thus there is some prime $\mathfrak{P}$ of N containing $\mathfrak{p}$ such that $((N/K)/\mathfrak{P}) = \sigma$. We claim that $\mathfrak{p} \in \tilde{\mathcal{S}}_{M/K}$. To see why, let $\mathfrak{P}' = \mathfrak{P} \cap \mathcal{O}_M$. Then, for $\alpha \in \mathcal{O}_M$, we have

$$\alpha \equiv \sigma(\alpha) \equiv \alpha^{N(\mathfrak{p})} \bmod \mathfrak{P}'.$$

The first congruence follows from $\sigma_{|M} = 1$, and the second follows by the definition of the Artin symbol (see Lemma 5.19). Thus $\mathcal{O}_M/\mathfrak{P}' \simeq \mathcal{O}_K/\mathfrak{p}$, so that $f_{\mathfrak{P}'|\mathfrak{p}} = 1$. This shows that $\mathfrak{p} \in \tilde{\mathcal{S}}_{M/K}$ as claimed.

The Density Theorem implies that there are infinitely many such $\mathfrak{p}$'s. Thus our hypothesis $\tilde{\mathcal{S}}_{M/K} \dot{\subset} \mathcal{S}_{L/K}$ allows us to assume that $\mathfrak{p} \in \mathcal{S}_{L/K}$, i.e., $((L/K)/\mathfrak{p}) = 1$. But Exercise 5.9 tells us that $((L/K)/\mathfrak{p}) = ((N/K)/\mathfrak{P})|_L$. Since $\sigma = ((N/K)/\mathfrak{P})$, we see that $\sigma_{|L} = 1$ as desired.

To prove (i), first note $L \subset M$ easily implies $\mathcal{S}_{M/K} \dot{\subset} \mathcal{S}_{L/K}$ (see Exercise 8.12). To prove the other direction, let L' be the Galois closure of L over K. It is a standard fact that a prime of K splits completely in L if and only if it splits completely in L' (see Exercises 8.13–8.15 or Marcus [77, Corollary to Theorem 31]). This implies that $\mathcal{S}_{L/K} = \mathcal{S}_{L'/K}$. Since M is Galois over K, we've already observed that $\tilde{\mathcal{S}}_{M/K} = \mathcal{S}_{M/K}$. Thus our hypothesis $\mathcal{S}_{M/K} \dot{\subset} \mathcal{S}_{L/K}$ can be written $\tilde{\mathcal{S}}_{M/K} \dot{\subset} \mathcal{S}_{L'/K}$, so that by part (ii) we obtain $L' \subset M$, which obviously implies $L \subset M$. This completes the proofs of Proposition 8.20 and Theorem 8.19. Q.E.D.

Theorem 8.19 is closely related to Corollary 8.7. The reason is that if $K \subset L$ is Abelian, then the set $\mathcal{S}_{L/K}$ of primes that split completely is, up to a finite set, exactly the prime ideals in $\ker(\Phi_{L/K,\mathfrak{m}})$, where $\mathfrak{m}$ is any modulus divisible by all of the ramified primes. Thus we don't need the whole kernel of the Artin map to determine the extension—just the primes in it will suffice! In particular, this shows that Theorem 8.19 is relevant to our question of which primes p are of the form $x^2 + ny^2$. To see why, consider the situation of Theorem 5.1. Here, K is an imaginary quadratic field of discriminant $d_K = -4n$ (which means that n satisfies (5.2)). Then, by Theorem 5.26,

$$p = x^2 + ny^2 \iff p \text{ splits completely in the Hilbert class field of } K$$

whenever p is an odd prime not dividing n. Thus Theorem 8.19 shows that the primes represented by $x^2 + ny^2$ characterize the Hilbert class field of $\mathbb{Q}(\sqrt{-n})$ uniquely. In §9 we will give a version of this result that holds for arbitrary n.

C. Norms and Ideles

Our discussion of class field theory has omitted several important topics. To give the reader a sense of what's been left out, we will say a few words about norms and ideles.

Given a finite extension $K \subset L$, there is the norm map $N_{L/K} : L^* \to K^*$, and $N_{L/K}$ can be extended to a map of ideals

$$N_{L/K} : I_L \longrightarrow I_K$$

(see Janusz [62, §I.8]). The importance of the norm map is that it gives a precise description of the kernel of the Artin map. Specifically, let L be an Abelian extension of K, and let $\mathfrak{m}$ be a modulus for which $P_{K,1}(\mathfrak{m}) \subset \ker(\Phi_{L/K,\mathfrak{m}})$. Then an important part of the Artin Reciprocity Theorem states that

$$\ker(\Phi_{L/K,\mathfrak{m}}) = N_{L/K}(I_L(\mathfrak{m}))P_{K,1}(\mathfrak{m}) \tag{8.21}$$

(see Janusz [62, Chapter V, Theorem 5.7]). Norms play an essential role in the proofs of the theorems of class field theory.

Class field theory can be presented without reference to ideles (as we have done above), but the idelic approach has some distinct advantages. Before we can see why, we need some definitions. Given a number field K, the *idele group* $\mathbf{I}_K$ is the restricted product

$$\mathbf{I}_K = \prod_{\mathfrak{p}}{}^* K_{\mathfrak{p}}^*,$$

where $\mathfrak{p}$ runs over all primes of K, finite and infinite, and $K_{\mathfrak{p}}$ is the completion of K at $\mathfrak{p}$. The symbol $\prod_{\mathfrak{p}}^*$ means that $\mathbf{I}_K$ consists of all tuples $(x_{\mathfrak{p}})$ such that $x_{\mathfrak{p}} \in \mathcal{O}_{K_{\mathfrak{p}}}$ for all but finitely many $\mathfrak{p}$. $\mathbf{I}_K$ is a locally compact topological group, and the multiplicative group K^* imbeds naturally in $\mathbf{I}_K$ as a discrete subgroup (see Neukirch [80, §V.2] for all of this). The quotient group

$$\mathbf{C}_K = \mathbf{I}_K / K^*$$

is called the *idele class group*.

We can now restate the theorems of class field theory using ideles. Given an Abelian extension L of K, there is an Artin map

$$\Phi_{L/K} : \mathbf{C}_K \longrightarrow \mathrm{Gal}(L/K)$$

which is continuous and surjective. This is the idele theoretic analog of the Artin Reciprocity Theorem. Note that $\ker(\Phi_{L/K})$ is a closed subgroup of finite index in $\mathbf{C}_K$. There is also an idelic version of the Existence Theorem, which asserts that there is a 1–1 correspondence between the Abelian extensions of K and the closed subgroups of finite index in $\mathbf{C}_K$. The nice feature of this approach is that it always uses the same group $\mathbf{C}_K$, unlike our situation, where we had to vary the modulus $\mathfrak{m}$ in $I_K(\mathfrak{m})$ as we moved from one Abelian extension to the next.

Norms also play an important role in the idelic theory. Given an Abelian extension L of K, there is a norm map

$$N_{L/K} : \mathbf{C}_L \longrightarrow \mathbf{C}_K,$$

and the idelic analog of (8.21) is that the kernel of the Artin map $\Phi_{L/K} : \mathbf{C}_K \to \mathrm{Gal}(L/K)$ is exactly $N_{L/K}(\mathbf{C}_L)$. Thus the subgroups of $\mathbf{C}_K$ of finite index are precisely the norm groups $N_{L/K}(\mathbf{C}_L)$.

Standard references for the idele theoretic formulation of class field theory are Neukirch [80] and Weil [104]. Neukirch also explains carefully the relation between the two approaches to class field theory.

D. Exercises

8.1. Let K be an imaginary quadratic field, and let $\mathfrak{m}$ be a modulus for K (which can be regarded as an ideal of $\mathcal{O}_K$). We want to show that $P_{K,1}(\mathfrak{m})$ has finite index in $I_K(\mathfrak{m})$.

(a) Show that the map $\alpha \mapsto \alpha\mathcal{O}_K$ induces a well-defined homomorphism

$$\phi : (\mathcal{O}_K/\mathfrak{m})^* \longrightarrow I_K(\mathfrak{m}) \cap P_K/P_{K,1}(\mathfrak{m}),$$

and then show that there is an exact sequence

$$\mathcal{O}_K^* \longrightarrow (\mathcal{O}_K/\mathfrak{m})^* \xrightarrow{\phi} I_K(\mathfrak{m}) \cap P_K/P_{K,1}(\mathfrak{m}) \longrightarrow 1.$$

Conclude that $I_K(\mathfrak{m}) \cap P_K/P_{K,1}(\mathfrak{m})$ is finite. Hint: see the proof of Theorem 7.24.

(b) Adapt the exact sequence (7.25) to show that $I_K(\mathfrak{m})/P_{K,1}(\mathfrak{m})$ is finite (recall that $C(\mathcal{O}_K)$ is finite by §2 and Theorem 7.7).

8.2. This problem is concerned with the Artin map of the cyclotomic extension $\mathbb{Q} \subset \mathbb{Q}(\zeta_m)$, where $\zeta_m = e^{2\pi i/m}$. We will assume that $m > 2$.

(a) Use Proposition 5.11 to prove that all finite ramified primes of this extension divide m. Thus the Artin map $\Phi_{m\infty}$ is defined.

(b) Show that $\Phi_{m\infty} : I_{\mathbb{Q}}(m\infty) \to \mathrm{Gal}(\mathbb{Q}(\zeta_m)/\mathbb{Q}) \simeq (\mathbb{Z}/m\mathbb{Z})^*$ is as described in (8.3). Hint: use Lemma 5.19.

(c) Conclude that $\ker(\Phi_{m\infty}) = P_{\mathbb{Q},1}(m\infty)$.

8.3. Let $\mathbb{Q} \subset \mathbb{Q}(\zeta_m)$ be as in the previous problem, and assume that $m > 2$.

(a) Show that $\mathbb{R} \cap \mathbb{Q}(\zeta_m) = \mathbb{Q}(\cos(2\pi/m))$, and then conclude that $[\mathbb{Q}(\cos(2\pi/m)) : \mathbb{Q}] = (1/2)\phi(m)$.

(b) Compute the Artin map $\Phi_m : I_{\mathbb{Q}}(m) \to \mathrm{Gal}(\mathbb{Q}(\cos(2\pi/m))/\mathbb{Q}) \simeq (\mathbb{Z}/m\mathbb{Z})^*/\{\pm 1\}$. Hint: use the previous exercise.

(c) Show that $\ker(\Phi_m) = P_{\mathbb{Q},1}(m)$.

8.4. Let $K \subset L$ be an Abelian extension, and let $\mathfrak{m}$ be a modulus for which the Artin map $\Phi_{\mathfrak{m}}$ is defined. If $\mathfrak{n}$ is another modulus and $\mathfrak{m} \mid \mathfrak{n}$, prove that

$$P_{K,1}(\mathfrak{m}) \subset \ker(\Phi_{\mathfrak{m}}) \Rightarrow P_{K,1}(\mathfrak{n}) \subset \ker(\Phi_{\mathfrak{n}}).$$

8.5. Prove that the conductor of the cyclotomic extension $\mathbb{Q} \subset \mathbb{Q}(\zeta_m)$ is given by

$$\mathfrak{f}(\mathbb{Q}(\zeta_m)/\mathbb{Q}) = \begin{cases} 1 & m \leq 2 \\ (m/2)\infty & m = 2n,\ n > 1 \text{ odd} \\ m\infty & \text{otherwise.} \end{cases}$$

Hint: when $m > 2$, use Theorem 8.5 and Exercise 8.2 to show that the conductor is of the form $n\infty$ for some n dividing m. Then use Corollary 8.7 to show that $\mathbb{Q}(\zeta_m) \subset \mathbb{Q}(\zeta_n)$, which implies that $\phi(m) \mid \phi(n)$. The formula for $\mathfrak{f}(\mathbb{Q}(\zeta_m)/\mathbb{Q})$ now follows from elementary arguments about the Euler ϕ-function.

8.6. This exercise is concerned with conductors.

(a) Given a modulus $\mathfrak{m}$ for a number field K, let $K_{\mathfrak{m}}$ denote the ray class field defined in the text. If L is an Abelian extension of K, then show that the conductor $\mathfrak{f}(L/K)$ is the greatest common divisor of all moduli $\mathfrak{m}$ for which $L \subset K_{\mathfrak{m}}$.

(b) If L is an Abelian extension of $\mathbb{Q}$, let m be the smallest positive integer for which $L \subset \mathbb{Q}(\zeta_m)$ (note that m exists by the Kronecker–Weber Theorem). Then show that

$$\mathfrak{f}(L/\mathbb{Q}) = \begin{cases} m & \text{if } L \subset \mathbb{R} \\ m\infty & \text{otherwise.} \end{cases}$$

8.7. In this exercise we will fill in some of the details omitted in the proof of quadratic reciprocity given in Theorem 8.12. Let p be an odd prime.

(a) If $K \subset L$ is an Abelian extension such that $\mathrm{Gal}(L/K)$ is a generalized ideal class group for the modulus $\mathfrak{m}$ of K, then prove that the same is true for any intermediate field $K \subset M \subset L$.

(b) If K is a quadratic field which ramifies only at p, then use Corollary 5.17 to show that $K = \mathbb{Q}(\sqrt{p^*})$, $p^* = (-1)^{(p-1)/2}p$.

(c) Show that the map $a \mapsto a\mathbb{Z}$ induces an isomorphism $(\mathbb{Z}/p\mathbb{Z})^* \xrightarrow{\sim} I_{\mathbb{Q}}(p\infty)/P_{\mathbb{Q},1}(p\infty)$.

8.8. This exercise will adapt the proof of Theorem 8.12 to prove $(2/p) = (-1)^{(p^2-1)/8}$.

(a) Let $H = \{\pm 1\}P_{\mathbb{Q},1}(8\infty)$. Show that via the Existence Theorem, H corresponds to $\mathbb{Q}(\sqrt{2})$. Hint: using the arguments of Theorem 8.12 and part (b) of Exercise 8.7, show that H corresponds to one of $\mathbb{Q}(i)$, $\mathbb{Q}(\sqrt{2})$ or $\mathbb{Q}(\sqrt{-2})$. Then use $-1 \in H$ to show that the corresponding field must be real.

(b) Construct an isomorphism $(\mathbf{Z}/8\mathbf{Z})^* \xrightarrow{\sim} I_{\mathbf{Q}}(8\infty)/P_{\mathbf{Q},1}(8\infty)$, and then use Weak Reciprocity to show that $(2/\cdot)$ induces a well-defined homomorphism on $(\mathbf{Z}/8\mathbf{Z})^*$ whose kernel is $\{\pm 1\}$.

(c) Show that $(2/p) = (-1)^{(p^2-1)/8}$.

8.9. In this exercise we will use Strong Reciprocity and the properties of the Hilbert symbol to prove cubic reciprocity. We will assume that the reader is familiar with p-adic fields. To list the properties of the Hilbert symbol, let K be a number field containing a primitive nth of unity, and let $\mathfrak{p}$ be a prime of K. The completion of K at $\mathfrak{p}$ will be denoted $K_{\mathfrak{p}}$. Then the Hilbert symbol $(\alpha,\beta/\mathfrak{p})_n$ is defined for $\alpha,\beta \in K_{\mathfrak{p}}^*$ and gives a map

$$\left(\frac{\cdot,\cdot}{\mathfrak{p}}\right)_n : K_{\mathfrak{p}}^* \times K_{\mathfrak{p}}^* \longrightarrow \mu_n,$$

where μ_n is the group of nth roots of unity. The Hilbert symbol has the following properties:

(i) $(\alpha\alpha',\beta/\mathfrak{p})_n = (\alpha,\beta/\mathfrak{p})_n(\alpha',\beta/\mathfrak{p})_n$.

(ii) $(\alpha',\beta\beta'/\mathfrak{p})_n = (\alpha,\beta/\mathfrak{p})_n(\alpha,\beta'/\mathfrak{p})_n$.

(iii) $(\alpha,\beta/\mathfrak{p})_n = (\beta,\alpha\mathfrak{p})_n^{-1}$.

(iv) $(\alpha,-\alpha/\mathfrak{p})_n = 1$.

(v) $(\alpha,1-\alpha/\mathfrak{p})_n = 1$.

For proofs of these properties of the Hilbert symbol, see Neukirch [80, §III.5].

Now let's specialize to the case $n = 3$ and $K = \mathbf{Q}(\omega)$, $\omega = e^{2\pi i/3}$. As we saw in (8.15), Strong Reciprocity shows that cubic reciprocity is equivalent to the assertion

$$\pi,\theta \text{ primary in } \mathcal{O}_K \Rightarrow \left(\frac{\pi,\theta}{\lambda}\right)_3 = 1$$

where $\lambda = 1-\omega$. Recall that π primary means that $\pi \equiv \pm 1 \bmod 3\mathcal{O}_K$. In §4 we saw that replacing π by $-\pi$ doesn't affect the statement of cubic reciprocity, so that we can assume that $\pi \equiv \theta \equiv 1 \bmod \lambda^2\mathcal{O}_K$ (note that λ^2 and 3 differ by a unit in $\mathcal{O}_K$). Let K_λ be the completion of K at λ, and let $\mathcal{O}_\lambda$ be the valuation ring of K_λ. We will use the properties of the cubic Hilbert symbol to show that

$$\alpha,\beta \equiv 1 \bmod \lambda^2\mathcal{O}_\lambda \Rightarrow \left(\frac{\alpha,\beta}{\lambda}\right)_3 = 1,$$

and then cubic reciprocity will be proved.

(a) If $\alpha \equiv 1 \bmod \lambda^4\mathcal{O}_\lambda$, then prove that $\alpha = u^3$ for some $u \in \mathcal{O}_\lambda$. Hint: if $\alpha \equiv u_n^3 \bmod \lambda^n\mathcal{O}_\lambda$ for $n \geq 4$, then show that $\alpha \equiv (u_n + a\lambda^{n-2})^3 \bmod \lambda^{n+1}\mathcal{O}_\lambda$ for an appropriately chosen $a \in \mathcal{O}_\lambda$.

(b) If $\alpha \in \mathcal{O}_\lambda^*$ and $\alpha \equiv \alpha' \bmod \lambda^4\mathcal{O}_\lambda$, then prove that for any $\beta \in K_\lambda^*$

$$\left(\frac{\alpha,\beta}{\lambda}\right)_3 = \left(\frac{\alpha',\beta}{\lambda}\right)_3.$$

Hint: use (a) and property (i) above. Remember that $(\alpha,\beta/\lambda)_3$ is a cube root of unity.

(c) Now assume that $\alpha \equiv \beta \equiv 1 \bmod \lambda^2\mathcal{O}_\lambda$, and write $\alpha = 1 + a\lambda^2$, $a \in \mathcal{O}_\lambda$. Then first, apply property (v) to $1 + a\beta\lambda^2$, and second, apply (b) to $1 + a\beta\lambda^2 \equiv 1 + a\lambda^2 \bmod \lambda^4\mathcal{O}_\lambda$. This proves that

$$1 = \left(\frac{1 + a\lambda^2, -a\beta\lambda^2}{\lambda}\right)_3.$$

From here, properties (ii) and (v) easily imply that $(\alpha,\beta/\lambda)_3 = 1$, which completes the proof of cubic reciprocity.

8.10. In this exercise we will study the properties of the Dirichlet density.

(a) Assuming that $\zeta_K(s) = \sum_{\mathfrak{a}\subset\mathcal{O}_K} N(\mathfrak{a})^{-s}$ converges absolutely for $\mathrm{Re}(s) > 1$, show that for $\mathcal{S} \subset \mathcal{P}_K$, the sum $\sum_{\mathfrak{p}\in\mathcal{S}} N(\mathfrak{p})^{-s}$ also converges absolutely for $\mathrm{Re}(s) > 1$.

(b) Use (a) to prove that properties (ii)–(iv) of the Dirichlet density follow from (i) and the definition.

8.11. Apply the Čebotarev Density Theorem to the cyclotomic extension $\mathbb{Q} \subset \mathbb{Q}(\zeta_m)$ to show that the primes in a fixed congruence class in $(\mathbb{Z}/m\mathbb{Z})^*$ have Dirichlet density $1/\phi(m)$. This proves Dirichlet's theorem that there are infinitely many primes in an arithmetic progression where the first term and common difference are relatively prime.

8.12. Let M be a finite extension of a number field K, and let $\tilde{\mathcal{S}}_{M/K}$ be the set of primes of $\mathcal{O}_K$ defined in Proposition 8.20.

(a) If M is Galois over K, then show that $\tilde{\mathcal{S}}_{M/K}$ equals the set $\mathcal{S}_{M/K}$ of Theorem 8.19.

(b) If L is a Galois extension of K and $L \subset M$, then show that $\tilde{\mathcal{S}}_{M/K} \dot{\subset} \mathcal{S}_{L/K}$.

(c) If $L \subset M$ are finite extensions of K, then show that $\mathcal{S}_{M/K} \subset \mathcal{S}_{L/K}$.

8.13. Let $K \subset N$ be a Galois extension, and let $\mathfrak{P}$ be a prime of $\mathcal{O}_N$. Set $\mathfrak{p} = \mathfrak{P} \cap \mathcal{O}_K$, $e = e_{\mathfrak{P}|\mathfrak{p}}$ and $f = f_{\mathfrak{P}|\mathfrak{p}}$. If $D_{\mathfrak{P}} \subset \mathrm{Gal}(N/K)$ is the decomposition group of $\mathfrak{P}$, we will denote the fixed field of $D_{\mathfrak{P}}$ by $N_{\mathfrak{P}}$. From Proposition 5.10, we know that $|D_{\mathfrak{P}}| = ef$, and Galois theory tells us that $[N : N_{\mathfrak{P}}] = |D_{\mathfrak{P}}|$. Let $\mathfrak{P}' = \mathfrak{P} \cap \mathcal{O}_{N_{\mathfrak{P}}}$.

(a) Prove that $e_{\mathfrak{P}'|\mathfrak{p}} = f_{\mathfrak{P}'|\mathfrak{p}} = 1$. Hint: by Proposition 5.10, the map $D_{\mathfrak{P}} \to \widetilde{G}$ is surjective, where $\widetilde{G}$ is the Galois group of $\mathcal{O}_K/\mathfrak{p} \subset \mathcal{O}_N/\mathfrak{P}$. Use $\mathcal{O}_K/\mathfrak{p} \subset \mathcal{O}_{N_{\mathfrak{P}}}/\mathfrak{P}' \subset \mathcal{O}_N/\mathfrak{P}$, and remember that the e's and f's are multiplicative (see Exercise 5.15).

(b) Given an intermediate field $K \subset M \subset N$, let $\mathfrak{P}_M = \mathfrak{P} \cap \mathcal{O}_M$. Then prove that

$$e_{\mathfrak{P}_M|\mathfrak{p}} = f_{\mathfrak{P}_M|\mathfrak{p}} = 1 \iff M \subset N_{\mathfrak{P}}.$$

Hint: if $M \subset N_{\mathfrak{P}}$, then apply (a). Conversely, show that the compositum $N_{\mathfrak{P}}M$ is the fixed field for the decomposition group of $\mathfrak{P}$ in $\mathrm{Gal}(N/M)$. By applying the result of (a) to $M \subset N_{\mathfrak{P}}M$ and computing degrees, one sees that $N_{\mathfrak{P}}M = M$, which implies $M \subset N_{\mathfrak{P}}$.

8.14. Let L and M be finite extensions of a number field K, and let $\mathfrak{p}$ be a prime of K that splits completely in L and M. Then prove that $\mathfrak{p}$ splits completely in LM. Hint: let N be a Galois extension of K containing both L and M, and let $\mathfrak{P}$ be a prime of N containing $\mathfrak{p}$. From Exercise 8.13 we get the intermediate field $K \subset N_{\mathfrak{P}} \subset N$. Then use part (b) of that exercise to show that L and M lie in $N_{\mathfrak{P}}$, which implies $LM \subset N_{\mathfrak{P}}$.

8.15. Let L be a finite extension of a number field K, and let L' be the Galois closure of L over K. The goal of this exercise is to prove that $\mathcal{S}_{L/K} = \mathcal{S}_{L'/K}$. By part (c) of Exercise 8.12, we have $\mathcal{S}_{L'/K} \subset \mathcal{S}_{L/K}$, so that it suffices to show that a prime of K that splits completely in L also splits completely in L'.

(a) Let $\sigma : L \to \mathbb{C}$ be an embedding which is the identity on K, and let $\mathfrak{p}$ be an ideal of K which splits completely in L. Then prove that $\mathfrak{p}$ splits completely in $\sigma(L)$.

(b) Since L' is the compositum of the $\sigma(L)$'s, use the previous exercise to show that $\mathfrak{p}$ splits completely in L'.

8.16. Let $K \subset M$ be a finite extension of number fields. Then prove that $K \subset M$ is a Galois extension if and only if $\tilde{\mathcal{S}}_{M/K} = \mathcal{S}_{M/K}$. Hint: one implication is covered in part (a) of Exercise 8.12, and the other implication is an easy consequence of Proposition 8.20.

§9. RING CLASS FIELDS AND $p = x^2 + ny^2$

In Theorem 5.1 we used the Hilbert class field to characterize $p = x^2 + ny^2$ when n is a positive, squarefree and $n \not\equiv 3 \bmod 4$. In §4, we also proved that for an odd prime p,

$$p = x^2 + 27y^2 \iff \begin{cases} p \equiv 1 \bmod 3 \text{ and } x^3 \equiv 2 \bmod p \\ \text{has an integer solution} \end{cases}$$

$$p = x^2 + 64y^2 \iff \begin{cases} p \equiv 1 \bmod 4 \text{ and } x^4 \equiv 2 \bmod p \\ \text{has an integer solution.} \end{cases}$$

These earlier results follow the format of Theorem 5.1 (note that both exponents are class numbers), yet *neither* is a corollary of the theorem, for 27 and 64 are not squarefree. In §9 we will use the theory developed in §§7 and 8 to overcome this limitation. Specifically, given an order $\mathcal{O}$ in an imaginary quadratic field K, we will construct a generalization of the Hilbert class field called the ring class field of $\mathcal{O}$. Then, using the ring class field of the order $\mathbf{Z}[\sqrt{-n}]$, where $n > 0$ is now arbitrary, we will prove a version of Theorem 5.1 that holds for all n (see Theorem 9.2 below). This, of course, is the main theorem of the whole book. The basic idea is that the criterion for $p = x^2 + ny^2$ is determined by a primitive element of the ring class field of $\mathbf{Z}[\sqrt{-n}]$. To see how this works in practice, we will describe the ring class fields of $\mathbf{Z}[\sqrt{-27}]$ and $\mathbf{Z}[\sqrt{-64}]$, and then Theorem 9.2 will give us class field theory proofs of Euler's conjectures for $p = x^2 + 27y^2$ or $x^2 + 64y^2$. To complete the circle of ideas, we will then explain how class field theory implies those portions of cubic and biquadratic reciprocity used in §4 in our earlier discussion of $x^2 + 27y^2$ and $x^2 + 64y^2$.

The remainder of the section will explore two other aspects of ring class fields. We will first use the Čebotarev Density Theorem to prove that a primitive positive definite quadratic form represents infinitely many prime numbers. Then, in a different direction, we will give a purely field-theoretic characterization of ring class fields and their subfields.

A. Solution of $p = x^2 + ny^2$ for all n

Before introducing ring class fields, we need some notation. If K is a number field, an ideal $\mathfrak{m}$ of $\mathcal{O}_K$ can be regarded as a modulus, and in §8 we defined the ideal groups $I_K(\mathfrak{m})$ and $P_{K,1}(\mathfrak{m})$. In this section, $\mathfrak{m}$ will usually be a principal ideal $\alpha\mathcal{O}_K$, and the above groups will be written $I_K(\alpha)$ and $P_{K,1}(\alpha)$.

To define a ring class field, let $\mathcal{O}$ be an order of conductor f in an imaginary quadratic field K. We know from Proposition 7.22 that the ideal class group $C(\mathcal{O})$ can be written

$$C(\mathcal{O}) \simeq I_K(f)/P_{K,\mathbf{Z}}(f) \tag{9.1}$$

(recall that $P_{K,\mathbb{Z}}(f)$ is generated by the principal ideals $\alpha\mathcal{O}_K$, where $\alpha \equiv a \bmod f\mathcal{O}_K$ for some integer a with $\gcd(a,f) = 1$). Furthermore, in §8 we saw that

$$P_{K,1}(f) \subset P_{K,\mathbb{Z}}(f) \subset I_K(f),$$

so that $C(\mathcal{O})$ is a generalized ideal class group of K for the modulus $f\mathcal{O}_K$ (see (8.1)). By the Existence Theorem (Theorem 8.6), this data determines a unique Abelian extension L of K, which is called the *ring class field* of the order $\mathcal{O}$. The basic properties of the ring class field L are, first, all primes of K ramified in L must divide $f\mathcal{O}_K$, and second, the Artin map and (9.1) give us isomorphisms

$$C(\mathcal{O}) \simeq I_K(f)/P_{K,\mathbb{Z}}(f) \simeq \mathrm{Gal}(L/K).$$

In particular the degree of L over K is the class number, i.e., $[L:K] = h(\mathcal{O})$. For an example of a ring class field, note that the ring class field of the maximal order $\mathcal{O}_K$ is the Hilbert class field of K (see Exercise 9.1). Later in this section we will give other examples of ring class fields.

We can now state the main theorem of the book:

Theorem 9.2. *Let $n > 0$ be an integer. Then there is a monic irreducible polynomial $f_n(x) \in \mathbb{Z}[x]$ of degree $h(-4n)$ such that if an odd prime p divides neither n nor the discriminant of $f_n(x)$, then*

$$p = x^2 + ny^2 \iff \begin{cases} (-n/p) = 1 \text{ and } f_n(x) \equiv 0 \bmod p \\ \text{has an integer solution.} \end{cases}$$

Furthermore, $f_n(x)$ may be taken to be the minimal polynomial of a real algebraic integer α for which $L = K(\alpha)$ is the ring class field of the order $\mathbb{Z}[\sqrt{-n}]$ in the imaginary quadratic field $K = \mathbb{Q}(\sqrt{-n})$.

Finally, if $f_n(x)$ is any monic integer polynomial of degree $h(-4n)$ for which the above equivalence holds, then $f_n(x)$ is irreducible over $\mathbb{Z}$ and is the minimal polynomial of a primitive element of the ring class field L described above.

Remark. This theorem generalizes Theorem 5.1, and the last part of the theorem shows that knowing $f_n(x)$ is *equivalent* to knowing the ring class field of $\mathbb{Z}[\sqrt{-n}]$.

Proof. Before proceeding with the proof, we will first prove the following general fact about ring class fields:

Lemma 9.3. *Let L be the ring class field of an order $\mathcal{O}$ in an imaginary quadratic field K. Then L is a Galois extension of $\mathbb{Q}$, and its Galois group can be written as a semidirect product*

$$\mathrm{Gal}(L/\mathbb{Q}) \simeq \mathrm{Gal}(L/K) \rtimes (\mathbb{Z}/2\mathbb{Z})$$

where the nontrivial element of $\mathbb{Z}/2\mathbb{Z}$ *acts on* $\mathrm{Gal}(L/K)$ *by sending* σ *to its inverse* σ^{-1}.

Proof. In the case of the Hilbert class field, this lemma was proved in §6 (see the discussion following (6.3)). To do the general case, we first need to show that $\tau(L) = L$, where τ denotes complex conjugation. Let $\mathfrak{m}$ denote the modulus $f\mathcal{O}_K$, and note that $\tau(\mathfrak{m}) = \mathfrak{m}$. Since $\ker(\Phi_{L/K,\mathfrak{m}}) = P_{K,\mathbb{Z}}(f)$, an easy computation shows that

$$\ker(\Phi_{\tau(L)/K,\mathfrak{m}}) = \tau(\ker(\Phi_{L/K,\mathfrak{m}})) = \tau(P_{K,\mathbb{Z}}(f)) = P_{K,\mathbb{Z}}(f)$$

(see Exercise 9.2), and thus $\ker(\Phi_{\tau(L)/K,\mathfrak{m}}) = \ker(\Phi_{L/K,\mathfrak{m}})$. Then $\tau(L) = L$ follows from Corollary 8.7.

As we noticed in the proof of Lemma 5.28, this implies that L is Galois over $\mathbb{Q}$, so that we have an exact sequence

$$1 \longrightarrow \mathrm{Gal}(L/K) \longrightarrow \mathrm{Gal}(L/\mathbb{Q}) \longrightarrow \mathrm{Gal}(K/\mathbb{Q}) (\simeq \mathbb{Z}/2\mathbb{Z}) \longrightarrow 1.$$

Since $\tau \in \mathrm{Gal}(L/\mathbb{Q})$, $\mathrm{Gal}(L/\mathbb{Q})$ is the semidirect product $\mathrm{Gal}(L/K) \rtimes (\mathbb{Z}/2\mathbb{Z})$, where the nontrivial element of $\mathbb{Z}/2\mathbb{Z}$ acts by conjugation by τ. However, for a prime $\mathfrak{p}$ of K, Lemma 5.19 implies that

$$\tau\left(\frac{L/K}{\mathfrak{p}}\right)\tau^{-1} = \left(\frac{L/K}{\tau(\mathfrak{p})}\right) = \left(\frac{L/K}{\overline{\mathfrak{p}}}\right)$$

(see Exercise 6.3). Thus, under the isomorphism $I_K(f)/P_{K,\mathbb{Z}}(f) \simeq \mathrm{Gal}(L/K)$, conjugation by τ in $\mathrm{Gal}(L/K)$ corresponds to the usual action of τ on $I_K(f)$. But if $\mathfrak{a}$ is any ideal in $I_K(f)$, then $\mathfrak{a}\overline{\mathfrak{a}} = N(\mathfrak{a})\mathcal{O}_K$ lies in $P_{K,\mathbb{Z}}(f)$ since $N(\mathfrak{a})$ is prime to f. Thus $\overline{\mathfrak{a}}$ gives the *inverse* of $\mathfrak{a}$ in the quotient $I_K(f)/P_{K,\mathbb{Z}}(f)$, and the lemma is proved. Q.E.D.

We can now proceed with the proof of Theorem 9.2. Let L be the ring class field of $\mathbb{Z}[\sqrt{-n}]$. We start by relating $p = x^2 + ny^2$ to the behavior of p in L:

Theorem 9.4. *Let* $n > 0$ *be an integer, and* L *be the ring class field of the order* $\mathbb{Z}[\sqrt{-n}]$ *in the imaginary quadratic field* $K = \mathbb{Q}(\sqrt{-n})$. *If* p *is an odd prime not dividing* n, *then*

$$p = x^2 + ny^2 \iff p \text{ splits completely in } L.$$

Proof. Let $\mathcal{O} = \mathbb{Z}[\sqrt{-n}]$. The discriminant of $\mathcal{O}$ is $-4n$, and then $-4n = f^2 d_K$ by (7.3), where f is the conductor of $\mathcal{O}$. Let p be an odd prime not dividing n. Then $p \nmid f^2 d_K$, which implies that p is unramified in K. We will prove the following equivalences:

$$\begin{aligned} p = x^2 + ny^2 &\iff p\mathcal{O}_K = \mathfrak{p}\bar{\mathfrak{p}},\ \mathfrak{p} \neq \bar{\mathfrak{p}},\ \text{and } \mathfrak{p} = \alpha\mathcal{O}_K,\ \alpha \in \mathcal{O} \\ &\iff p\mathcal{O}_K = \mathfrak{p}\bar{\mathfrak{p}},\ \mathfrak{p} \neq \bar{\mathfrak{p}},\ \text{and } \mathfrak{p} \in P_{K,\mathbf{Z}}(f) \\ &\iff p\mathcal{O}_K = \mathfrak{p}\bar{\mathfrak{p}},\ \mathfrak{p} \neq \bar{\mathfrak{p}},\ \text{and } ((L/K)/\mathfrak{p}) = 1 \\ &\iff p\mathcal{O}_K = \mathfrak{p}\bar{\mathfrak{p}},\ \mathfrak{p} \neq \bar{\mathfrak{p}},\ \text{and } \mathfrak{p} \text{ splits completely in } L \\ &\iff p \text{ splits completely in } L, \end{aligned}$$

and Theorem 9.4 will follow.

To prove the first equivalence, suppose that $p = x^2 + ny^2 = (x + \sqrt{-n}y) \times (x - \sqrt{-n}y)$. If we set $\mathfrak{p} = (x + \sqrt{-n}y)\mathcal{O}_K$, then $p\mathcal{O}_K = \mathfrak{p}\bar{\mathfrak{p}}$ is the prime factorization of $p\mathcal{O}_K$ in $\mathcal{O}_K$. Note that $x + \sqrt{-n}y \in \mathcal{O}$, and $\mathfrak{p} \neq \bar{\mathfrak{p}}$ since p is unramified in K. Conversely, if $p\mathcal{O}_K = \mathfrak{p}\bar{\mathfrak{p}}$, where $\mathfrak{p} = (x + \sqrt{-n}y)\mathcal{O}_K$, then it follows easily that $p = x^2 + ny^2$.

Since $p \nmid f$, the second equivalence follows from Proposition 7.22, and the next two equivalences are equally straightforward: the isomorphism $I_K(f)/P_{K,\mathbf{Z}}(f) \simeq \mathrm{Gal}(L/K)$ given by the Artin map shows that $\mathfrak{p} \in P_{K,\mathbf{Z}}(f)$ if and only if $((L/K)/\mathfrak{p}) = 1$, and then Lemma 5.21 shows that $((L/K)/\mathfrak{p}) = 1$ if and only if $\mathfrak{p}$ splits completely in L. Finally, recall from Lemma 9.3 that L is Galois over $\mathbf{Q}$. Thus, the proof of the last equivalence is identical to the proof of the last equivalence of (5.27). This completes the proof of Theorem 9.4. Q.E.D.

The next step is to prove the main equivalence of Theorem 9.2. By Lemma 9.3, the ring class field L is Galois over $\mathbf{Q}$, and thus Proposition 5.29 enables us to find a real algebraic integer α such that $L = K(\alpha)$. Let $f_n(x) \in \mathbf{Z}[x]$ be the minimal polynomial of α over K. Since $\mathcal{O}$ has discriminant $-4n$, the degree of $f_n(x)$ is $[L:K] = h(\mathcal{O}) = h(-4n)$. Then, combining Theorem 9.4 with the last part of Proposition 5.29, we have

$$\begin{aligned} p = x^2 + ny^2 &\iff p \text{ splits completely in } L \\ &\iff \begin{cases} (-n/p) = 1 \text{ and } f_n(x) \equiv 0 \bmod p \\ \text{has an integer solution,} \end{cases} \end{aligned}$$

whenever p is an odd prime dividing neither n nor the discriminant of $f_n(x)$. This proves the main equivalence of Theorem 9.2.

The final part of the theorem is concerned with the "uniqueness" of $f_n(x)$. Of course, there are infinitely many real algebraic integers which are primitive elements of the extension $K \subset L$, and correspondingly there are infinitely many $f_n(x)$'s. So the best we could hope for in the way of uniqueness is that these are *all* of the possible $f_n(x)$'s. This is almost what the last part of the statement of Theorem 9.2 asserts—the $f_n(x)$'s that can occur are exactly the monic integer polynomials which are minimal polynomials of primitive elements (not necessarily real) of L over K.

To prove this assertion, let $f_n(x)$ be a monic integer polynomial of degree $h(-4n)$ which satisfies the equivalence of Theorem 9.2. Then let $g(x) \in K[x]$ be an irreducible factor of $f_n(x)$ over K, and let $M = K(\alpha)$ be the field generated by a root of $g(x)$. Note that α is an algebraic integer. If we can show that $L \subset M$, then

$$h(-4n) = [L:K] \leq [M:K] = \deg(g(x)) \leq \deg(f_n(x)) = h(-4n),$$

which will prove that $L = M = K(\alpha)$ and that $f_n(x)$ is the minimal polynomial of α over K (and hence over $\mathbf{Q}$).

It remains to prove $L \subset M$. Since L is Galois over $\mathbf{Q}$ by Lemma 9.3, Proposition 8.20 tells us that $L \subset M$ if and only if $\tilde{\mathcal{S}}_{M/\mathbf{Q}} \dot{\subset} \mathcal{S}_{L/\mathbf{Q}}$, where:

$$\mathcal{S}_{L/\mathbf{Q}} = \{p \text{ prime} : p \text{ splits completely in } L\}$$

$$\tilde{\mathcal{S}}_{M/\mathbf{Q}} = \{p \text{ prime} : \text{there is an ideal } \mathfrak{P} \text{ of } M \text{ with } f_{\mathfrak{P}|p} = 1\}.$$

Let's first study $\mathcal{S}_{L/\mathbf{Q}}$. By Theorem 9.4, this is the set of primes p represented by $x^2 + ny^2$. Since $f_n(x)$ satisfies the equivalence of Theorem 9.2, it follows that $\mathcal{S}_{L/\mathbf{Q}}$ is (with finitely many exceptions) the set of primes p which split completely in K and for which $f_n(x) \equiv 0 \bmod p$ has a solution.

To prove $\tilde{\mathcal{S}}_{M/\mathbf{Q}} \dot{\subset} \mathcal{S}_{L/\mathbf{Q}}$, suppose that $p \in \tilde{\mathcal{S}}_{M/\mathbf{Q}}$. Then $f_{\mathfrak{P}|p} = 1$ for some prime $\mathfrak{P}$ of M, and if we set $\mathfrak{p} = \mathfrak{P} \cap \mathcal{O}_K$, then $1 = f_{\mathfrak{P}|p} = f_{\mathfrak{P}|\mathfrak{p}} f_{\mathfrak{p}|p}$. Thus $f_{\mathfrak{p}|p} = 1$, which implies that p splits completely in K (since it's unramified). Note also that $f_n(x) \equiv 0 \bmod \mathfrak{P}$ has a solution in $\mathcal{O}_M$ since $\alpha \in \mathcal{O}_M$ and $g(\alpha) = f_n(\alpha) = 0$. But $f_{\mathfrak{P}|p} = 1$ implies that $\mathbf{Z}/p\mathbf{Z} \simeq \mathcal{O}_M/\mathfrak{P}$, and hence $f_n(x) \equiv 0 \bmod p$ has an integer solution. By the above description of $\mathcal{S}_{L/\mathbf{Q}}$, it follows that $p \in \mathcal{S}_{L/\mathbf{Q}}$. This proves $\tilde{\mathcal{S}}_{M/\mathbf{Q}} \dot{\subset} \mathcal{S}_{L/\mathbf{Q}}$ and completes the proof of Theorem 9.2. Q.E.D.

There are also versions of Theorems 9.2 and 9.4 that characterize which primes are represented by the form $x^2 + xy + ((1-D)/4)y^2$, where $D \equiv 1 \bmod 4$ is negative (see Exercise 9.3).

B. The Ring Class Fields of $\mathbf{Z}[\sqrt{-27}]$ and $\mathbf{Z}[\sqrt{-64}]$

Theorem 9.2 shows how the ring class field solves our basic problem of determining when $p = x^2 + ny^2$, and the last part of the theorem points out that our problem is in fact *equivalent* to finding the appropriate ring class field. To see how this works in practice, we will next use Theorem 9.2 to give new proofs of Euler's conjectures for when a prime is represented by $x^2 + 27y^2$ or $x^2 + 64y^2$ (proved in §4 as Theorems 4.15 and 4.23). The first step, of course, is to determine the ring class fields involved:

Proposition 9.5.

(i) *The ring class field of the order* $\mathbf{Z}[\sqrt{-27}] \subset K = \mathbf{Q}(\sqrt{-3})$ *is* $L = K(\sqrt[3]{2})$.

(ii) *The ring class field of the order* $\mathbf{Z}[\sqrt{-64}] \subset K = \mathbf{Q}(i)$ *is* $L = K(\sqrt[4]{2})$.

Proof. To prove (i), let L be the ring class field of $\mathbf{Z}[\sqrt{-27}]$. Although L is defined abstractly by class field theory, we still know the following facts about L:

(i) L is a cubic Galois extension of $K = \mathbf{Q}(\sqrt{-3})$ since $[L:K] = h(-4\cdot 27) = 3$.

(ii) L is Galois over $\mathbf{Q}$ with group $\mathrm{Gal}(L/\mathbf{Q})$ isomorphic to the symmetric group S_3. This follows from Lemma 9.3 since S_3 is the semidirect product $(\mathbf{Z}/3\mathbf{Z}) \rtimes (\mathbf{Z}/2\mathbf{Z})$ with $\mathbf{Z}/2\mathbf{Z}$ acting nontrivially.

(iii) All primes of K that ramify in L must divide $6\mathcal{O}_K$. To see this, note that $\mathbf{Z}[\sqrt{-27}] = \mathbf{Z}[3\sqrt{-3}]$ is an order of conductor 6 (since $\mathcal{O}_K = \mathbf{Z}[(-1+\sqrt{-3})/2]$), so that L corresponds to a generalized ideal class group for the modulus $6\mathcal{O}_K$. By the Existence Theorem (Theorem 8.6), the ramification must divide the modulus.

We will show that only four fields satisfy these conditions. To see this, first note that K contains a primitive cube root of unity, and hence any cubic Galois extension of K is of the form $K(\sqrt[3]{u})$ for some $u \in K$. (This is a standard result of Galois theory—see Artin [2, Corollary to Theorem 25].) However, the fact that $\mathrm{Gal}(L/\mathbf{Q}) \simeq S_3$ allows us to assume that u is an ordinary integer. More precisely, we have:

Lemma 9.6. *If M is a cubic extension of $K = \mathbf{Q}(\sqrt{-3})$ with* $\mathrm{Gal}(M/\mathbf{Q}) \simeq S_3$, *then $M = K(\sqrt[3]{m})$ for some cubefree positive integer m.*

Proof. The idea is to modify the classical proof that $M = K(\sqrt[3]{u})$ for some $u \in K$. We know that M is Galois over $\mathbf{Q}$ and that complex conjugation τ is in $\mathrm{Gal}(M/\mathbf{Q})$. Furthermore, if σ is a generator of $\mathrm{Gal}(L/K) \simeq \mathbf{Z}/3\mathbf{Z}$, then $\mathrm{Gal}(L/\mathbf{Q}) \simeq S_3$ implies that $\tau\sigma\tau = \sigma^{-1}$.

By Proposition 5.29, we can find a real algebraic integer α such that $M = K(\alpha)$. Then define $u_i \in M$ by

$$u_i = \alpha + \omega^i\sigma^{-1}(\alpha) + \omega^{2i}\sigma^{-2}(\alpha), \qquad i = 0,1,2.$$

The u_i's are algebraic integers satisfying $\sigma(u_i) = \omega^i u_i$, and note that $\tau(u_i) = u_i$ since α is real and $\tau\sigma\tau = \sigma^{-1}$. Thus the u_i's are all real. Then u_0 is fixed by both σ and τ, which implies that $u_0 \in \mathbf{Z}$. Similar arguments show that u_1^3 and u_2^3 are also integers. If $u_1 \neq 0$, we claim that $M = K(u_1)$. This is easy to see, for $[M:K]=3$, and thus $M \neq K(u_1)$ could only happen when $u_1 \in K$. Since u_1 is real, this would force u_1 to be an integer, which would contradict $\sigma(u_1)=\omega u_1$ and $u_1 \neq 0$. This proves our claim, and if we set $m = u_1^3 \in \mathbf{Z}$, it follows that $M = K(u_1) = K(\sqrt[3]{m})$. We may assume that m is positive and cubefree, and we are done.

If $u_2 \neq 0$, we are done by a similar argument. The remaining case to consider is when $u_1 = u_2 = 0$. However, in this situation a simple application of Cramer's rule shows that our original α would lie in K and hence be rational (since we chose α to be real in the first place). The details of this argument are left to the reader (see Exercise 9.4), and this completes the proof of Lemma 9.6. Q.E.D.

Once we know $L = K(\sqrt[3]{m})$ for some cubefree integer m, the next step is to use the ramification of $K \subset L$ to restrict m. Specifically, it is easy to show that *any* prime of $\mathcal{O}_K$ dividing m ramifies in $K(\sqrt[3]{m})$ (see Exercise 9.5). However, by (iii) above, we know that all ramified primes divide $6\mathcal{O}_K$, and consequently 2 and 3 are the only integer primes that can divide m. Since m is also positive and cubefree, it must be one of the following eight numbers:

$$2,\ 3,\ 4,\ 6,\ 9,\ 12,\ 18,\ 36,$$

and this in turn implies that L must be one of the following four fields:

$$K(\sqrt[3]{2}),\ K(\sqrt[3]{3}),\ K(\sqrt[3]{6}),\ K(\sqrt[3]{12}) \tag{9.7}$$

(see Exercise 9.6). All four of these fields satisfy conditions (i)–(iii) above, so that we will need something else to decide which one is the ring class field L.

Surprisingly, the extra ingredient is none other than Theorem 9.2. More precisely, each field listed in (9.7) gives a different candidate for the polynomial $f_{27}(x)$ that characterizes $p = x^2 + 27y^2$, and then numerical computations can determine which one is the correct field. To illustrate what this means, suppose that L were $K(\sqrt[3]{3})$, the second field in (9.7). This would imply that $f_{27}(x) = x^3 - 3$, which has discriminant -3^5 (see Exercise 9.7). If Theorem 9.2 held with this particular $f_{27}(x)$, then the congruence $x^3 \equiv 3 \bmod 31$ would have a solution since $31 = 2^2 + 27 \cdot 1^2$ is of the form $x^2 + 27y^2$. Using a computer, it is straightforward to show that there are no solutions, so that $K(\sqrt[3]{3})$ can't be the ring class field in question. Similar arguments (also using $p = 31$) suffice to rule out the third and fourth fields given in (9.7) (see Exercise 9.8), and it follows that $L = K(\sqrt[3]{2})$ as claimed.

The second part of the proposition, which concerns the ring class field of the order $\mathbf{Z}[\sqrt{-64}] \subset K = \mathbf{Q}(i)$, is easier to prove than the first, for in this case one can show that $K(\sqrt[4]{2})$ is the *unique* field satisfying the analogs of conditions (i)—(iii) above (see Exercise 9.9). Q.E.D.

Another example of a ring class field is given in Exercise 9.10, where we will show that the field $K(\sqrt[3]{3})$ from (9.7) is the ring class field of the order $\mathbf{Z}[9\omega]$ of conductor 9 in $K = \mathbf{Q}(\sqrt{-3})$.

If we combine Theorem 9.2 with the explicit ring class fields of Proposition 9.5, then we get the following characterizations of when $p = x^2 + 27y^2$ and $p = x^2 + 64y^2$ (proved earlier as Theorems 4.15 and 4.23):

Theorem 9.8.

(i) *If $p > 3$ is prime, then*

$$p = x^2 + 27y^2 \iff \begin{cases} p \equiv 1 \bmod 3 \textit{ and } x^3 \equiv 2 \bmod p \\ \textit{has an integer solution.} \end{cases}$$

(ii) *If p is an odd prime, then*

$$p = x^2 + 64y^2 \iff \begin{cases} p \equiv 1 \bmod 4 \textit{ and } x^4 \equiv 2 \bmod p \\ \textit{has an integer solution.} \end{cases}$$

Proof. By Proposition 9.5, the ring class field of $\mathbf{Z}[\sqrt{-27}]$ is $L = K(\sqrt[3]{2})$, where $K = \mathbf{Q}(\sqrt{-3})$. Since $\sqrt[3]{2}$ is a real algebraic integer, the polynomial $f_{27}(x)$ of Theorem 9.2 may be taken to be $x^3 - 2$. Then the main equivalence of Theorem 9.2 is exactly what we need, once once checks that the condition $(-27/p) = 1$ is equivalent to the congruence $p \equiv 1 \bmod 3$. The final detail to check is that the discriminant of $x^3 - 2$ is $-2^2 \cdot 3^3$ (see Exercise 9.7), so that the only excluded primes are 2 and 3, and then (i) follows. The proof of (ii) is similar and is left to the reader (see Exercise 9.11). Q.E.D.

Besides allowing us to prove Theorem 9.8, the ring class fields determined in Proposition 9.5 have other uses. For example, if we combine them with Weak Reciprocity from §8, we then get the following partial results concerning cubic and biquadratic reciprocity:

Theorem 9.9.

(i) *If a primary prime π of $\mathbf{Z}[\omega]$, $\omega = e^{2\pi i/3}$, is relatively prime to 6, then*

$$\left(\frac{2}{\pi}\right)_3 = \left(\frac{\pi}{2}\right)_3.$$

(ii) *If $p \equiv 1 \bmod 4$ is prime and $p = a^2 + b^2$, then $\pi = a + bi$ is prime in $\mathbf{Z}[i]$, and*

$$\left(\frac{2}{\pi}\right)_4 = i^{ab/2}.$$

Remark. Notice that these are *exactly* the portions of cubic and biquadratic reciprocity used in our discussion of $p = x^2 + 27y^2$ and $x^2 + 64y^2$ in §4 (see Theorems 4.15 and 4.23).

Proof. We will prove (i) and leave the proof of (ii) as an exercise (see Exercise 9.12). The basic idea is to combine Weak Reciprocity (Theorem 8.11) with the explicit description of the ring class field given in Proposition 9.5.

If $K = \mathbf{Q}(\omega)$, then $\mathcal{O}_K$ is the ring $\mathbf{Z}[\omega]$ from §4. Thus $L = K(\sqrt[3]{2})$ is the ring class field of the order of conductor 6, and hence corresponds to a subgroup of $I_K(6)$ containing $P_{K,1}(6)$. This shows that the conductor $\mathfrak{f}$ divides $6\mathcal{O}_K$. Then Weak Reciprocity tells us that the cubic Legendre symbol $(2/\cdot)_3$ induces a well-defined homomorphism

$$I_K(6)/P_{K,1}(6) \longrightarrow \mu_3$$

where μ_3 is the group of cube roots of unity. However, the map sending $\alpha \in \mathcal{O}_K$ to the principal ideal $\alpha\mathcal{O}_K$ induces a homomorphism

$$(\mathcal{O}_K/6\mathcal{O}_K)^* \longrightarrow I_K(6)/P_{K,\mathbf{Z}}(6)$$

(this is similar to what we did in §7—see part (c) of Exercise 9.21). Combining these two maps, the Legendre symbol $(2/\cdot)_3$ induces a well-defined homomorphism

$$(\mathcal{O}_K/6\mathcal{O}_K)^* \longrightarrow \mu_3. \tag{9.10}$$

Recall that π is primary by assumption, which means that $\pi \equiv \pm 1 \bmod 3\mathcal{O}_K$. Replacing π by $-\pi$ affects neither $(2/\pi)_3$ nor $(\pi/2)_3$, so that we can assume $\pi \equiv 1 \bmod 3\mathcal{O}_K$. Now consider the isomorphism

$$(\mathcal{O}_K/6\mathcal{O}_K)^* \simeq (\mathcal{O}_K/2\mathcal{O}_K)^* \times (\mathcal{O}_K/3\mathcal{O}_K)^*. \tag{9.11}$$

By (9.10), $(2/\cdot)_3$ is a homomorphism on $(\mathcal{O}_K/6\mathcal{O}_K)^*$, and the condition $\pi \equiv 1 \bmod 3\mathcal{O}_K$ means we are restricting this homomorphism to the subgroup $(\mathcal{O}_K/2\mathcal{O}_K)^* \times \{1\}$ relative to (9.11). But the cubic Legendre symbol $(\cdot/2)_3$ can also be regarded as a homomorphism on this subgroup, and we thus need only show that these homomorphisms are equal.

To prove this, first note that $(\mathcal{O}_K/2\mathcal{O}_K)^* \times \{1\}$ is cyclic of order 3 ($\mathcal{O}_K/2\mathcal{O}_K$ is a field with four elements), and the class of $\theta = 1 + 3\omega$ in $(\mathcal{O}_K/6\mathcal{O}_K)^*$ is a generator. Thus, to show that the two homomorphisms are equal, it suffices to prove that

$$\left(\frac{2}{\theta}\right)_3 = \left(\frac{\theta}{2}\right)_3.$$

Using (4.10), this is straightforward to check—see Exercise 9.12 for the details. Theorem 9.9 is proved. Q.E.D.

C. Primes Represented by Positive Definite Quadratic Forms

As an application of ring class fields, we will prove the classic theorem that a primitive positive definite quadratic form $ax^2 + bxy + cy^2$ represents infinitely many prime numbers. The basic idea is to compute the Dirichlet density (in the sense of §8) of the set of primes represented by $ax^2 + bxy + cy^2$, for once we show that the density is positive, there must be infinitely many primes represented. Here is the precise statement of what we will prove:

Theorem 9.12. *Let $ax^2 + bxy + cy^2$ be a primitive positive definite quadratic form of discriminant $D < 0$, and let $\mathcal{S}$ be the set of primes represented by $ax^2 + bxy + cy^2$. Then the Dirichlet density $\delta(\mathcal{S})$ exists and is given by the formula*

$$\delta(\mathcal{S}) = \begin{cases} \dfrac{1}{h(D)} & \text{if } ax^2 + bxy + cy^2 \text{ is properly equivalent to its opposite} \\ \dfrac{1}{2h(D)} & \text{otherwise.} \end{cases}$$

In particular, $ax^2 + bxy + cy^2$ represents infinitely many prime numbers.

Proof. Let $\mathcal{O}$ be the order of the discriminant D, and let $K = \mathbb{Q}(\sqrt{D})$. By (7.3), we have $D = f^2 d_K$, where f is the conductor of $\mathcal{O}$. As in the statement of the theorem, let $\mathcal{S} = \{p \text{ prime} : p = ax^2 + bxy + cy^2\}$. We need to compute the Dirichlet density of $\mathcal{S}$.

The first step is to relate $\mathcal{S}$ to the generalized ideal class group $I_K(f)/P_{K,\mathbb{Z}}(f)$. From Theorem 7.7 we have the isomorphism $C(D) \simeq C(\mathcal{O})$, so that the class $[ax^2 + bxy + cy^2] \in C(D)$ corresponds to the class $[\mathfrak{a}_0] \in C(\mathcal{O})$ for some proper $\mathcal{O}$-ideal $\mathfrak{a}_0$. Then part (iii) of Theorem 7.7 tells us that

$$\mathcal{S} = \{p \text{ prime} : p = N(\mathfrak{b}),\ \mathfrak{b} \in [\mathfrak{a}_0]\}. \tag{9.13}$$

We need to state this in terms of the maximal order $\mathcal{O}_K$. By Corollary 7.17 we may assume that $\mathfrak{a}_0$ is prime to f, and from here on we will consider only primes p not dividing f. Under the map $\mathfrak{a} \mapsto \mathfrak{a}\mathcal{O}_K$, we know that $\mathfrak{b} \in [\mathfrak{a}_0] \in C(\mathcal{O})$ corresponds to $\mathfrak{b}\mathcal{O}_K \in [\mathfrak{a}_0\mathcal{O}_K] \in I_K(f)/P_{K,\mathbb{Z}}(f)$ (Proposition 7.22), and furthermore, $\mathfrak{b}$ and $\tilde{\mathfrak{b}} = \mathfrak{b}\mathcal{O}_K$ have the same norm when prime to f (Proposition 7.20). Thus (9.13) implies

$$\mathcal{S} \doteq \{p \text{ prime} : p \nmid f,\ p = N(\tilde{\mathfrak{b}}),\ \tilde{\mathfrak{b}} \in [\mathfrak{a}_0\mathcal{O}_K]\}.$$

Since p is prime, the equation $p = N(\tilde{\mathfrak{b}})$ forces $\tilde{\mathfrak{b}}$ to be prime, so that this description of $\mathcal{S}$ can be written

$$(9.14) \qquad \mathcal{S} \doteq \{p \text{ prime} : p \nmid f,\ p = N(\mathfrak{p}),\ \mathfrak{p} \text{ prime}, \mathfrak{p} \in [\mathfrak{a}_0\mathcal{O}_K]\}.$$

If L is the ring class field of $\mathcal{O}$, then Artin Reciprocity gives us an isomorphism

$$(9.15) \qquad I_K(f)/P_{K,\mathbf{Z}}(f) \simeq \mathrm{Gal}(L/K).$$

Under this isomorphism, the class of $\mathfrak{a}_0\mathcal{O}_K$ maps to an element $\sigma_0 \in \mathrm{Gal}(L/K)$, which we can regard as an element of $\mathrm{Gal}(L/\mathbf{Q})$. Letting $\langle\sigma_0\rangle$ denote its conjugacy class in $\mathrm{Gal}(L/\mathbf{Q})$, we claim that

$$(9.16) \qquad \mathcal{S} \doteq \left\{p \text{ prime} : p \text{ unramified in } L,\ \left(\frac{L/\mathbf{Q}}{p}\right) = \langle\sigma_0\rangle\right\}.$$

The right hand side of (9.16) will be denoted $\mathcal{S}'$, so that we must prove $\mathcal{S} \doteq \mathcal{S}'$.

To show $\mathcal{S}' \dot{\subset} \mathcal{S}$, let $p \in \mathcal{S}'$. Thus $((L/\mathbf{Q})/p) = \langle\sigma_0\rangle$, which means that $((L/\mathbf{Q})/\mathfrak{P}) = \sigma_0$ for some prime $\mathfrak{P}$ of L containing p. Then $\mathfrak{p} = \mathfrak{P} \cap \mathcal{O}_K$ is a prime of K containing p, and we claim that $p = N(\mathfrak{p})$. To see this, note that for any $\alpha \in \mathcal{O}_L$,

$$(9.17) \qquad \sigma_0(\alpha) \equiv \alpha^p \bmod \mathfrak{P}$$

since $\sigma_0 = ((L/\mathbf{Q})/\mathfrak{P})$. But $\sigma_0 \in \mathrm{Gal}(L/K)$, so that when $\alpha \in \mathcal{O}_K$, the above congruence reduces to

$$\alpha \equiv \alpha^p \bmod \mathfrak{p}.$$

This implies $\mathcal{O}_K/\mathfrak{p} \simeq \mathbf{Z}/p\mathbf{Z}$, and $N(\mathfrak{p}) = p$ follows. This fact and (9.17) then imply that σ_0 is the Artin symbol $((L/K)/\mathfrak{p})$. Since $[\mathfrak{a}_0\mathcal{O}_K] \in I_K(f)/P_{K,\mathbf{Z}}(f)$ corresponds to $\sigma_0 \in \mathrm{Gal}(L/K)$ under the isomorphism (9.15), it follows that $\mathfrak{p}$ is in the class of $\mathfrak{a}_0\mathcal{O}_K$. Then (9.14) implies that $p \in \mathcal{S}$, at least when $p \nmid f$, and $\mathcal{S}' \dot{\subset} \mathcal{S}$ follows. The opposite inclusion is straightforward and is left to the reader (see Exercise 9.14). This completes the proof of (9.16).

From (9.16), the Čebotarev Density Theorem shows that $\mathcal{S}$ has Dirichlet density

$$\delta(\mathcal{S}) = \frac{|\langle\sigma_0\rangle|}{[L:\mathbf{Q}]}.$$

However, since $\sigma_0 \in \mathrm{Gal}(L/K)$, Lemma 9.3 implies that $\langle\sigma_0\rangle = \{\sigma_0, \sigma_0^{-1}\}$ (see Exercise 9.15). Since $[L:\mathbf{Q}] = 2h(D)$, we see that

$$\delta(\mathcal{S}) = \begin{cases} \dfrac{1}{2h(D)}, & \sigma_0 \text{ has order} \leq 2 \\[2ex] \dfrac{1}{h(D)}, & \text{otherwise.} \end{cases}$$

Now σ_0 has order ≤ 2 if and only if $ax^2 + bxy + cy^2$ has order ≤ 2 in $C(D)$, and this last statement means that $ax^2 + bxy + cy^2$ is properly equivalent to its opposite. This completes the proof of Theorem 9.12. Q.E.D.

As an example of what the theorem says, consider forms of discriminant -56. The class number is 4, and we know the reduced forms from §2. Then Theorem 9.12 implies that

$$\delta(\{p \text{ prime} : p = x^2 + 14y^2\}) = \tfrac{1}{8}$$

$$\delta(\{p \text{ prime} : p = 2x^2 + 7y^2\}) = \tfrac{1}{8}$$

$$\delta(\{p \text{ prime} : p = 3x^2 \pm 2xy + 5y^2\}) = \tfrac{1}{4}.$$

Notice that these densities sum to 1/2, which is the density of primes for which $(-56/p) = 1$. This example is no accident, for given any negative discriminant, the densities of primes represented by the reduced forms (counted properly) always sum to 1/2 (see Exercise 9.17).

A weaker form of Theorem 9.12, which asserts that $ax^2 + bxy + cy^2$ represents infinitely many primes, was first stated by Dirichlet in 1840, though his proof applied only to a restricted class of discriminants (see [27, Vol. I, pp. 497–502]). A complete proof was given by Weber in 1882 [101], and in 1954 Briggs [10] found an "elementary" proof (in the sense of the "elementary" proofs of the prime number theorem due to Erdös and Selberg).

D. Ring Class Fields and Generalized Dihedral Extensions

We will conclude §9 by asking if there is an intrinsic characterization of ring class fields. We know that they are Abelian extensions of K, but which ones? The remarkable fact is that there is a purely field-theoretic way to characterize ring class fields and their subfields. The key idea is to work with the Galois group over $\mathbb{Q}$. We used this strategy in §6 in dealing with the genus field, and here it will be similarly successful. For the genus field, we wanted $\text{Gal}(L/\mathbb{Q})$ to be Abelian, while in the present case we will allow slightly more complicated Galois groups. The crucial notion is when an extension of K is *generalized dihedral* over $\mathbb{Q}$. To define this, let K be an imaginary quadratic field, and let L be an Abelian extension of K which is Galois over $\mathbb{Q}$. As we saw in the proof of Lemma 9.3, complex conjugation τ is an automorphism of L, and the Galois group $\text{Gal}(L/K)$ can be written as a semidirect product

$$\text{Gal}(L/\mathbb{Q}) \simeq \text{Gal}(L/K) \rtimes (\mathbb{Z}/2\mathbb{Z}),$$

where the nontrivial element of $\mathbb{Z}/2\mathbb{Z}$ acts on $\text{Gal}(L/K)$ via conjugation by τ. We say that L is *generalized dihedral* over $\mathbb{Q}$ if this action sends every element in $\text{Gal}(L/K)$ to its inverse.

In Lemma 9.3 we proved that every ring class field L is generalized dihedral over $\mathbf{Q}$, and it is easy to show that every subfield of L containing K is also generalized dihedral over $\mathbf{Q}$ (see Exercise 9.18). The unexpected result, due to Bruckner [11], is that this gives *all* extensions of K which are generalized dihedral over $\mathbf{Q}$:

Theorem 9.18. *Let K be an imaginary quadratic field. Then an Abelian extension L of K is generalized dihedral over $\mathbf{Q}$ if and only if L is contained in a ring class field of K.*

Proof. By the above discussion, we know that any extension of K contained in a ring class field is generalized dihedral over $\mathbf{Q}$. To prove the converse, fix an Abelian extension L of K which is generalized dihedral over $\mathbf{Q}$. By Artin Reciprocity, there is an ideal $\mathfrak{m}$ and a subgroup $P_{K,1}(\mathfrak{m}) \subset H \subset I_K(\mathfrak{m})$ such that the Artin map induces an isomorphism

$$(9.19) \qquad I_K(\mathfrak{m})/H \xrightarrow{\sim} \mathrm{Gal}(L/K).$$

We saw in §8 that all of this remains true when $\mathfrak{m}$ is enlarged, so that we may assume that $\mathfrak{m} = f\mathcal{O}_K$ for some integer f, and we can also assume that f is divisible by the discriminant d_K of K (this will be useful later in the proof). To prove the theorem, it suffices to show that $P_{K,\mathbf{Z}}(f) \subset H$, for this will imply that L lies in the ring class field of the order of conductor f in $\mathcal{O}_K$. From the definition of $P_{K,\mathbf{Z}}(f)$, this means that we have to prove the following for elements $u \in \mathcal{O}_K$:

$$(9.20) \qquad c \in \mathbf{Z},\ c \text{ prime to } f,\ u \equiv c \bmod f \Rightarrow u\mathcal{O}_K \in H.$$

The first step is to use the fact that $P_{K,1}(f\mathcal{O}_K) \subset H$: if $\alpha, \beta \in \mathcal{O}_K$ are prime to f, then we claim that

$$(9.21) \qquad \alpha \equiv \beta \bmod f\mathcal{O}_K \Rightarrow (\alpha\mathcal{O}_K \in H \iff \beta\mathcal{O}_K \in H).$$

To prove this, pick an element $\gamma \in \mathcal{O}_K$ such that $\alpha\gamma \equiv 1 \bmod f\mathcal{O}_K$. Then $\beta\gamma \equiv 1 \bmod f\mathcal{O}_K$ also holds, so that $\alpha\gamma\mathcal{O}_K$ and $\beta\gamma\mathcal{O}_K$ both lie in $P_{K,1}(f\mathcal{O}_K) \subset H$, and (9.21) follows immediately. One consequence of (9.21) is that (9.20) is equivalent to the simpler statement

$$(9.22) \qquad c \in \mathbf{Z}, c \text{ prime to } f \Rightarrow c\mathcal{O}_K \in H.$$

So we need to see how (9.22) follows from L being generalized dihedral over $\mathbf{Q}$. Under the isomorphism (9.19), we know that conjugation by τ on $\mathrm{Gal}(L/K)$ corresponds to the usual action of τ on $I_K(f)$. Then L being generalized dihedral over $\mathbf{Q}$ means that for $\mathfrak{a} \in I_K(f)$, the class of $\bar{\mathfrak{a}}$ gives the inverse of $\mathfrak{a}$ in $I_K(f)/H$, which in turn means that $\mathfrak{a}\bar{\mathfrak{a}} \in H$. Since $\mathfrak{a}\bar{\mathfrak{a}} = N(\mathfrak{a})\mathcal{O}_K$ by Lemma 7.14, we see that for any ideal $\mathfrak{a} \in I_K(f)$, we have

$$(9.23) \qquad N(\mathfrak{a})\mathcal{O}_K \in H.$$

It remains to prove that (9.23) implies (9.22). Note first that it suffices to prove (9.22) when c is a prime p not dividing f. Recall that $d_K \mid f$, so that p is unramified in K. There are two cases to consider, depending on whether or not p splits in K. If p splits, then $p = N(\mathfrak{p})$, where $\mathfrak{p}$ is a prime factor of $p\mathcal{O}_K$. Then, by (9.23), we have $p\mathcal{O}_K = N(\mathfrak{p})\mathcal{O}_K \in H$, as desired. If p doesn't split, then $(d_K/p) = -1$ by Corollary 5.17. Let q be a prime such that $q \equiv -p \bmod f$ (such primes exist by Dirichlet's theorem). We claim that q splits completely in K. The proof will use the character χ from Lemma 1.14. Recall that this lemma states that the Legendre symbol $(d_K/\cdot)$ induces a well defined homomorphism $\chi : (\mathbf{Z}/d_K\mathbf{Z})^* \to \{\pm 1\}$, and since $d_K < 0$, it also tells us that $\chi([-1]) = -1$. Since $d_K \mid f$, we have $q \equiv -p \bmod d_K$, and thus

$$\left(\frac{d_K}{q}\right) = \chi([q]) = \chi([-p]) = \chi([-1])\chi([p]) = -\left(\frac{d_K}{p}\right) = 1.$$

Hence q splits completely in K. The argument for the split case implies that $q\mathcal{O}_K \in H$, and then $q \equiv -p \bmod f\mathcal{O}_K$ and (9.21) imply that $p\mathcal{O}_K = (-p)\mathcal{O}_K \in H$. This proves (9.22) and completes the proof of Theorem 9.18. Q.E.D.

In Exercises 9.19–9.24, we will explore some other aspects of ring class fields, including a computation of the conductor (in the sense of class field theory) of a ring class field. For further discussion of ring class fields, see Bruckner [11], Cohn [19, §15.I] and Cohn [21, Chapter 8].

E. Exercises

9.1. Prove that the Hilbert class field of an imaginary quadratic field is the ring class field of the maximal order.

9.2. Let $\mathcal{O}$ be the order of conductor f in the imaginary quadratic field K, and let L be the ring class field of $\mathcal{O}$. Let $\mathfrak{m} = f\mathcal{O}_K$, and let τ denote complex conjugation.
(a) Show that $\tau(\mathfrak{m}) = \mathfrak{m}$ and that $\tau(P_{K,\mathbf{Z}}(f)) = P_{K,\mathbf{Z}}(f)$.
(b) Show that $\ker(\Phi_{\tau(L)/K,\mathfrak{m}}) = \tau(\ker(\Phi_{L/K,\mathfrak{m}}))$.
(c) Using $\ker(\Phi_{L/K,\mathfrak{m}}) = P_{K,\mathbf{Z}}(f)$, conclude that $\ker(\Phi_{\tau(L)/K,\mathfrak{m}}) = \ker(\Phi_{L/K,\mathfrak{m}})$.

9.3. Formulate and prove versions of Theorems 9.2 and 9.4 for primes represented by the principal form $x^2 + xy + ((1-D)/4)y^2$ when $D \equiv 1 \bmod 4$ is negative.

9.4. Let u_i, $i = 0, 1, 2$ be as in the proof of Lemma 9.6. If $u_1 = u_2 = 0$, then use Cramer's rule to prove that $\alpha \in K$.

9.5. Let $L = K(\sqrt[3]{m})$ be a cubic extension of K where m is a cubefree integer and K is an imaginary quadratic field. If $\mathfrak{p}$ is any prime of K dividing m, then prove that $\mathfrak{p}$ ramifies in L.

9.6. Verify that if $K = \mathbf{Q}(\sqrt{-3})$ and $L = K(\sqrt[3]{m})$, where m is a cubefree integer of the form $2^a 3^b$, then L is one of the four fields listed in (9.7).

9.7. Prove that the discriminant of the cubic polynomial $x^3 - a$ is $-27a^2$.

9.8. Use the arguments outlined in the proof of Proposition 9.5 to show that none of the fields $K(\sqrt[3]{3})$, $K(\sqrt[3]{6})$ and $K(\sqrt[3]{12})$ can be the ring class field of the order $\mathbf{Z}[\sqrt{-27}]$. Hint: use $31 = 2^2 + 27 \cdot 1^2$.

9.9. Prove part (ii) of Proposition 9.5 using the hints given in the text.

9.10. This exercise is concerned with the order $\mathbf{Z}[9\omega]$ of conductor 9 in the field $K = \mathbf{Q}(\omega)$, $\omega = e^{2\pi i/3}$.

(a) Prove that $L = K(\sqrt[3]{3})$ is the ring class field of $\mathbf{Z}[9\omega]$. Hint: adapt the proof of Proposition 9.5.

(b) Use Exercise 9.3 to prove that for primes $p \geq 5$, $p = x^2 + xy + 61y^2$ if and only if $p \equiv 1 \bmod 3$ and 3 is a cubic residue modulo p.

(c) Use (b) to prove that for primes $p \geq 5$, $4p = x^2 + 243y^2$ if and only if $p \equiv 1 \bmod 3$ and 3 is a cubic residue modulo p. Note that this result, conjectured by Euler, was proved earlier in Exercise 4.15 using the supplementary laws of cubic reciprocity.

9.11. Prove part (ii) of Theorem 9.8.

9.12. This exercise is concerned with the proof of Theorem 9.9.

(a) Let $\theta = 1 + 3\omega$. To prove that $(2/\theta)_3 = (\theta/2)_3$, first use (4.10) to show

$$\left(\frac{2}{\theta}\right)_3 = \left(\frac{2}{1+3\omega}\right)_3 \equiv 2^{(N(1+3\omega)-1)/3} \equiv 4 \bmod (1+3\omega)\mathcal{O}_K$$

$$\left(\frac{\theta}{2}\right)_3 = \left(\frac{1+3\omega}{2}\right)_3 \equiv \left(\frac{1+\omega}{2}\right)_3 \equiv (1+\omega)^{(N(2)-1)/3}$$

$$\equiv 1 + \omega \bmod 2\mathcal{O}_K,$$

and then note that $1 + \omega + \omega^2 = 0$ and $4 - \omega^2 = -(1+2\omega)(1+3\omega)$.

(b) Prove part (ii) of Theorem 9.9.

9.13. Let $K = \mathbf{Q}(\omega)$, $\omega = e^{2\pi i/3}$. In this exercise we will use the ring class field $K(\sqrt[3]{3})$ from Exercise 9.10 to prove the supplementary laws of cubic reciprocity. Let $p \equiv 1 \bmod 3$ be prime. In Exercise 4.15 we saw that $4p = a^2 + 27b^2$, which gave us the factorization $p = \pi\overline{\pi}$ where $\pi = (a + \sqrt{-27}b)/2$ is primary. We can assume that $a \equiv 1 \bmod 3$.

(a) Prove that $(\omega/\pi)_3 = \omega^{2(a+2)/3}$. Hint: use (4.10).

(b) Adapt the proof of Theorem 9.9 to prove that $(3/\pi)_3 = \omega^{2b}$. Hint: use Exercise 9.10.

(c) Use $3 = -\omega^2(1-\omega)^2$ to prove that $(1-\omega/\pi)_3 = \omega^{b+(a+2)/3}$.

(d) Show that the results of (a) and (c) imply the supplementary laws for cubic reciprocity as stated in (4.13).

9.14. Let $\mathcal{S}$ and $\mathcal{S}'$ be the two sets of primes defined in the proof of Theorem 9.12. Prove that $\mathcal{S} \dot{\subset} \mathcal{S}'$. Hint: use (9.14).

9.15. Let K be an imaginary quadratic field, and let $K \subset L$ be an Abelian extension which is generalized dihedral over $\mathbf{Q}$. If $\sigma \in \mathrm{Gal}(L/K) \subset \mathrm{Gal}(L/\mathbf{Q})$, then prove that the conjugacy class $\langle\sigma\rangle$ of σ in $\mathrm{Gal}(L/\mathbf{Q})$ is the set $\{\sigma, \sigma^{-1}\}$.

9.16. In this exercise we will use (8.16) to give a different proof of Theorem 9.12. We will use the notation of the proof of Theorem 9.12. Thus $\mathcal{O}$ is the order of conductor f in an imaginary quadratic field K, and L is the ring class field of $\mathcal{O}$. Let

$$\mathcal{S} = \{p \text{ prime} : p = ax^2 + bxy + cy^2\}.$$

(a) If $ax^2 + bxy + cy^2$ gives us the class $[\mathfrak{a}_0\mathcal{O}_K] \in I_K(f)/P_{K,\mathbf{Z}}(f)$, show that

$$\mathcal{S} \doteq \{p \text{ primes} : p \nmid f,\ p\mathcal{O}_K = \mathfrak{p}\overline{\mathfrak{p}},\ \mathfrak{p} \in [\mathfrak{a}_0\mathcal{O}_K]\}.$$

Hint: use (9.14).

(b) Use the Čebotarev Density Theorem to show that

$$\mathcal{S}'' = \{\mathfrak{p} \in \mathcal{P}_K : \mathfrak{p} \in [\mathfrak{a}_0\mathcal{O}_K]\}$$

has Dirichlet density $\delta(\mathcal{S}'') = 1/h(D)$. Then use (8.16) to show that $\delta(\mathcal{S}'' \cap \mathcal{P}_{K,1}) = 1/h(D)$. Recall that $\mathcal{P}_{K,1} = \{\mathfrak{p} \in \mathcal{P}_K : N(\mathfrak{p})$ is prime$\}$.

(c) Show that the mapping $\mathfrak{p} \mapsto N(\mathfrak{p})$ from $\mathcal{S}'' \cap \mathcal{P}_{K,1}$ to $\mathcal{S}$ is either two-to-one or one-to-one, depending on whether or not $\mathfrak{a}_0\mathcal{O}_K$ has order ≤ 2 in the class group. Then use (b) to prove Theorem 9.12.

9.17. Fix a negative discriminant D.

(a) Show that the sum of the densities of the primes represented by the reduced forms of discriminant D with middle coefficient $b > 0$ is always 1/2.

(b) To explain the result of (a), first use Lemma 2.5 to show that the primes represented by the forms listed in (a) are, up to a finite set, exactly the primes for which $(D/p) = 1$. Then use the Čebotarev Density Theorem to show that this set has density 1/2.

9.18. Let K be an imaginary quadratic field. Use Lemma 9.3 to prove that any intermediate field between K and a ring class field of K is generalized dihedral over $\mathbb{Q}$.

9.19. An imaginary quadratic field K has infinitely many ring class fields associated with it. In this exercise we will work out the relation between the different ring class fields.

(a) If $\mathcal{O}_1$ and $\mathcal{O}_2$ are orders in K, then we get ring class fields L_1 and L_2. Prove that

$$\mathcal{O}_1 \subset \mathcal{O}_2 \Rightarrow L_2 \subset L_1.$$

(b) If f_i is the conductor of $\mathcal{O}_i$, then prove that $\mathcal{O}_1 \subset \mathcal{O}_2$ if and only if $f_2 \mid f_1$, and conclude that the result of (a) can be stated in terms of conductors as follows:

$$f_2 \mid f_1 \Rightarrow L_2 \subset L_1.$$

In Exercise 9.24, we will see that the converse of this implication is false.

(c) Show that the Hilbert class field is contained in the ring class field of any order, and conclude that $h(d_K) \mid h(f^2 d_K)$. This fact was proved earlier in Theorem 7.24.

9.20. Let L be the ring class field of an order $\mathcal{O}$ in an imaginary quadratic field K. Such a field has two "conductors" associated to it: first, there is the conductor f of the order $\mathcal{O}$, and second, there is the class field theory conductor $\mathfrak{f}(L/K)$ of L as an Abelian extension of K. There should be a close relation between these conductors, and the obvious guess would be that

$$\mathfrak{f}(L/K) = f\mathcal{O}_K.$$

In Exercises 9.20–9.23, we will show that the answer is a bit more complicated: the conductor is given by the formula

$$\mathfrak{f}(L/K) = \begin{cases} \mathcal{O}_K, & f = 2 \text{ or } 3, K = \mathbb{Q}(\sqrt{-3}) \\ \mathcal{O}_K, & f = 2,\ K = \mathbb{Q}(i) \\ (f/2)\mathcal{O}_K, & f = 2f',\ f' \text{ odd, 2 splits completely in } K \\ f\mathcal{O}_K, & \text{otherwise.} \end{cases}$$

To begin the proof, let f be a positive integer, and let K be an imaginary quadratic field. Assume that $f = 2f'$, where f' is odd and 2 splits completely in K. Let L and L' be the ring class fields of K corresponding to the orders of conductor f and f' respectively. Then prove that

$$\mathfrak{f}(L/K) = \mathfrak{f}(L'/K).$$

Hint: first show that $L' \subset L$, and then use Theorem 7.24 to conclude that $L' = L$.

9.21. Let L be the ring class field of the order of conductor f in an imaginary quadratic field K, and assume that $\mathfrak{f}(L/K) \neq f\mathcal{O}_K$.

(a) Show that $f\mathcal{O}_K = \mathfrak{p}\mathfrak{m}$, where $\mathfrak{p}$ is prime and $\mathfrak{f}(L/K) \mid \mathfrak{m}$. We will fix $\mathfrak{p}$ and $\mathfrak{m}$ for the rest of this exercise.

(b) Prove that $I_K(f) \cap P_{K,1}(\mathfrak{m}) \subset P_{K,\mathbb{Z}}(f)$.

(c) Show that there is an exact sequence

$$\mathcal{O}_K^* \longrightarrow (\mathcal{O}_K/f\mathcal{O}_K)^* \xrightarrow{\phi} P_K \cap I_K(f)/P_{K,1}(f) \longrightarrow 1,$$

where P_K is the group of all principal ideals and ϕ is the map which sends $[\alpha] \in (\mathcal{O}_K/f\mathcal{O}_K)^*$ to $[\alpha\mathcal{O}_K] \in P_K \cap I_K(f)/P_{K,1}(f)$. Hint: This is similar to what we did in (7.27).

(d) Consider the natural maps

$$\pi : (\mathcal{O}_K/f\mathcal{O}_K)^* \longrightarrow (\mathcal{O}_K/\mathfrak{m})^*$$
$$\beta : (\mathbb{Z}/f\mathbb{Z})^* \longrightarrow (\mathcal{O}_K/f\mathcal{O}_K)^*.$$

Show that $\ker(\pi) \subset \mathcal{O}_K^* \cdot \mathrm{Im}(\beta)$. Hint: use (b) and the exact sequence of (c) to show that $\phi^{-1}(I_K(f) \cap P_{K,1}(\mathfrak{m})) = \mathcal{O}_K^* \cdot \ker(\pi)$ and $\phi^{-1}(P_{K,\mathbb{Z}}(f)) = \mathcal{O}_K^* \cdot \mathrm{Im}(\beta)$.

9.22. In this exercise we will assume that $\mathcal{O}_K^* = \{\pm 1\}$ (by Exercise 5.9, this excludes the fields $\mathbb{Q}(\sqrt{-3})$ and $\mathbb{Q}(i)$). Let K, f and L be as in the previous exercise, and assume in addition that if $f = 2f'$, f' odd, then 2 doesn't split completely in K. Our goal is to prove that

$$\mathfrak{f}(L/K) = f\mathcal{O}_K.$$

We will argue by contradiction. Suppose that $\mathfrak{f}(L/K) \neq f\mathcal{O}_K$. Then Exercise 9.21 implies that $f\mathcal{O}_K = \mathfrak{p}\mathfrak{m}$, where $\mathfrak{p}$ is prime and $\mathfrak{f}(L/K) \mid \mathfrak{m}$. Furthermore, if π and β are the natural maps

$$\pi : (\mathcal{O}_K/f\mathcal{O}_K)^* \longrightarrow (\mathcal{O}_K/\mathfrak{m})^*$$
$$\beta : (\mathbf{Z}/f\mathbf{Z})^* \longrightarrow (\mathcal{O}_K/f\mathcal{O}_K)^*$$

then Exercise 9.21 also implies that $\ker(\pi) \subset \mathcal{O}_K^* \cdot \mathrm{Im}(\beta)$, and since $\mathcal{O}_K^* = \{\pm 1\}$, we see that

$$\ker(\pi) \subset \mathrm{Im}(\beta).$$

We will show that this inclusion leads to a contradiction.

(a) Prove that

$$|\ker(\pi)| = \begin{cases} N(\mathfrak{p}), & \mathfrak{p} \mid \mathfrak{m} \\ N(\mathfrak{p}) - 1, & \mathfrak{p} \nmid \mathfrak{m}, \end{cases}$$

where p is the unique integer prime contained in $\mathfrak{p}$. Hint: use Exercise 7.29.

(b) Note that $N(\mathfrak{p}) = p$ or p^2. Suppose first that $N(\mathfrak{p}) = p$.
   (i) Show that $\mathfrak{m} = m\bar{\mathfrak{p}}$ for some integer m.
   (ii) Use (i) to show that the map $(\mathbf{Z}/f\mathbf{Z})^* \to (\mathcal{O}_K/\mathfrak{m})^*$ is injective, and conclude that $\ker(\pi) \cap \mathrm{Im}(\beta) = \{1\}$.
   (iii) Since $\ker(\pi) \subset \mathrm{Im}(\beta)$, (ii) implies that $\ker(\pi) = \{1\}$. Use (a) to show that $p = 2$, 2 splits completely in K, and $f = 2m$ where m is odd. This contradicts our assumption on f.

(c) It remains to consider the case when $N(\mathfrak{p}) = p^2$. Here, $f = pm$ and $\mathfrak{m} = m\mathcal{O}_K$.
   (i) Show that $\ker(\pi) \cap \mathrm{Im}(\beta) \simeq \ker(\theta)$, where $\theta : (\mathbf{Z}/f\mathbf{Z})^* \to (\mathbf{Z}/m\mathbf{Z})^*$ is the natural map.
   (ii) Since $\ker(\pi) \subset \mathrm{Im}(\beta)$, (i) implies that $|\ker(\pi)| \le |\ker(\theta)|$, and we know $|\ker(\pi)|$ from (a). Now compute $|\ker(\theta)|$ and use this to show that $|\ker(\pi)| \le |\ker(\theta)|$ is impossible. Again we have a contradiction.

9.23. Recall the formula for the conductor $\mathfrak{f}(L/K)$ stated in Exercise 9.20.

(a) Using Exercises 9.20 and 9.22, prove the desired formula when $\mathcal{O}_K^* = \{\pm 1\}$.

(b) Adapt the proof of Exercise 9.22 to the case $\mathcal{O}_K^* \neq \{\pm 1\}$, and prove the formula for $\mathfrak{f}(L/K)$ for all K.

9.24. Use the conductor formula from Exercise 9.20 to give infinitely many examples where $\mathfrak{f}(L/K) \neq f\mathcal{O}_K$. Also show that the converse of part (b) of Exercise 9.19 is not true in general (i.e., $L_2 \subset L_1$ need not imply $f_2 \mid f_1$).

CHAPTER THREE

COMPLEX MULTIPLICATION

§10. ELLIPTIC FUNCTIONS AND COMPLEX MULTIPLICATION

In Chapter Two we solved our problem of when a prime p can be written in the form $x^2 + ny^2$. The criterion from Theorem 9.2 states that, with finitely many exceptions,

$$p = x^2 + ny^2 \iff \begin{cases} (-n/p) = 1 \text{ and } f_n(x) \equiv 0 \bmod p \\ \text{has an integer solution.} \end{cases}$$

The key ingredient is the polynomial $f_n(x)$, which we know is the minimal polynomial of a primitive element of the ring class field of $\mathbf{Z}[\sqrt{-n}]$. But the proof of Theorem 9.2 doesn't explain how to find such a primitive element, so that we have only an abstract solution of the problem of $p = x^2 + ny^2$. In this chapter, we will use modular functions and the theory of complex multiplication to give a systematic method for finding $f_n(x)$.

In §10 we will study elliptic functions and introduce the idea of complex multiplication. A key role is played by the j-invariant of a lattice, and we will show that if $\mathcal{O}$ is an order in an imaginary quadratic field K, then its j-invariant $j(\mathcal{O})$ is an algebraic number. But before we can get to the real depth of the subject, we need to learn about modular functions. Thus §11 will present a brief but complete account of the main properties of modular functions, including the modular equation. Then we will prove that $j(\mathcal{O})$ is not only an algebraic integer, but also that it generates (over K) the ring class field of $\mathcal{O}$. This theorem, often called the "First Main Theorem" of complex multiplication, is the main result of §11. In §12 we will compute $j(\mathcal{O})$ in some special cases, and in §13 we will complete our study of $j(\mathcal{O})$ by describing an algorithm for computing its minimal polynomial (the so-called "class equation"). When applied to the order $\mathbf{Z}[\sqrt{-n}]$, this theory will give us an algorithm for constructing the polynomial $f_n(x)$ that solves $p = x^2 + ny^2$. Finally, in §14 we will discuss elliptic curves and primality testing.

Before we can begin our discussion of complex multiplication, we need to learn some basic facts about elliptic functions and j-invariants.

A. Elliptic Functions and the Weierstrass $\wp$-Function

To start, we define a *lattice* to be an additive subgroup L of $\mathbf{C}$ which is generated by two complex numbers ω_1 and ω_1 which are linearly independent over $\mathbf{R}$. We express this by writing $L = [\omega_1, \omega_2]$. Then an *elliptic function* for L is a function $f(z)$ defined on $\mathbf{C}$, except for isolated singularities, which satisfies the following two conditions:

(i) $f(z)$ is meromorphic on $\mathbf{C}$.

(ii) $f(z+\omega) = f(z)$ for all $\omega \in L$.

If $L = [\omega_1, \omega_2]$, note that the second condition is equivalent to

$$f(z+\omega_1) = f(z+\omega_2) = f(z).$$

Thus an elliptic function is a doubly periodic meromorphic function, and elements of L are often referred to as periods.

One of the most important elliptic functions is the Weierstrass $\wp$-function, which is defined as follows: given a complex number z not in the lattice L, we set

$$\wp(z;L) = \frac{1}{z^2} + \sum_{\omega \in L - \{0\}} \left(\frac{1}{(z-\omega)^2} - \frac{1}{\omega^2} \right).$$

When working with a fixed lattice L, we will usually write $\wp(z)$ instead of $\wp(z;L)$. Here are some basic properties of the $\wp$-function:

Theorem 10.1. *Let $\wp(z)$ be the Weierstrass $\wp$-function for the lattice L.*

(i) *$\wp(z)$ is an elliptic function for L whose singularities consist of double poles at the points of L.*

(ii) *$\wp(z)$ satisfies the differential equation*

$$\wp'(z)^2 = 4\wp(z)^3 - g_2(L)\wp(z) - g_3(L),$$

where the constants $g_2(L)$ and $g_3(L)$ are defined by

$$
\begin{aligned}
g_2(L) &= 60 \sum_{\omega \in L-\{0\}} \frac{1}{\omega^4} \\
g_3(L) &= 140 \sum_{\omega \in L-\{0\}} \frac{1}{\omega^6}.
\end{aligned}
$$

(iii) $\wp(z)$ *satisfies the addition law*

$$
\wp(z+w) = -\wp(z) - \wp(w) + \frac{1}{4}\left(\frac{\wp'(z)-\wp'(w)}{\wp(z)-\wp(w)}\right)^2
$$

whenever $z, w \notin L$ *and* $z+w \notin L$.

Proof. The first step is to prove the following lemma:

Lemma 10.2. *If L is a lattice and $r > 2$, then the series*

$$
G_r(L) = \sum_{\omega \in L-\{0\}} \frac{1}{\omega^r}
$$

converges absolutely.

Proof. If $L = [\omega_1, \omega_2]$, then we need to show that the series

$$
\sum_{\omega \in L-\{0\}} \frac{1}{|\omega|^r} = \sum_{m,n}{}' \frac{1}{|m\omega_1 + n\omega_2|^r}
$$

converges, where $\sum'_{m,n}$ denotes summation over all ordered pairs $(m,n) \neq (0,0)$ of integers. If we let $M = \min\{|x\omega_1 + y\omega_2| : x^2 + y^2 = 1\}$, then it is easy to see that for all $x, y \in \mathbf{R}$,

$$
|x\omega_1 + y\omega_2| \geq M\sqrt{x^2+y^2}
$$

(see Exercise 10.1), and it follows that

$$
\sum_{m,n}{}' \frac{1}{|m\omega_1 + n\omega_2|^r} \leq \frac{1}{M^r} \sum_{m,n}{}' \frac{1}{(m^2+n^2)^{r/2}}.
$$

By comparing the sum on the right to the integral

$$
\int\!\!\int_{x^2+y^2 \geq 1} \frac{1}{(x^2+y^2)^{r/2}}\, dx\, dy,
$$

it is easy to show that the sum in question converges when $r > 2$ (see Exercise 10.1). Q.E.D.

We can now show that $\wp(z)$ is holomorphic outside L. Namely, if Ω is a compact subset of $\mathbf{C}$ missing L, it suffices to show that the sum in

$$\wp(z) = \frac{1}{z^2} + \sum_{\omega \in L - \{0\}} \left(\frac{1}{(z-\omega)^2} - \frac{1}{\omega^2} \right)$$

converges absolutely and uniformly on Ω. Pick a number R such that $|z| \leq R$ for all $z \in \Omega$. Now suppose that $z \in \Omega$ and that $\omega \in L$ satisfies $|\omega| \geq 2R$. Then $|z - \omega| \geq \frac{1}{2}|\omega|$, and one sees that

$$\left| \frac{1}{(z-\omega)^2} - \frac{1}{\omega^2} \right| \leq \left| \frac{z(2\omega - z)}{\omega^2 (z-\omega)^2} \right| \leq \frac{R(2|\omega| + \frac{1}{2}|\omega|)}{|\omega|^2 (\frac{1}{4}|\omega|^2)} = \frac{10R}{|\omega|^3}.$$

Since $|\omega| \geq 2R$ holds for all but finitely many elements of L, it follows from Lemma 10.2 that the sum in the $\wp$-function converges absolutely and uniformly on Ω. Thus $\wp(z)$ is holomorphic on $\mathbf{C} - L$ and has a double pole at the origin.

Notice that since $(-z-\omega)^2 = (z - (-\omega))^2$, the identity $\wp(-z) = \wp(z)$ follows immediately from absolute convergence. Thus the $\wp$-function is an even function.

To show that $\wp(z)$ is periodic is a bit trickier. We first differentiate the series for $\wp(z)$ to obtain

$$\wp'(z) = -2 \sum_{\omega \in L} \frac{1}{(z-\omega)^3}.$$

Arguing as above, this series converges absolutely, and it follows easily that $\wp'(z)$ is an elliptic function for L (see Exercise 10.2). Now suppose that $L = [\omega_1, \omega_2]$. The functions $\wp(z)$ and $\wp(z + \omega_i)$ have the same derivative (since $\wp'(z)$ is periodic), and hence they differ by a constant, say $\wp(z) = \wp(z + \omega_i) + C$. Evaluating this at $-\omega_i/2$ (which is not in L), we obtain

$$\wp(-\omega_i/2) = \wp(-\omega_i/2 + \omega_i) + C = \wp(\omega_i/2) + C.$$

Since $\wp(z)$ is an even function, C must be zero, and periodicity is proved. It follows that the poles of $\wp(z)$ are all double poles and lie exactly on the points of L, and (i) is proved.

Turning to (ii), we will first compute the Laurent expansion of $\wp(z)$ about the origin:

Lemma 10.3. *Let $\wp(z)$ be the $\wp$-function for the lattice L, and let $G_r(L)$ be the constants defined in Lemma 10.2. Then, in a neighborhood of the origin, we have*

$$\wp(z) = \frac{1}{z^2} + \sum_{n=1}^{\infty} (2n+1) G_{2n+2}(L) z^{2n}.$$

Proof. For $|x| < 1$, we have the series expansion

$$\frac{1}{(1-x)^2} = 1 + \sum_{n=1}^{\infty}(n+1)x^n$$

(see Exercise 10.3). Thus, if $|z| < |\omega|$, we can put $x = z/\omega$ in the above series, and it follows easily that

$$\frac{1}{(z-\omega)^2} - \frac{1}{\omega^2} = \sum_{n=1}^{\infty}\frac{n+1}{\omega^{n+2}}z^n.$$

Summing over all $\omega \in L - \{0\}$ and using absolute convergence, we obtain

$$\wp(z) = \frac{1}{z^2} + \sum_{n=1}^{\infty}(n+1)G_{n+2}(L)z^n.$$

Since the $\wp(z)$ is an even function, all of the odd coefficients must vanish, giving us the desired Laurent expansion. Q.E.D.

From this lemma, we see that

$$\wp'(z) = \frac{-2}{z^3} + \sum_{n=1}^{\infty}2n(2n+1)G_{2n+2}(L)z^{2n-1},$$

and then one computes the first few terms of $\wp(z)^3$ and $\wp'(z)^2$ as follows:

$$\wp(z)^3 = \frac{1}{z^6} + \frac{9G_4(L)}{z^2} + 15G_6(L) + \cdots$$

$$\wp'(z)^2 = \frac{4}{z^6} - \frac{24G_4(L)}{z^2} - 80G_6(L) + \cdots,$$

where $+\cdots$ indicates terms involving positive powers of z (see Exercise 10.4). Now consider the elliptic function

$$F(z) = \wp'(z)^2 - 4\wp(z)^3 - 60G_4(L)\wp(z) - 140G_6(L).$$

Using the above expansions, it is easy to see that $F(z)$ vanishes at the origin, and then by periodicity, $F(z)$ vanishes at all points of L. But it is also holomorphic on $\mathbf{C} - L$, so that $F(z)$ is holomorphic on all of $\mathbf{C}$. An easy argument using Liouville's Theorem shows that $F(z)$ is constant (see Exercise 10.5), so that $F(z)$ is identically zero. Since $g_2(L)$ and $g_3(L)$ were defined to be $60G_4(L)$ and $140G_6(L)$ respectively, the proof of (ii) is complete.

In order to prove (iii), we will need the following lemma:

Lemma 10.4. *If $z, w \notin L$, then $\wp(z) = \wp(w)$ if and only if $z \equiv \pm w \bmod L$.*

Proof. The $\Leftarrow$ direction of the proof is trivial since $\wp(z)$ is an even function. To argue the other way, suppose that $L = [\omega_1, \omega_2]$, and fix a number $-1 < \delta < 0$. Let $\mathbf{P}$ denote the parallelogram $\{s\omega_1 + t\omega_2 : \delta \leq s, t \leq \delta + 1\}$, and let Γ be its boundary oriented counterclockwise. Note that every complex number is congruent modulo L to a number in $\mathbf{P}$ (see Exercise 10.6).

Fix w and consider the function $f(z) = \wp(z) - \wp(w)$. By adjusting δ, we can arrange that $f(z)$ has no zeros or poles on Γ. Then it is well known that

$$\frac{1}{2\pi i}\int_\Gamma \frac{f'(z)}{f(z)}\,dz = Z - P,$$

where Z (resp. P) is the number of zeros (resp. poles) of $f(z)$ in $\mathbf{P}$, counting multiplicity. Since $f'(z)/f(z)$ is periodic, the integrals on opposite sides of Γ cancel, and thus $\int_\Gamma (f'(z)/f(z))dz = 0$. This shows that $Z = P$. However, P is easy to compute: from the definition of $\mathbf{P}$, it's obvious that 0 is the only pole of $f(z) = \wp(z) - \wp(w)$ in $\mathbf{P}$. It's a double pole, and thus $Z = P = 2$, so that $f(z)$ has two zeros (counting multiplicity) in $\mathbf{P}$.

There are now two cases to consider. If $w \not\equiv -w \bmod L$, then modulo L, w and $-w$ give rise to two distinct points of $\mathbf{P}$, both of which are zeros of $f(z) = \wp(z) - \wp(w)$. Since $Z = 2$, these are all of the zeros, and their multiplicity is one, i.e., $\wp'(w) \neq 0$. If $w \equiv -w \bmod L$, then $2w \in L$. Since $\wp'(z)$ is an odd function (being the derivative of an even function), we obtain

$$\wp'(w) = \wp'(w - 2w) = \wp'(-w) = -\wp'(w),$$

which forces $\wp'(w) = 0$. Thus modulo L, w gives rise to a zero of $f(z)$ of multiplicity ≥ 2 in $\mathbf{P}$, and again $Z = 2$ implies that these are all. This proves the lemma. Q.E.D.

The proof of Lemma 10.4 yields the following useful corollary:

Corollary 10.5. *If $w \notin L$, then $\wp'(w) = 0$ if and only if $2w \in L$.* Q.E.D.

Now we can finally prove the addition theorem. Fix $w \notin L$, and consider the elliptic function

$$G(z) = \wp(z + w) + \wp(z) + \wp(w) - \frac{1}{4}\left(\frac{\wp'(z) - \wp'(w)}{\wp(z) - \wp(w)}\right)^2.$$

If we can show that $G(z)$ is holomorphic on $\mathbf{C}$ and vanishes at the origin, then as in (ii), Liouville's Theorem will imply that $G(z)$ vanishes identically, and the addition theorem will be proved.

Using Lemma 10.4, we see that the possible singularities of $G(z)$ come from three sources: L, $L + \{w\}$ and $L - \{w\}$. By periodicity, it suffices to consider $G(0)$, $G(w)$ and $G(-w)$. Let's begin with $G(0)$. Using the Laurent expansions for $\wp(z)$ and $\wp'(z)$, one sees that

$$\frac{1}{4}\left(\frac{\wp'(z)-\wp'(w)}{\wp(z)-\wp(w)}\right)^2 = \frac{1}{4}\left(\frac{-2/z^3-\wp'(w)+\cdots}{1/z^2-\wp(w)+\cdots}\right)^2 = \frac{1}{z^2}+2\wp(w)+\cdots,$$

where as usual, $+\cdots$ means terms involving positive powers of z. Hence

$$G(z) = \wp(z+w)+\wp(w)+\frac{1}{z^2}+\cdots-\frac{1}{z^2}-2\wp(w)-\cdots,$$

and it follows that $G(0)=0$.

To simplify the remainder of the argument, we will assume that $2w \notin L$. Turning to $G(w)$, we use L'Hospital's Rule to obtain

$$G(w) = \wp(2w)+2\wp(w)-\frac{1}{4}\left(\frac{\wp''(w)}{\wp'(w)}\right)^2.$$

Since $2w \notin L$, Corollary 10.5 shows that $\wp'(w) \neq 0$, and thus $G(w)$ is defined. It remains to consider $G(-w)$. We begin with some Laurent expansions about $z=-w$:

$$\wp(z+w) = \frac{1}{(z+w)^2}+\cdots$$

$$\wp(z) = \wp(-w)+\wp'(-w)(z+w)+\cdots = \wp(w)-\wp'(w)(z+w)+\cdots.$$

where $+\cdots$ now refers to higher powers of $z+w$. Since $\wp'(w)\neq 0$, these formulas make it easy to show that $G(-w)$ is defined (see Exercise 10.7). This shows that $G(z)$ is holomorphic and vanishes at 0, so that $G(z)$ vanishes everywhere.

To complete the proof, we need to consider the case $2w \in L$. We leave this to the reader (see Exercise 10.7). We have now proved all three parts of Theorem 10.1. Q.E.D.

There are many more results connected with the Weierstrass $\wp$-function, and we refer the reader to Chandrasekharan [16, Chapter III], Lang [73, Chapter 1] or Whittaker and Watson [109, Chapter XX] for more details.

B. The j-Invariant of a Lattice

Elliptic functions depend on which lattice is being used, but sometimes different lattices can have basically the same elliptic functions. We say that two lattices L and L' are *homothetic* if there is a nonzero complex number λ such that $L' = \lambda L$. Note that homothety is an equivalence relation. It is easy to check how homothety affects elliptic functions: if $f(z)$ is an elliptic function for L, then $f(\lambda z)$ is an elliptic function for λL. Furthermore, the $\wp$-function transforms as follows:

$$\wp(\lambda z; \lambda L) = \lambda^{-2}\wp(z; L).$$

Thus we would like to classify lattices up to homothety, and this is where the j-invariant comes in.

Given a lattice L, we have the constants $g_2(L)$ and $g_3(L)$ which appear in the differential equation for $\wp(z)$. It is customary to set

$$\Delta(L) = g_2(L)^3 - 27g_3(L)^2.$$

The number $\Delta(L)$ is closely related to the discriminant of the polynomial $x^3 - g_2(L)x - g_3(L)$ that appears in the differential equation for $\wp(z)$. In fact, if e_1, e_2 and e_3 are the roots of this polynomial, then one can show that

$$\Delta(L) = 16(e_1 - e_2)^2(e_1 - e_3)^2(e_2 - e_3)^2 \tag{10.6}$$

(see Exercise 10.8). An important fact is that $\Delta(L)$ never vanishes, i.e.,

Proposition 10.7. *If L is a lattice, then $\Delta(L) \neq 0$.*

Proof. If $w \notin L$ and $2w \in L$, then Corollary 10.5 implies that $\wp'(w) = 0$. Then the differential equation from Theorem 10.1 tells us that

$$0 = \wp'(w)^2 = 4\wp(w)^3 - g_2(L)\wp(w) - g_3(L),$$

so that $\wp(w)$ is a root of $4x^3 - g_2(L)x - g_3(L)$. If $L = [\omega_1, \omega_2]$, this process gives three roots $\wp(\omega_1/2)$, $\wp(\omega_2/2)$ and $\wp((\omega_1+\omega_2)/2)$, which are distinct by Lemma 10.4 since $\pm\omega_1/2$, $\pm\omega_2/2$ and $\pm(\omega_1+\omega_2)/2$ are distinct modulo L. Thus the roots of $4x^3 - g_2(L)x - g_3(L)$ are distinct, and $\Delta(L) \neq 0$ by (10.6). Q.E.D.

The *j-invariant* $j(L)$ of the lattice L is defined to be the complex number

$$j(L) = 1728\frac{g_2(L)^3}{g_2(L)^3 - 27g_3(L)^2} = 1728\frac{g_2(L)^3}{\Delta(L)}. \tag{10.8}$$

Note that $j(L)$ is always defined since $\Delta(L) \neq 0$. The reason for the factor of 1728 will become clear in §11. The remarkable fact is that the j-invariant $j(L)$ characterizes the lattice L up to homothety:

Theorem 10.9. *If L and L' are lattices in $\mathbb{C}$, then $j(L) = j(L')$ if and only if L and L' are homothetic.*

Proof. It is easy to see that homothetic lattices have the same j-invariant. Namely, if $\lambda \in \mathbf{C}^*$, then the definition of $g_2(L)$ and $g_3(L)$ implies that

$$(10.10)\qquad \begin{aligned} g_2(\lambda L) &= \lambda^{-4} g_2(L) \\ g_3(\lambda L) &= \lambda^{-6} g_3(L), \end{aligned}$$

and $j(\lambda L) = j(L)$ follows easily.

Now suppose that L and L' are lattices such that $j(L) = j(L')$. We first claim that there is a complex number λ such that

$$(10.11)\qquad \begin{aligned} g_2(L') &= \lambda^{-4} g_2(L) \\ g_3(L') &= \lambda^{-6} g_3(L). \end{aligned}$$

When $g_2(L') \neq 0$ and $g_3(L') \neq 0$, we can pick a number λ such that

$$\lambda^4 = \frac{g_2(L)}{g_2(L')}.$$

Since $j(L) = j(L')$, some easy algebra shows that

$$\lambda^{12} = \left(\frac{g_3(L)}{g_3(L')}\right)^2,$$

so that

$$\lambda^6 = \pm\frac{g_3(L)}{g_3(L')}.$$

Replacing λ by $i\lambda$ if necessary, we can assume that the above sign is $+$, and then (10.11) follows. The proof when $g_2(L') = 0$ or $g_3(L') = 0$ is similar and is left to the reader (see Exercise 10.9).

To exploit (10.11), we need to learn more about the Laurent expansion of the $\wp$-function:

Lemma 10.12. *Let $\wp(z)$ be the $\wp$-function for the lattice L, and as in Lemma 10.3, let*

$$\wp(z) = \frac{1}{z^2} + \sum_{n=1}^{\infty} (2n+1) G_{2n+2}(L) z^{2n}$$

be its Laurent expansion. Then for $n \geq 1$, the coefficient $(2n+1)G_{2n+2}(L)$ of z^{2n} is a polynomial with rational coefficients, independent of L, in $g_2(L)$ and $g_3(L)$.

Proof. For simplicity, we will write the coefficients of the Laurent expansion as $a_n = (2n+1)G_{2n+2}(L)$. To get a relation among the a_n's, we differentiate the equation $\wp'(z)^2 = 4\wp(z)^3 - g_2(L)\wp(z) - g_3(L)$ to obtain

$$\wp''(z) = 6\wp(z)^2 - (1/2)g_2(L).$$

By substituting in the Laurent expansion for $\wp(z)$ and comparing the coefficients of z^{2n-2}, one easily sees that for $n \geq 3$,

$$2n(2n-1)a_n = 6\left(2a_n + \sum_{i=1}^{n-2} a_i a_{n-1-i}\right)$$

(see Exercise 10.10), and hence

$$(2n+3)(n-2)a_n = 3\sum_{i=1}^{n-2} a_i a_{n-1-i}.$$

Since $g_2(L) = 60G_4(L) = 20a_1$ and $g_3(L) = 140G_6(L) = 28a_2$, an easy induction shows that a_n is a polynomial with rational coefficients in $g_2(L)$ and $g_3(L)$. This proves the lemma. Q.E.D.

Now suppose that we have lattices L and L' such that (10.11) holds for some constant λ. We claim that $L' = \lambda L$. To see this, first note that by (10.10), we have $g_2(L') = g_2(\lambda L)$ and $g_3(L') = g_3(\lambda L)$. Then Lemma 10.12 implies that $\wp(z; L')$ and $\wp(z; \lambda L)$ have the same Laurent expansion about 0, so that the two functions agree in a neighborhood of the origin, and hence $\wp(z; L') = \wp(z; \lambda L)$ everywhere. Since the lattice is the set of poles of the $\wp$-function, this proves that $L' = \lambda L$, and the theorem is proved. Q.E.D.

Besides the notion of the j-invariant of a lattice, there is another way to think about the j-invariant which will be useful when we study modular functions. Given a complex number τ in the upper half plane $\mathfrak{h} = \{\tau \in \mathbf{C} : \mathrm{Im}(\tau) > 0\}$, we get the lattice $[1, \tau]$, and then the *j-function* $j(\tau)$ is defined by

$$j(\tau) = j([1, \tau]).$$

The analytic properties of $j(\tau)$ play an important role in the theory of complex multiplication and will be studied in detail in §11.

C. Complex Multiplication

We begin with the simple observation that orders in imaginary quadratic fields give rise to a natural class of lattices. Namely, let $\mathcal{O}$ be an order in the imaginary quadratic field K, and let $\mathfrak{a}$ be a proper fractional $\mathcal{O}$-ideal. We know from §7 that $\mathfrak{a} = [\alpha, \beta]$ for some $\alpha, \beta \in K$ (see Exercise 7.8). We can regard K as a subset of $\mathbf{C}$, and since K is imaginary quadratic, α and β are linearly independent over $\mathbf{R}$ (see Exercise 10.11). Thus $\mathfrak{a} = [\alpha, \beta]$ is a lattice in $\mathbb{C}$, and consequently the j-invariant $j(\mathfrak{a})$ is defined. These complex numbers, often called *singular moduli*, have some remarkable properties which will be explored in §11. For now, we have the more modest goal of trying to motivate the idea of complex multiplication.

In order to simplify our discussion of complex multiplication, we will fix the lattice L. As usual, $\wp(z; L)$ is written $\wp(z)$, and to simplify things further, $g_2(L)$ and $g_3(L)$ will be written g_2 and g_3. The basic idea of complex multiplication goes back to the addition law for the $\wp$-function, proved in part (iii) of Theorem 10.1. If we specialize to the case $z = w$, then L'Hospital's rule gives the following duplication formula for the $\wp$-function:

$$\wp(2z) = -2\wp(z) + \frac{1}{4}\left(\frac{\wp''(z)}{\wp'(z)}\right)^2. \tag{10.13}$$

However, the differential equation from Theorem 10.1 implies that

$$\wp'(z)^2 = 4\wp(z)^3 - g_2\wp(z) - g_3$$
$$\wp''(z) = 6\wp(z)^2 - (1/2)g_2,$$

and substituting these expressions into (10.13), we obtain

$$\wp(2z) = -2\wp(z) + \frac{(12\wp(z)^2 - g_2)^2}{16(4\wp(z)^3 - g_2\wp(z) - g_3)}.$$

Thus $\wp(2z)$ is a rational function in $\wp(z)$. More generally, one can show by induction that for any positive integer n, $\wp(nz)$ is a rational function in $\wp(z)$ (see Exercise 10.12). So the natural question to ask is whether there are any other complex numbers α for which $\wp(\alpha z)$ is a rational function in $\wp(z)$. The answer is rather surprising:

Theorem 10.14. *Let L be a lattice, and let $\wp(z)$ be the $\wp$ function for L. Then, for a number $\alpha \in \mathbb{C} - \mathbb{Z}$, the following statements are equivalent:*

(i) *$\wp(\alpha z)$ is a rational function in $\wp(z)$.*

(ii) *$\alpha L \subset L$.*

(iii) *There is an order $\mathcal{O}$ in an imaginary quadratic field K such that $\alpha \in \mathcal{O}$ and L is homothetic to a proper fractional $\mathcal{O}$-ideal.*

Furthermore, if these conditions are satisfied, then $\wp(\alpha z)$ can be written in the form

$$\wp(\alpha z) = \frac{A(\wp(z))}{B(\wp(z))}$$

where $A(x)$ and $B(x)$ are relatively prime polynomials such that

$$\deg(A(x)) = \deg(B(x)) + 1 = [L : \alpha L] = N(\alpha).$$

Proof. (i) $\Rightarrow$ (ii). If $\wp(\alpha z)$ is a rational function in $\wp(z)$, then there are polynomials $A(x)$ and $B(x)$ such that

$$B(\wp(z))\wp(\alpha z) = A(\wp(z)). \tag{10.15}$$

Since $\wp(z)$ and $\wp(\alpha z)$ have double poles at the origin, it follows from (10.15) that

$$\deg(A(x)) = \deg(B(x)) + 1. \tag{10.16}$$

Now let $\omega \in L$. Then (10.15) and (10.16) show that $\wp(\alpha z)$ has a pole at ω, which means that $\wp(z)$ has a pole at $\alpha\omega$. Since the poles of $\wp(z)$ are exactly the period lattice L, this implies that $\alpha\omega \in L$, and $\alpha L \subset L$ follows.

(ii) $\Rightarrow$ (i). If $\alpha L \subset L$, it follows that $\wp(\alpha z)$ is meromorphic and has L as a lattice of periods. Furthermore, note that $\wp(\alpha z)$ is an even function since $\wp(z)$ is. Then the following theorem immediately implies that $\wp(\alpha z)$ is a rational function in $\wp(z)$:

Lemma 10.17. *Any even elliptic function for L is a rational function in* $\wp(z)$.

Proof. This proof of this assertion is covered in Exercise 10.13. Q.E.D.

(ii) $\Rightarrow$ (iii). Suppose that $\alpha L \subset L$. Replacing L by λL for suitable λ, we can assume that $L = [1, \tau]$ for some $\tau \in \mathbf{C} - \mathbf{R}$. Then $\alpha L \subset L$ means that $\alpha = a + b\tau$ and $\alpha\tau = c + d\tau$ for some integers a, b, c and d. Taking the quotient of the two equations, we obtain

$$\tau = \frac{c + d\tau}{a + b\tau},$$

which gives us the quadratic equation

$$b\tau^2 + (a - d)\tau - c = 0.$$

Since τ is not real, we must have $b \neq 0$, and then $K = \mathbf{Q}(\tau)$ is an imaginary quadratic field. It follows that

$$\mathcal{O} = \{\beta \in K : \beta L \subset L\}$$

is an order of K for which L is a proper fractional $\mathcal{O}$-ideal, and since α is obviously in $\mathcal{O}$, we are done.

(iii) $\Rightarrow$ (ii). This implication is trivial.

Finally, to prove the last statement of the proposition, suppose that

$$\wp(\alpha z) = \frac{A(\wp(z))}{B(\wp(z))}. \tag{10.18}$$

By (10.16), we know that $\deg(A(x)) = \deg(B(x)) + 1$, and in Corollary

11.27, we will show that $N(\alpha) = [L : \alpha L]$. It remains to prove that the degree of $A(x)$ is the index $[L : \alpha L]$.

Fix $z \in \mathbf{C}$ such that $2z \notin (1/\alpha)L$, and consider the polynomial $A(x) - \wp(\alpha z)B(x)$. This polynomial has the same degree as $A(x)$, and z can be chosen so that it has distinct roots (see Exercise 10.14). Then consider the lattices $L \subset (1/\alpha)L$, and let $\{w_i\}$ be coset representatives of L in $(1/\alpha)L$. We claim that

(10.19)

$$\text{The } \wp(z + w_i) \text{ are distinct and give all roots of } A(x) - \wp(\alpha z)B(x).$$

This will imply $\deg(A(x)) = [(1/\alpha)L : L] = [L : \alpha L]$, and the theorem will be proved.

To prove (10.19), we first show that the $\wp(z + w_i)$ are distinct. If not, we would have $\wp(z + w_i) = \wp(z + w_j)$ for some $i \neq j$. Then Lemma 10.4 implies that $z + w_i \equiv \pm(z + w_j) \bmod L$. The plus sign implies $w_i \equiv w_j \bmod L$, which contradicts $i \neq j$, and the minus sign implies $2z \equiv w_j - w_i \bmod L$, which contradicts $2z \notin (1/\alpha)L$. Thus the $\wp(z + w_i)$ are distinct.

From (10.18), we see that $A(\wp(z + w_i)) = \wp(\alpha(z + w_i))B(\wp(z + w_i))$. But $w_i \in (1/\alpha)L$, so that $\alpha(z + w_i) \equiv \alpha z \bmod L$, and hence $\wp(\alpha(z + w_i)) = \wp(\alpha z)$. This shows that the $\wp(z + w_i)$ are roots of $A(x) - \wp(\alpha z)B(x)$. To see that all roots arise this way, let u be another root. Note that $B(u) \neq 0$ since $B(u) = 0$ implies $A(u) = 0$, which is impossible since $A(x)$ and $B(x)$ are relatively prime. By adapting the argument of Lemma 10.4, it is easy to see that $u = \wp(w)$ for some complex number w (see Exercise 10.14). Then

$$\wp(\alpha z) = \frac{A(u)}{B(u)} = \frac{A(\wp(w))}{B(\wp(w))} = \wp(\alpha w),$$

and using Lemma 10.4 again, we see that $\alpha w \equiv \pm \alpha z \bmod L$. Changing w to $-w$ if necessary (which doesn't affect $u = \wp(w) = \wp(-w)$), we can assume that $w \equiv z \bmod (1/\alpha)L$. Working modulo L, this means $w \equiv z + w_i \bmod L$ for some i, and thus $u = \wp(w) = \wp(z + w_i)$ is one of the known roots. This proves (10.19), and we are done with Theorem 10.14. Q.E.D.

This theorem shows if an elliptic function has multiplication by some $\alpha \in \mathbf{C} - \mathbf{R}$, then it has mutiplication by an entire order $\mathcal{O}$ in an imaginary quadratic field. Notice that all of the elements of $\mathcal{O} - \mathbf{Z}$ are genuinely complex, i.e., not real. This accounts for the name *complex multiplication*.

One important consequence of Theorem 10.14 is that complex multiplication is an intrinsic property of the lattice. So rather than talk about elliptic functions with complex multiplication, it makes more sense to talk about lattices with complex multiplication. Since changing the lattice by a constant multiple doesn't affect the complex multiplications, we will work with homothety classes of lattices.

Using Theorem 10.14, we can relate homothety classes of lattices and ideal class groups of orders as follows. Fix an order $\mathcal{O}$ in an imaginary quadratic field, and consider those lattices $L \subset \mathbf{C}$ which have $\mathcal{O}$ as their full ring of complex multiplications. By Theorem 10.14, we can assume that L is a proper fractional $\mathcal{O}$-ideal, and conversely, every proper fractional $\mathcal{O}$-ideal is a lattice with $\mathcal{O}$ as its ring of complex multiplications. Furthermore, two proper fractional $\mathcal{O}$-ideals are homothetic as lattices if and only if they determine the same class in the ideal class group $C(\mathcal{O})$ (see Exercise 10.15). We have thus proved the following:

Corollary 10.20. *Let $\mathcal{O}$ be an order in an imaginary quadratic field. Then there is a one-to-one correspondence between the ideal class group $C(\mathcal{O})$ and the homothety classes of lattices with $\mathcal{O}$ as their full ring of complex multiplications.* Q.E.D.

It follows that the class number $h(\mathcal{O})$ tells us the number of homothety classes of lattices having $\mathcal{O}$ as their full ring of complex multiplications.

Here are some examples. First, consider all lattices which have complex multiplication by $\sqrt{-3}$. This means that we are dealing with an order $\mathcal{O}$ containing $\sqrt{-3}$ in the field $K = \mathbf{Q}(\sqrt{-3})$. Then $\mathcal{O}$ must be either $\mathbf{Z}[\sqrt{-3}]$ or $\mathbf{Z}[\omega]$, $\omega = e^{2\pi i/3}$, and since both of these have class number 1, the only lattices are $[1, \sqrt{-3}]$ and $[1, \omega]$. Thus, up to homothety, there are only two lattices with complex multiplication by $\sqrt{-3}$. Next, consider complex multiplication by $\sqrt{-5}$. Here, $K = \mathbf{Q}(\sqrt{-5})$, and the only order containing $\sqrt{-5}$ is the maximal order $\mathcal{O}_K = \mathbf{Z}[\sqrt{-5}]$. The class number is $h(-20) = 2$, and since we know the reduced forms of discriminant -20, the results of §7 show that up to homothety, the only lattices with complex multiplication by $\sqrt{-5}$ are $[1, \sqrt{-5}]$ and $[2, 1+\sqrt{-5}]$ (see Exercise 10.16).

The discussion so far has concentrated on the elliptic functions and their lattices. Since our ultimate goal involves the j-invariant of the lattices, we need to indicate how complex multiplication influences the j-invariant. Let's start with the simplest case, complex multiplication by $i = \sqrt{-1}$. Up to a multiple, the only possible lattice is $L = [1, i]$. To compute $j(L) = j(i)$, note that $iL = L$, so that by the homogeneity (10.11) of $g_3(L)$,

$$g_3(L) = g_3(iL) = i^{-6} g_3(L) = -g_3(L).$$

This implies that $g_3(L) = 0$, and then the formula (10.8) for the j-invariant tells us that $j(i) = 1728$. Similarly, one can show that if $L = [1, \omega]$, $\omega = e^{2\pi i/3}$, then $g_2(L) = 0$, which tells us that $j(\omega) = 0$ (see Exercise 10.17).

A more interesting example is given by complex multiplication by $\sqrt{-2}$. By the above methods, the only lattice involved is $[1, \sqrt{-2}]$, up to homothety. We will follow the exposition of Stark [97] and show that

$$j(\sqrt{-2}) = 8000.$$

Since $N(\sqrt{-2}) = 2$, Theorem 10.14 tells us that

$$\wp(\sqrt{-2}z) = \frac{A(\wp(z))}{B(\wp(z))}$$

where $A(x)$ is quadratic and $B(x)$ is linear. Dividing $B(x)$ into $A(x)$, we can write this as

$$\wp(\sqrt{-2}z) = a\wp(z) + b + \frac{1}{c\wp(z) + d}, \tag{10.21}$$

where a and c are nonzero complex numbers. To exploit this identity, we will use the Laurent expansion of $\wp(z)$ at $z = 0$. The differential equation for $\wp(z)$ shows that the first few terms of the Laurent expansion are

$$\wp(z) = \frac{1}{z^2} + \frac{g_2}{20}z^2 + \frac{g_3}{28}z^4 + \frac{g_2^2}{1200}z^6 + \cdots$$

(this follows easily from the proof of Lemma 10.12—see Exercise 10.18). To simplify this expansion, first note that g_2 and g_3 are nonzero, for otherwise there would be complex multiplication by i or ω, which can't happen for $L = [1, \sqrt{-2}]$ (see Exercise 10.19). Then, replacing L by a suitable multiple, the homogeneity of g_2 and g_3 allows us to assume that $g_2 = 20g$ and $g_3 = 28g$ for some number g (see Exercise 10.19). With this choice of lattice, the expansion for $\wp(z)$ can be written

$$\wp(z) = \frac{1}{z^2} + gz^2 + gz^4 + \frac{g^2}{3}z^6 + \cdots,$$

and it follows that the expansion for $\wp(\sqrt{-2}z)$ is

$$\wp(\sqrt{-2}z) = \frac{-1}{2z^2} - 2gz^2 + 4gz^4 - \frac{8g^2}{3}z^6 + \cdots.$$

Now the constants a and b in (10.21) are the unique constants such that $\wp(\sqrt{-2}z) - a\wp(z) - b$ is zero when $z = 0$. Comparing the above expansions for $\wp(z)$ and $\wp(\sqrt{-2}z)$, we see that $a = -1/2$ and $b = 0$. Then (10.21) tells us the remarkable fact that $(\wp(\sqrt{-2}z) + \frac{1}{2}\wp(z))^{-1}$ is a linear polynomial in $\wp(z)$. Using the above expansions, one computes that

$$\begin{aligned}\left(\wp(\sqrt{-2}z) + \frac{1}{2}\wp(z)\right)^{-1} &= \left(-\frac{3g}{2}z^2 + \frac{9g}{2}z^4 - \frac{5g^2}{2}z^6 + \cdots\right)^{-1} \\ &= -\frac{2}{3gz^2} - \frac{2}{g} - \frac{2}{3g}\left(9 - \frac{5g}{3}\right)z^2 + \cdots\end{aligned} \tag{10.22}$$

(see Exercise 10.19). By (10.21), this expression is linear in $\wp(z)$. Looking at the behavior at $z = 0$, it follows that the bottom line of (10.22) must equal

$$-\frac{2}{3g}\wp(z) - \frac{2}{g},$$

and then comparing the coefficients of z^2 implies that

$$-\frac{2}{3g}\left(9-\frac{5g}{3}\right)=-\frac{2}{3g}g.$$

Solving this equation for g yields $g=\frac{27}{8}$, so that

$$g_2=20g=\frac{5\cdot 27}{2}$$
$$g_3=28g=\frac{7\cdot 27}{2},$$

and thus

$$j(\sqrt{-2})=1728\frac{g_2^3}{g_2^3-27g_3^2}=8000=20^3.$$

By a similar computation, one can also show that

$$j\left(\frac{1+\sqrt{-7}}{2}\right)=-3375=(-15)^3$$

(see Exercise 10.20). In §12 we will explain why these numbers are cubes.

Besides allowing us to compute $j(\sqrt{-2})$ and $j((1+\sqrt{-7})/2)$, the Laurent series of the $\wp$-function can be used to give an elementary proof that the j-invariant of a lattice with complex multiplication is an algebraic number:

Theorem 10.23. *Let $\mathcal{O}$ be an order in an imaginary quadratic field, and let $\mathfrak{a}$ be a proper fractional $\mathcal{O}$-ideal. Then $j(\mathfrak{a})$ is an algebraic number of degree at most $h(\mathcal{O})$.*

Proof. By Lemma 10.12, the Laurent expansion of $\wp(z)$ can be written

$$\wp(z)=\frac{1}{z^2}+\sum_{n=1}^{\infty}a_n(g_2,g_3)z^{2n},$$

where each $a_n(g_2,g_3)$ is a polynomial in g_2 and g_3 with rational coefficients. To emphasize the dependence on g_2 and g_3, we will write $\wp(z)$ as $\wp(z;g_2,g_3)$.

By assumption, for any $\alpha\in\mathcal{O}$, $\wp(\alpha z)$ is a rational function in $\wp(z)$, say

$$\wp(\alpha z;g_2,g_3)=\frac{A(\wp(z;g_2,g_3))}{B(\wp(z;g_2,g_3))}. \tag{10.24}$$

We then have the Laurent expansion

$$\wp(\alpha z;g_2,g_3)=\frac{1}{\alpha^2z^2}+\sum_{n=1}^{\infty}a_n(g_2,g_3)\alpha^{2n}z^{2n},$$

which means that (10.24) can be regarded as an identity in the field $\mathbb{C}((z))$ of formal meromorphic Laurent series. Recall that $\mathbb{C}((z))$ is the field of fractions of the formal power series ring $\mathbb{C}[[z]]$, so that an element of $\mathbb{C}((z))$ is a series of the form $\sum_{n=-M}^{\infty} b_n z^n$, $b_n \in \mathbb{C}$.

Now let σ be any automorphism of $\mathbb{C}$. Then σ induces an automorphism of $\mathbb{C}((z))$ by acting on the coefficients. Thus, if we apply σ to (10.24), we obtain the identity

$$\wp(\sigma(\alpha)z;\sigma(g_2),\sigma(g_3)) = \frac{A^\sigma(\wp(z;\sigma(g_2),\sigma(g_3)))}{B^\sigma(\wp(z;\sigma(g_2),\sigma(g_3)))}, \tag{10.25}$$

where $A^\sigma(x)$ (resp. $B^\sigma(x)$) is the polynomial obtained by applying σ to the coefficients of $A(x)$ (resp. $B(x)$). This follows because $a_n(g_2,g_3)$ is a polynomial in g_2 and g_3 with rational coefficients. We don't know much about $\sigma(g_2)$ and $\sigma(g_3)$, but $g_2^3 - 27g_3^2 \neq 0$ implies $\sigma(g_2)^3 - 27\sigma(g_3)^2 \neq 0$. In §11, we will prove that this condition on $\sigma(g_2)$ and $\sigma(g_3)$ guarantees that there is a lattice L such that

$$\begin{aligned} g_2(L) &= \sigma(g_2) \\ g_3(L) &= \sigma(g_3) \end{aligned}$$

(see Corollary 11.7). Thus the formal Laurent series $\wp(z;\sigma(g_2),\sigma(g_3))$ is the Laurent series of the $\wp$-function $\wp(z;L)$, and then (10.25) tells us that $\wp(z;L)$ has complex multiplication by $\sigma(\alpha)$. This holds for any $\alpha \in \mathcal{O}$, so that if $\mathcal{O}'$ is the ring of all complex multiplications of L, then we have proved that

$$\mathcal{O} = \sigma(\mathcal{O}) \subset \mathcal{O}'.$$

If we work with σ^{-1} and interchange $\mathfrak{a}$ and L, the above argument shows that $\mathcal{O}' \subset \mathcal{O}$, which shows that $\mathcal{O}$ is the ring of all complex multiplications of both $\mathfrak{a}$ and L.

Now consider j-invariants. The above formulas for $g_2(L)$ and $g_3(L)$ imply that

$$j(L) = \sigma(j(\mathfrak{a})). \tag{10.26}$$

Since L has $\mathcal{O}$ as its ring of complex multiplications, Corollary 10.20 implies that there are only $h(\mathcal{O})$ possibilities for $j(L)$. By (10.26), there are thus at most $h(\mathcal{O})$ possibilities for $\sigma(j(\mathfrak{a}))$. Since σ was an *arbitrary* automorphism of $\mathbb{C}$, it follows that $j(\mathfrak{a})$ must be an algebraic number, and in fact the degree of its minimal polynomial over $\mathbb{Q}$ is at most $h(\mathcal{O})$. This proves the theorem. Q.E.D.

In §11 we will prove the stronger result that $j(\mathfrak{a})$ is an algebraic *integer* and that the degree of its minimal polynomial *equals* the class number

$h(\mathcal{O})$. But we thought it worthwhile to show what can be done by elementary means. Furthermore, the method of proof used above (the action of an automorphism on the coefficients of a Laurent expansion) is similar to some of the arguments to be given in §11.

For a more classical introduction to complex multiplication, the reader should consult the recent book [9] by Borwein and Borwein.

D. Exercises

10.1. This exercise is concerned with the proof of Lemma 10.2.

(a) If $L = [\omega_1, \omega_2]$ is a lattice, let $M = \min\{|x\omega_1 + y\omega_2| : x^2 + y^2 = 1\}$. Show that $M > 0$ and that $|x\omega_1 + y\omega_2| \geq M\sqrt{x^2+y^2}$ for all $x, y \in \mathbf{R}$.

(b) Show that the integral $\int\int_{x^2+y^2 \geq 1}(x^2+y^2)^{-r/2}\,dx\,dy$ converges when $r > 2$.

(c) Show that the series $\sum'_{m,n}(m^2+n^2)^{-r/2}$ converges when $r > 2$. Hint: compare the series to the integral in part (b).

10.2. In the proof of Theorem 10.1, we proved that $\wp'(z) = -2\sum_{\omega\in L}(z-\omega)^{-3}$.

(a) Show that this series converges absolutely for $z \notin L$.

(b) Using (a), show that $\wp'(z+\omega) = \wp'(z)$ for $Z \in L$.

10.3. Show that for $|x| < 1$, $(1-x)^{-2} = \sum_{n=0}^{\infty}(n+1)x^n$. Hint: differentiate the standard identity $(1-x)^{-1} = \sum_{n=0}^{\infty} x^n$.

10.4. Use Lemma 10.3 to show that

$$\wp(z)^3 = \frac{1}{z^6} + \frac{9G_4(L)}{z^2} + 15G_6(L) + \cdots$$

$$\wp'(z)^2 = \frac{4}{z^6} - \frac{24G_4(L)}{z^2} - 80G_6(L) + \cdots,$$

where $+\cdots$ indicates terms involving positive powers of z.

10.5. Use Liouville's Theorem to show that a holomorphic elliptic function $f(z)$ must be constant. Hint: consider $|f(z)|$ on the parallelogram $\{s\omega_1 + t\omega_2 : 0 \leq s,t \leq 1\}$. Exercise 10.6 will be useful.

10.6. Let $L = [\omega_1, \omega_2]$ be a lattice. For a fixed $\alpha \in \mathbf{C}$, consider the parallelogram $\mathbf{P} = \{\alpha + s\omega_1 + t\omega_2 : 0 \leq s,t \leq 1\}$. Show that if $z \in \mathbf{C}$, then there is $z' \in \mathbf{P}$ such that $z \equiv z' \bmod \mathbf{P}$. Note that the parallelogram used in Lemma 10.4 corresponds to $\alpha = \delta\omega_1 + \delta\omega_2$.

10.7. As in the proof of the addition theorem, let

$$G(z)=\wp(z+w)+\wp(z)+\wp(w)-\frac{1}{4}\left(\frac{\wp'(z)-\wp'(w)}{\wp(z)-\wp(w)}\right)^2.$$

(a) If $2w\notin L$, complete the argument begun in the text to show that $G(-w)$ is defined.

(b) Prove the addition law when $2w\in L$. Hint: take a sequence of points w_i converging to w such that $2w_i\notin L$ for all i.

10.8. Let $4x^3-g_2x-g_3$ be a cubic polynomial with roots e_1, e_2 and e_3.

(a) Show that $e_1+e_2+e_3=0$, $e_1e_2+e_1e_3+e_2e_3=-g_2/4$ and $e_1e_2e_3=g_3/4$.

(b) Using (a), show that $g_2^3-27g_3^2=16(e_1-e_2)^2(e_1-e_3)^2(e_2-e_3)^2$.

10.9. Let L and L' be lattice such that $j(L)=j(L')$. If $g_2(L')=0$ or $g_3(L')=0$, prove that there is a complex number λ such that (10.11) holds. Hint: by Proposition 10.7, they can't both be zero.

10.10. Let the Laurent expansion of the $\wp$-function about 0 be $\wp(z)=z^{-2}+\sum_{n=1}^{\infty}a_nz^{2n}$, where $a_n=(2n+1)G_{2n+2}$ is as in Lemma 10.3.

(a) Use the differential equation for the $\wp$-function to show that $\wp''(z)=6\wp(z)^2-(1/2)g_2(L)$.

(b) Use (a) to show that for $n\geq 3$,

$$2n(2n-1)a_n=6\left(2a_n+\sum_{i=1}^{n-2}a_ia_{n-1-i}\right).$$

10.11. Let K be an imaginary quadratic field, which can be regarded as a subfield of $\mathbf{C}$.

(a) If $\mathcal{O}$ is an order in K and $\mathfrak{a}=[\alpha,\beta]$ is a proper fractional $\mathcal{O}$-ideal, then show that α and β are linearly independent over $\mathbf{R}$. Thus $\mathfrak{a}\subset\mathbf{C}$ is a lattice.

(b) Conversely, let $L\subset\mathbf{C}$ be a lattice which is contained in K. Show that L is a proper fractional $\mathcal{O}$-ideal for some order $\mathcal{O}$ of K.

10.12. Let L be a lattice, and let n be a positive integer.

(a) Prove that $\wp(nz)$ is a rational function in $\wp(z)$. Hint: use the addition law and induction on n. For a quicker proof, use Lemma 10.17.

(b) Adapt the proof of Theorem 10.14 to show that the numerator of the rational function of part (a) has degree n^2 and the denominator has degree n^2-1.

10.13. In this exercise we will see how to express elliptic functions for a given lattice L in terms of $\wp(z)$ and $\wp'(z)$.

(a) Let $f(z)$ be an even elliptic function which is holomorphic on $\mathbf{C}-L$. Prove that $f(z)$ is a polynomial in $\wp(z)$. Hint: show that there is a polynomial $A(x)$ such that the Laurent expansion of $f(z)-A(\wp(z))$ has only terms of nonnegative degree. Then use Exercise 10.5.

(b) Let $f(z)$ be an even elliptic function that has a pole of order m at $w \in \mathbf{C}$. We will assume that $w \notin L$.

(i) If $2w \notin L$, prove that $(\wp(z)-\wp(w))^m f(z)$ is holomorphic at w. Hint: use Corollary 10.5.

(ii) If $2w \in L$, prove that m is even. Hint: $f(z) = f(2w-z)$.

(iii) If $2w \in L$, prove that $(\wp(z)-\wp(w))^{m/2} f(z)$ is holomorphic at w. Hint: use the proof of Lemma 10.4 to show that $\wp''(w) \neq 0$.

(c) Show that an even elliptic function $f(z)$ is a rational function in $\wp(z)$. This will prove Lemma 10.17. Hint: write $L = [\omega_1, \omega_2]$, and consider the parallelogram $\mathbf{P} = \{s\omega_1 + t\omega_2 : 0 \leq s,t \leq 1\}$. Note that only finitely many poles of $f(z)$ lie in $\mathbf{P}$. Now use part (b) to find a polynomial $B(x)$ such that $B(\wp(z))f(z)$ is holomorphic on $\mathbf{C}-L$ (use Exercise 10.6). Then the claim follows easily by part (a).

(d) Show that all elliptic functions for L are rational functions in $\wp(z)$ and $\wp'(z)$. Hint:

$$f(z) = \frac{f(z)+f(-z)}{2} + \left(\frac{f(z)-f(-z)}{2\wp'(z)}\right)\wp'(z).$$

10.14. This exercise is concerned with the proof of Theorem 10.14.

(a) Let $A(x)$ and $B(x)$ be relatively prime polynomials. Prove that there are only finitely many complex numbers λ such that the polynomial $A(x)-\lambda B(x)$ has a multiple root. Hint: show that every multiple root is a root of $A(x)B'(x)-A'(x)B(x)$.

(b) Adapt the proof of Lemma 10.4 to show that for any complex number u, the equation $u = \wp(w)$ always has a solution.

10.15. Let $\mathfrak{a}$ and $\mathfrak{b}$ be two proper fractional $\mathcal{O}$-ideals, where $\mathcal{O}$ is an order in an imaginary quadratic field. Prove that $\mathfrak{a}$ and $\mathfrak{b}$ determine the same class in the ideal class group $C(\mathcal{O})$ if and only if they are homothetic as lattices in $\mathbf{C}$.

10.16. In this exercise we study lattices with complex multiplication by a fixed $\alpha \in \mathbf{C}$.

(a) Verify that up to a multiple, the only lattices with complex multiplication by $\sqrt{-5}$ are $[1, \sqrt{-5}]$ and $[2, 1+\sqrt{-5}]$.

(b) Determine, up to a multiple, all lattices with complex multiplication by $\sqrt{-14}$. Hint: see the example following Theorem 5.25.

(c) Let K be an imaginary quadratic field of discriminant d_K, and let $\alpha \in \mathcal{O}_K - \mathbf{Z}$. Show that up to homothety, the number of lattices given with complex multiplication by α is given by

$$\sum_{f=1}^{[\mathcal{O}_K:\mathbf{Z}[\alpha]]} h(f^2 d_K).$$

10.17. Let $\omega = e^{2\pi i/3}$, and let L be the lattice $[1,\omega]$. Show that $g_2(L) = j(\omega) = 0$.

10.18. Use the proof of Lemma 10.12 to show that in a neighborhood of $z = 0$, the Laurent expansion of the $\wp$-function is

$$\wp(z) = \frac{1}{z^2} + \frac{g_2}{20}z^2 + \frac{g_3}{28}z^4 + \frac{g_2^2}{1200}z^6 + \cdots.$$

10.19. This exercise is concerned with the computation $j(\sqrt{-2}) = 8000$.

(a) If L is a lattice with $g_2(L) = 0$, then prove that L is a multiple of $[1,\omega]$, $\omega = e^{2\pi i/3}$. Hint: use Theorem 10.9 and Exercise 10.17.

(b) Similarly, show that if $g_3(L) = 0$, then L is a multiple of $[1,i]$.

(c) If L is a lattice with $g_2 g_3 \neq 0$, then show that there is a nonzero complex number λ such that for some $g \in \mathbf{C}$, $\lambda^{-4} g_2 = 20g$ and $\lambda^{-6} g_3 = 28g$. Hint: use (10.10).

(d) Verify the computations made in (10.22).

10.20. Show that $j((1+\sqrt{-7})/2) = -3375$.

§11. MODULAR FUNCTIONS AND RING CLASS FIELDS

In §10 we studied complex multiplication, and we saw that for an order $\mathcal{O}$ in an imaginary quadratic field, the j-invariant $j(\mathfrak{a})$ of a proper fractional $\mathcal{O}$-ideal $\mathfrak{a}$ is an algebraic number. This suggests a strong connection with number theory, and the goal of §11 is to unravel this connection by relating $j(\mathfrak{a})$ to the ring class field of $\mathcal{O}$ introduced in §9. The precise statement of this relation is the "First Main Theorem" of complex multiplication, which is the main result of this section:

Theorem 11.1. *Let $\mathcal{O}$ be an order in an imaginary quadratic field K, and let $\mathfrak{a}$ be a proper fractional $\mathcal{O}$-ideal. Then the j-invariant $j(\mathfrak{a})$ is an algebraic integer and $K(j(\mathfrak{a}))$ is the ring class field of the order $\mathcal{O}$.*

For a fixed order $\mathcal{O}$, we will prove in §13 that the $j(\mathfrak{a})$'s are all conjugate and hence are roots of the same irreducible polynomial over $\mathbf{Q}$. This polynomial is called the *class equation* of $\mathcal{O}$ and will be studied in detail in §13.

Of special interest is the case when $\mathcal{O} = \mathbf{Z}[\sqrt{-n}]$. Here, Theorem 11.1 implies that $j(\mathcal{O}) = j(\sqrt{-n})$ is an algebraic integer and is a primitive element of the ring class field of $\mathbf{Z}[\sqrt{-n}]$. It is elementary to see that $j(\sqrt{-n})$ is real (see Exercise 11.1), and thus, by Theorem 9.2, the class equation of $\mathbf{Z}[\sqrt{-n}]$ can be used to characterize primes of the form $p = x^2 + ny^2$.

Before we can prove Theorem 11.1, we need to learn about modular functions and the modular equation. The first step is to study the j-function $j(\tau)$ in detail.

A. The j-Function

The j-invariant $j(L)$ of a lattice L was defined in §10 in terms of the constants $g_2(L)$ and $g_3(L)$. Given τ in the upper half plane $\mathfrak{h}$, we get the lattice $[1,\tau]$, and then the j-function $j(\tau)$ is defined by

$$j(\tau) = j([1,\tau]).$$

We also define $g_2(\tau)$ and $g_3(\tau)$ by

$$g_2(\tau) = g_2([1,\tau]) = 60\sum_{m,n}{}' \frac{1}{(m+n\tau)^4}$$

$$g_3(\tau) = g_3([1,\tau]) = 140\sum_{m,n}{}' \frac{1}{(m+n\tau)^6},$$

where $\sum'_{m,n}$ denotes summation over all ordered pairs of integers $(m,n) \neq (0,0)$. By (10.8), it follows that $j(\tau)$ is given by the formula

$$j(\tau) = 1728\frac{g_2(\tau)^3}{\Delta(\tau)},$$

where $\Delta(\tau) = g_2(\tau)^3 - 27g_3(\tau)^2$.

The properties of $j(\tau)$ are closely related to the action of the group $\mathrm{SL}(2,\mathbf{Z})$ on the upper half plane $\mathfrak{h}$. This action is defined as follows: if $z \in \mathfrak{h}$ and $\gamma = \begin{pmatrix} a & b \\ c & d \end{pmatrix}$, then

$$\gamma\tau = \frac{a\tau + b}{c\tau + d}.$$

It is easy to check that $\gamma\tau \in \mathfrak{h}$ (see Exercise 11.2), and we say that $\gamma\tau$ and τ are $\mathrm{SL}(2,\mathbf{Z})$-equivalent. Then the j-function has the following properties:

Theorem 11.2.

(i) *$j(\tau)$ is a holomorphic function on $\mathfrak{h}$.*

(ii) *If τ and τ' lie in $\mathfrak{h}$, then $j(\tau) = j(\tau')$ if and only if $\tau' = \gamma\tau$ for some $\gamma \in \mathrm{SL}(2,\mathbf{Z})$. In particular, $j(\tau)$ is $\mathrm{SL}(2,\mathbf{Z})$-invariant.*

(iii) *$j : \mathfrak{h} \to \mathbf{C}$ is surjective.*

(iv) *For $\tau \in \mathfrak{h}$, $j'(\tau) \neq 0$, except in the following cases:*

  (a) *$\tau = \gamma i$, $\gamma \in \mathrm{SL}(2,\mathbf{Z})$, where $j'(\tau) = 0$ but $j''(\tau) \neq 0$.*

  (b) *$\tau = \gamma\omega$, $\omega = e^{2\pi i/3}$, $\gamma \in \mathrm{SL}(2,\mathbf{Z})$, where $j'(\tau) = j''(\tau) = 0$ but $j'''(\tau) \neq 0$.*

Proof. To prove (i), recall from Proposition 10.7 that $\Delta(\tau)$ never vanishes. Thus it suffices to show that $g_2(\tau)$ and $g_3(\tau)$ are holomorphic. For $g_2(\tau)$, this works as follows. By Lemma 10.2, the sum defining $g_2(\tau)$ converges absolutely, but we still must show that the convergence is uniform on compact subsets of $\mathfrak{h}$. To see this, first note that $g_2(\tau+1) = g_2(\tau)$ (this follows from absolute convergence). Thus it suffices to show that convergence is uniform when τ satisfies $|\mathrm{Re}(\tau)| \leq 1/2$ and $\mathrm{Im}(\tau) \geq \epsilon$, where $\epsilon < 1$ is an arbitrary positive number. In this case it is easy to show that

$$|m + n\tau| \geq \frac{\epsilon}{2}\sqrt{m^2+n^2}$$

(see Exercise 11.3), and then uniform convergence is immediate. The proof for $g_3(\tau)$ is similar, so that $g_2(\tau)$, $g_3(\tau)$, $\Delta(\tau)$ and $j(\tau)$ are all holomorphic on $\mathfrak{h}$.

Turning to (ii), we need to recall the following fact from in §7: if $\tau, \tau' \in \mathfrak{h}$, then

$$[1,\tau] \text{ and } [1,\tau'] \text{ are homothetic} \iff \tau' = \gamma\tau \text{ for some } \gamma \in \mathrm{SL}(2,\mathbf{Z}).$$

See (7.8) for the proof (in §7, we assumed that τ and τ' lay in an imaginary quadratic field, but the proof given for (7.8) holds for arbitrary $\tau, \tau' \in \mathfrak{h}$). From Theorem 10.9, we also know that

$$j(\tau) = j(\tau') \iff [1,\tau] \text{ and } [1,\tau'] \text{ are homothetic.}$$

Combining these two equivalences, (ii) is immediate.

Before we can prove (iii), we need to compute the limits of $g_2(\tau)$ and $g_3(\tau)$ as $\mathrm{Im}(\tau) \to \infty$. To study $g_2(\tau)$, write

$$g_2(\tau) = 60\sum_{m,n}{}' \frac{1}{(m+n\tau)^4} = 60\left(2\sum_{m=1}^{\infty}\frac{1}{m^4} + \sum_{\substack{m,n=-\infty \\ n\neq 0}}^{\infty} \frac{1}{(m+n\tau)^4}\right).$$

Using the uniform convergence proved in (i), we see that

$$\lim_{\mathrm{Im}(\tau)\to\infty} g_2(\tau) = 120\sum_{m=1}^{\infty}\frac{1}{m^4},$$

and then the well known formula $\sum_{m=1}^{\infty} 1/m^4 = \pi^4/90$ (see Serre [88, §VII.4.1]) implies that

$$\lim_{\mathrm{Im}(\tau)\to\infty} g_2(\tau) = \frac{4}{3}\pi^4.$$

The case of $g_3(\tau)$ is similar. Here, the key formula is $\sum_{m=1}^{\infty} 1/m^6 = \pi^6/945$ (see Serre [88, §VII.4.1]), and one obtains

$$\lim_{\mathrm{Im}(\tau)\to\infty} g_3(\tau) = \frac{8}{27}\pi^6.$$

These limits imply that

$$\lim_{\mathrm{Im}(\tau)\to\infty} \Delta(\tau) = \left(\frac{4}{3}\pi^4\right)^3 - 27\left(\frac{8}{27}\pi^6\right)^2 = 0,$$

and it follows easily that

$$\lim_{\mathrm{Im}(\tau)\to\infty} j(\tau) = \infty. \tag{11.3}$$

We will also need the following lemma:

Lemma 11.4. *Every $\tau \in \mathfrak{h}$ is* SL$(2,\mathbf{Z})$*-equivalent to a point τ' which satisfies $|\mathrm{Re}(\tau')| \le 1/2$ and $|\mathrm{Im}(\tau')| \ge 1/2$.*

Proof. If $|\mathrm{Im}(\tau)| \ge 1/2$, then there is an integer m such that $\tau' = \tau + m$ satisfies the desired inequalities. Since $\tau + m = \left(\begin{smallmatrix} 1 & m \\ 0 & 1\end{smallmatrix}\right)\tau$, we are done in this case.

If $|\mathrm{Im}(\tau)| \le 1/2$, set $\tau = x + iy$. Pick relatively prime integers c and d such that $|cx+d| \le y$, and then choose a and b so that $\gamma = \left(\begin{smallmatrix} a & b \\ c & d\end{smallmatrix}\right) \in \mathrm{SL}(2,\mathbf{Z})$. In (7.9) we showed that

$$\mathrm{Im}(\gamma\tau) = \frac{\mathrm{Im}(\tau)}{|c\tau + d|^2},$$

so that

$$\mathrm{Im}(\gamma\tau) = \frac{y}{(cx+d)^2 + y^2} \ge \frac{y}{y^2+y^2} = \frac{1}{2y} \ge 1.$$

Now proceed as in the previous paragraph. Q.E.D.

This lemma is related to the idea of finding a *fundamental domain* for the action of SL$(2,\mathbf{Z})$ on $\mathfrak{h}$. We won't use this concept in the text, but there is an interesting relation between fundamental domains and reduced forms (in the sense of Theorem 2.9). See Exercise 11.4 for the details.

We can now show that the j-function is surjective. Since it's holomorphic and nonconstant, its image is an open subset of $\mathbf{C}$. If we can show that the image is closed, surjectivity will follow. So take a sequence of points $j(\tau_k)$ which converges to some $w \in \mathbf{C}$. We need to show that $w = j(\tau)$ for some $\tau \in \mathfrak{h}$. By Lemma 11.4, we can assume that each τ_k lies in the region $R = \{\tau \in \mathfrak{h} : |\mathrm{Re}(\tau)| \leq 1/2, |\mathrm{Im}(\tau)| \geq 1/2\}$. If the imaginary parts of the τ_k's were unbounded, then by the limit (11.3), the $j(\tau_k)$'s would have a subsequence which converged to ∞. This is clearly impossible. But once the imaginary parts are bounded, the τ_k's lie in a compact subset of $\mathfrak{h}$. Then they have a subsequence converging to some $\tau \in \mathfrak{h}$, and it follows by continuity that $j(\tau) = w$, as desired.

The proof of (iv) will use the following lemma:

Lemma 11.5. *If $\tau, \tau' \in \mathfrak{h}$, then there exist neighborhoods U of τ and V of τ' such that the set $\{\gamma \in \mathrm{SL}(2,\mathbf{Z}) : \gamma(U) \cap V \neq \emptyset\}$ is finite.*

Proof. This lemma says that $\mathrm{SL}(2,\mathbf{Z})$ acts properly discontinuously on $\mathfrak{h}$, and the proof is given in Exercise 11.5. Q.E.D.

Corollary 11.6. *If $\tau \in \mathfrak{h}$, then τ has a neighborhood U such that for all $\gamma \in \mathrm{SL}(2,\mathbf{Z})$,*

$$\gamma(U) \cap U \neq \emptyset \iff \gamma\tau = \tau.$$

Proof. See Exercise 11.5. Q.E.D.

Now suppose that $j'(\tau) = 0$. Then τ has a neighborhood U such that for w sufficiently close to $j(\tau)$, there are $\tau' \neq \tau'' \in U$ such that $j(\tau') = j(\tau'') = w$. By (ii), $\tau'' = \gamma\tau'$ for some $\gamma \neq \pm I$, where $I = \left(\begin{smallmatrix} 1 & 0 \\ 0 & 1 \end{smallmatrix}\right)$. Thus $\gamma(U) \cap U \neq \emptyset$. By shrinking U and using Corollary 11.6, it follows that $\gamma\tau = \tau$, $\gamma \neq \pm I$. This is a very strong restriction on τ. To see why, let $\gamma = \left(\begin{smallmatrix} a & b \\ c & d \end{smallmatrix}\right)$. Then $\gamma\tau = \tau$ implies that

$$[1,\tau] = (c\tau + d)[1,\tau]$$

(see the proof of (7.8)), and since $\gamma \neq \pm I$, an easy argument shows that $c \neq 0$ (see Exercise 11.6). Thus $\alpha = c\tau + d \notin \mathbf{Z}$, so that by Theorem 10.14, the lattice $[1,\tau]$ has complex multiplication by an order $\mathcal{O}$ in an imaginary quadratic field. Furthermore, $\alpha[1,\tau] = [1,\tau]$ implies that $\alpha \in \mathcal{O}^*$. However, we know that $\mathcal{O}^* = \{\pm 1\}$ unless $\mathcal{O} = \mathcal{O}_K$ for $K = \mathbf{Q}(i)$ or $\mathbf{Q}(\omega)$, $\omega = e^{2\pi i/3}$ (see Exercise 11.6). Both of these orders have class number 1, so that $[1,\tau]$ is homothetic to either $[1,i]$ or $[1,\omega]$. Thus $j'(\tau) = 0$ implies that τ is $\mathrm{SL}(2,\mathbf{Z})$-equivalent to either i or ω.

When τ is SL(2, **Z**)-equivalent to i, we may assume that $\tau = i$, and we need to show that $j'(i) = 0$ and $j''(i) \neq 0$. To prove the former, note that

$$j(\tau) - 1728 = 1728 \frac{27g_3(\tau)^2}{\Delta(\tau)}.$$

In §10 we proved that $g_3(i) = 0$, and $j'(i) = 0$ follows immediately. Now suppose that $j''(i) = 0$. Then i is at least a triple zero of $j(\tau) - 1728$, so that for w sufficiently near 1728, there are distinct points τ, τ' and τ'' near i such that $j(\tau) = j(\tau') = j(\tau'') = w$. Then $\tau' = \gamma_1\tau$, $\tau'' = \gamma_2\tau$, where $\pm I$, $\pm\gamma_1$ and $\pm\gamma_2$ are all distinct elements of SL(2, **Z**). By Corollary 11.6, $\gamma_1 i = \gamma_2 i = i$, so that at least 6 elements of SL(2, **Z**) fix i. Since only 4 elements of SL(2, **Z**) fix i (see Exercise 11.6), we see that $j''(i) \neq 0$. The case when $\tau = \omega$ is similar and is left to the reader (see Exercise 11.6). Theorem 11.2 is proved. Q.E.D.

The surjectivity of the j-function implies the following result which was used in §10:

Corollary 11.7. *Let g_2 and g_3 be arbitrary complex numbers such that $g_2^3 - 27g_3^2 \neq 0$. Then there is a lattice L such that $g_2(L) = g_2$ and $g_3(L) = g_3$.*

Proof. Since the j-function is surjective and $g_2^3 - 27g_3^2 \neq 0$, there is some $\tau \in \mathfrak{h}$ such that

$$j(\tau) = 1728 \frac{g_2^3}{g_2^3 - 27g_3^2}.$$

Arguing as in the proof of (10.11), this equation implies that there is a nonzero complex number λ such that

$$g_2 = \lambda^{-4} g_2(\tau)$$
$$g_3 = \lambda^{-6} g_3(\tau).$$

Using (10.10), it follows that $L = \lambda[1, \tau]$ is the desired lattice. Q.E.D.

Since $j(\tau)$ is invariant under SL(2, **Z**), we see that

$$j(\tau + 1) = j\left(\begin{pmatrix} 1 & 1 \\ 0 & 1 \end{pmatrix} \tau\right) = j(\tau).$$

This implies that $j(\tau)$ is a holomorphic function in $q = q(\tau) = e^{2\pi i\tau}$, defined in the region $0 < |q| < 1$. Consequently $j(\tau)$ has a Laurent expansion

$$j(\tau) = \sum_{n=-\infty}^{\infty} c_n q^n,$$

which is called the *q-expansion* of $j(\tau)$. The following theorem will be used often in what follows:

Theorem 11.8. *The q-expansion of $j(\tau)$ is*

$$j(\tau) = \frac{1}{q} + 744 + 196884q + \cdots = \frac{1}{q} + \sum_{n=1}^{\infty} c_n q^n,$$

where the coefficients c_n are integers for all $n \geq 0$.

Proof. We will prove this in §12 using the Weber functions and the Weierstrass σ-function. More standard proofs may be found in Apostol [1, §1.15] or Lang [73, §4.1]. Q.E.D.

This theorem is the reason that the factor 1728 appears in the definition of the j-invariant: it's exactly the factor needed to guarantee that all of the coefficients of the q-expansion are integers without any common multiple.

B. Modular Functions for $\Gamma_0(m)$

One can define modular functions for any subgroup of $\mathrm{SL}(2,\mathbf{Z})$, but we will concentrate on the subgroups $\Gamma_0(m)$ of $\mathrm{SL}(2,\mathbf{Z})$, which are defined as follows: if m is a positive integer, then

$$\Gamma_0(m) = \left\{ \begin{pmatrix} a & b \\ c & d \end{pmatrix} \in \mathrm{SL}(2,\mathbf{Z}) : c \equiv 0 \bmod m \right\}.$$

Note that $\Gamma_0(1) = \mathrm{SL}(2,\mathbf{Z})$. Then a *modular function for* $\Gamma_0(m)$ is a complex-valued function $f(\tau)$ defined on the upper half plane $\mathfrak{h}$, except for isolated singularities, which satisfies the following three conditions:

(i) $f(\tau)$ is meromorphic on $\mathfrak{h}$.
(ii) $f(\tau)$ is invariant under $\Gamma_0(m)$.
(iii) $f(\tau)$ is meromorphic at the cusps.

By (ii), we mean that $f(\gamma\tau) = f(\tau)$ for all $\tau \in \mathfrak{h}$ and $\gamma \in \Gamma_0(m)$. To explain (iii), some more work is needed. Suppose that $f(\tau)$ satisfies (i) and (ii), and that $\gamma \in \mathrm{SL}(2,\mathbf{Z})$. We claim that $f(\gamma\tau)$ has period m. To see this, note that $\tau + m = U\tau$, where $U = \left(\begin{smallmatrix} 1 & m \\ 0 & 1 \end{smallmatrix}\right)$. An easy calculation shows that $\gamma U \gamma^{-1} \in \Gamma_0(m)$, and we then obtain

$$f(\gamma(\tau + m)) = f(\gamma U\tau) = f(\gamma U \gamma^{-1} \gamma\tau) = f(\gamma\tau)$$

since $f(\tau)$ is $\Gamma_0(m)$-invariant. It follows that if $q = q(\tau) = e^{2\pi i\tau}$, then $f(\gamma\tau)$ is a holomorphic function in $q^{1/m}$, defined for $0 < |q^{1/m}| < 1$. Thus $f(\gamma\tau)$

has a Laurent expansion

$$f(\gamma\tau) = \sum_{n=-\infty}^{\infty} a_n q^{n/m},$$

which by abuse of notation we will call the q-expansion of $f(\gamma\tau)$. Then $f(\tau)$ is *meromorphic at the cusps* if for all $\gamma \in \mathrm{SL}(2,\mathbf{Z})$, the q-expansion of $f(\gamma\tau)$ has only finitely many nonzero negative coefficients.

The basic example of a such a function is given by $j(\tau)$. It is holomorphic on $\mathfrak{h}$, invariant under $\mathrm{SL}(2,\mathbf{Z})$, and Theorem 11.8 implies that it is meromorphic at the cusps. Thus $j(\tau)$ is a modular function for $\mathrm{SL}(2,\mathbf{Z}) = \Gamma_0(1)$. The remarkable fact is that modular functions for both $\mathrm{SL}(2,\mathbf{Z})$ and $\Gamma_0(m)$ are easily described in terms of the j-function:

Theorem 11.9. *Let m be a positive integer.*

(i) *$j(\tau)$ is a modular function for* $\mathrm{SL}(2,\mathbf{Z})$, *and every modular function for* $\mathrm{SL}(2,\mathbf{Z})$ *is a rational function in $j(\tau)$.*

(ii) *$j(\tau)$ and $j(m\tau)$ are modular functions for $\Gamma_0(m)$, and every modular function for $\Gamma_0(m)$ is a rational function of $j(\tau)$ and $j(m\tau)$.*

Proof. Note that (i) is a special case of (ii). It is stated separately not only because of its independent interest, but also because it's what we must prove first.

Before beginning the proof, let's make a comment about q-expansions. Our definition requires checking the q-expansion of $f(\gamma\tau)$ for all $\gamma \in \mathrm{SL}(2,\mathbf{Z})$. Since $f(\tau)$ is $\Gamma_0(m)$-invariant, we actually need only consider the q-expansions of $f(\gamma_i\tau)$, where the γ_i's are right coset representatives of $\Gamma_0(m) \subset \mathrm{SL}(2,\mathbf{Z})$. So there are only finitely many q-expansions to check. The nicest case is when $f(\tau)$ is a modular function for $\mathrm{SL}(2,\mathbf{Z})$, for here we need only consider the q-expansion of $f(\tau)$.

We can now prove (i). We've seen that $j(\tau)$ is a modular function for $\mathrm{SL}(2,\mathbf{Z})$, so we need only show that every modular function $f(\tau)$ for $\mathrm{SL}(2,\mathbf{Z})$ is a rational function in $j(\tau)$. We will begin by studying some special cases. We say that a modular function $f(\tau)$ is *holomorphic at* ∞ if its q-expansion involves only nonnegative powers of q.

Lemma 11.10.

(i) *A holomorphic modular function for* $\mathrm{SL}(2,\mathbf{Z})$ *which is holomorphic at* ∞ *is constant.*

(ii) *A holomorphic modular function for* $\mathrm{SL}(2,\mathbf{Z})$ *is a polynomial in $j(\tau)$.*

Proof. To prove (i), let $f(\tau)$ be the modular function in question. Since $f(\tau)$ is holomorphic at ∞, we know that $f(\infty) = \lim_{\mathrm{Im}(\tau)\to\infty} f(\tau)$ exists as a complex number. We will show that $f(\mathfrak{h} \cup \{\infty\})$ is compact. By the maximum modulus principle, this will imply that $f(\tau)$ is constant.

Let $f(\tau_k)$ be a sequence of points in the image. We need to find a subsequence that converges to a point of the form $f(\tau)$ for some $\tau \in \mathfrak{h}$. Since $f(\tau)$ is SL(2, Z)-invariant, we can assume that the τ_k's lie in the region $R = \{\tau \in \mathfrak{h} : |\mathrm{Re}(\tau)| \leq 1/2,\ |\mathrm{Im}(\tau)| \geq 1/2\}$ (see Lemma 11.4). If the imaginary parts of the τ_k's are unbounded, then by the above limit, a subsequence converges to $f(\infty)$. If the imaginary parts are bounded, then the τ_k's lie in a compact subset of $\mathfrak{h}$, and the desired subsequence is easily found. This proves (i).

Turning to (ii), let $f(\tau)$ be a holomorphic modular function for SL(2, Z). Its q-expansion has only finitely many terms with negative powers of q. Since the q-expansion of $j(\tau)$ begins with $1/q$, one can find a polynomial $A(x)$ such that $f(\tau) - A(j(\tau))$ is holomorphic at ∞. Since it is also holomorphic on $\mathfrak{h}$, it is constant by (i). Thus $f(\tau)$ is a polynomial in $j(\tau)$, and the lemma is proved. Q.E.D.

To treat the general case, let $f(\tau)$ be an arbitrary modular function for SL(2, Z), possibly with poles on $\mathfrak{h}$. If we can find a polynomial $B(x)$ such that $B(j(\tau))f(\tau)$ is holomorphic on $\mathfrak{h}$, then the lemma will imply that $f(\tau)$ is a rational function in $j(\tau)$. Since $f(\tau)$ has a meromorphic q-expansion, it follows that $f(\tau)$ has only finitely many poles in the region $R = \{\tau \in \mathfrak{h} : |\mathrm{Re}(\tau)| \leq 1/2,\ |\mathrm{Im}(\tau)| \geq 1/2\}$, and since $f(\tau)$ is SL(2, Z)-invariant, Lemma 11.4 implies that every pole of $f(\tau)$ is SL(2, Z)-equivalent to one in R. Thus, if $B(j(\tau))f(\tau)$ has no poles in R, then it is holomorphic on $\mathfrak{h}$.

So suppose that $f(\tau)$ has a pole of order m at $\tau_0 \in R$. If $j'(\tau_0) \neq 0$, then $(j(\tau) - j(\tau_0))^m f(\tau)$ is holomorphic at τ_0. In this way we can find a polynomial $B(x)$ such that $B(j(\tau))f(\tau)$ has no poles in R, except possibly for those where $j'(\tau_0) = 0$. When this happens, part (iv) of Theorem 11.2 allows us to assume that $\tau_0 = i$ or $\omega = e^{2\pi i/3}$. When $\tau_0 = i$, we claim that m is even. To see this, note that in a neighborhood of i, $f(\tau)$ can be written in the form

$$f(\tau) = \frac{g(\tau)}{(\tau - i)^m},$$

where $g(\tau)$ is holomorphic and $g(i) \neq 0$. Now $\left(\begin{smallmatrix} 0 & 1 \\ -1 & 0 \end{smallmatrix}\right) \in \mathrm{SL}(2, \mathbf{Z})$ fixes i, so that

$$f(\tau) = f(-1/\tau) = \frac{g(-1/\tau)}{(-1/\tau - i)^m}.$$

Comparing these two expressions for $f(\tau)$, we see that

$$g(-1/\tau) = \frac{1}{(i\tau)^m} g(\tau).$$

Evaluating this at $\tau = i$ implies that $g(i) = (-1)^m g(i)$, and since $g(i) \neq 0$, it follows that m is even. By Theorem 11.2, $j(\tau) - 1728$ has a zero of order 2 at i, and hence $(j(\tau) - 1728)^{m/2} f(\tau)$ is holomorphic at i. The argument for $\tau_0 = \omega$ is similar and is left to the reader (see Exercise 11.7). This completes the proof of part (i) of Theorem 11.9.

To prove part (ii), it is trivial to show that $j(\tau)$ is a modular function for $\Gamma_0(m)$. As for $j(m\tau)$, it is certainly holomorphic, and to check its invariance properties, let $\gamma = \left(\begin{smallmatrix} a & b \\ c & d \end{smallmatrix}\right) \in \Gamma_0(m)$. Then

$$j(m\gamma\tau) = j\left(\frac{m(a\tau + b)}{c\tau + d}\right) = j\left(\frac{a \cdot m\tau + bm}{c/m \cdot m\tau + d}\right).$$

Since $\gamma \in \Gamma_0(m)$, it follows that $\gamma' = \left(\begin{smallmatrix} a & bm \\ c/m & d \end{smallmatrix}\right) \in \mathrm{SL}(2, \mathbf{Z})$. Thus

$$j(m\gamma\tau) = j(\gamma' m\tau) = j(m\tau),$$

which proves that $j(m\tau)$ is $\Gamma_0(m)$-invariant.

In order to show that $j(m\tau)$ is meromorphic at the cusps, we first relate $\Gamma_0(m)$ to the set of matrices

$$C(m) = \left\{ \begin{pmatrix} a & b \\ 0 & d \end{pmatrix} : ad = m,\ a > 0,\ 0 \leq b < d,\ \gcd(a, b, d) = 1 \right\}.$$

The matrix $\sigma_0 = \left(\begin{smallmatrix} m & 0 \\ 0 & 1 \end{smallmatrix}\right) \in C(m)$ has two properties of interest: first, $\sigma_0\tau = m\tau$, and second,

$$\Gamma_0(m) = (\sigma_0^{-1}\mathrm{SL}(2, \mathbf{Z})\sigma_0) \cap \mathrm{SL}(2, \mathbf{Z})$$

(see Exercise 11.8). Note that these two properties account for the $\Gamma_0(m)$-invariance of $j(m\tau)$ proved above. More generally, we have the following lemma:

Lemma 11.11. *For $\sigma \in C(m)$, the set*

$$(\sigma_0^{-1}\mathrm{SL}(2, \mathbf{Z})\sigma) \cap \mathrm{SL}(2, \mathbf{Z})$$

is a right coset of $\Gamma_0(m)$ in $\mathrm{SL}(2, \mathbf{Z})$. This induces a one-to-one correspondence between right cosets of $\Gamma_0(m)$ and elements of $C(m)$.

Proof. See Exercise 11.8. Q.E.D.

This lemma implies that $[\mathrm{SL}(2, \mathbf{Z}) : \Gamma_0(m)] = |C(m)|$. One can also compute the number of elements in $C(m)$: one gets the formula:

$$|C(m)| = m \prod_{p|m} \left(1 + \frac{1}{p}\right)$$

(see Exercise 11.9), and thus the index of $\Gamma_0(m)$ is $\mathrm{SL}(2,\mathbf{Z})$ is $m\prod_{p|m}(1+1/p)$.

We can now compute some q-expansions. Fix $\gamma \in \mathrm{SL}(2,\mathbf{Z})$, and choose $\sigma \in C(m)$ so that γ lies in the right coset corresponding to σ in Lemma 11.11. This means that $\sigma_0\gamma = \tilde{\gamma}\sigma$ for some $\tilde{\gamma} \in \mathrm{SL}(2,\mathbf{Z})$, and hence $j(m\gamma\tau) = j(\sigma_0\gamma\tau) = j(\tilde{\gamma}\sigma\tau) = j(\sigma\tau)$ since $j(\tau)$ is $\mathrm{SL}(2,\mathbf{Z})$-invariant. Hence

$$j(m\gamma\tau) = j(\sigma\tau). \tag{11.12}$$

Suppose that $\sigma = \left(\begin{smallmatrix} a & b \\ 0 & d \end{smallmatrix}\right)$. We know from Theorem 11.8 that the q-expansion of $j(\tau)$ is

$$j(\tau) = \frac{1}{q} + \sum_{n=1}^{\infty} c_n z^n, \qquad c_n \in \mathbf{Z},$$

and since $\sigma\tau = (a\tau + b)/d$, it follows that

$$q(\sigma\tau) = e^{2\pi i(a\tau+b)/d} = e^{2\pi i b/d} q^{a/d}.$$

If we set $\zeta_m = e^{2\pi i/m}$, we can write this as

$$q(\sigma\tau) = \zeta_m^{ab}(q^{1/m})^{a^2}$$

since $ad = m$. This gives us the q-expansion

$$j(m\gamma\tau) = j(\sigma\tau) = \frac{\zeta_m^{-ab}}{(q^{1/m})^{a^2}} + \sum_{n=1}^{\infty} c_n \zeta_m^{abn}(q^{1/m})^{a^2 n}, \qquad c_n \in \mathbf{Z}. \tag{11.13}$$

There are only finitely many negative exponents, which shows that $j(m\tau)$ is meromorphic at the cusps, and thus $j(m\tau)$ is a modular function for $\Gamma_0(m)$.

The next step is to introduce the modular equation $\Phi_m(X,Y)$. Let the right cosets of $\Gamma_0(m)$ in $\mathrm{SL}(2,\mathbf{Z})$ be $\Gamma_0(m)\gamma_i$, $i = 1,\ldots,|C(m)|$. Then consider the polynomial in X

$$\Phi_m(X,\tau) = \prod_{i=1}^{|C(m)|} (X - j(m\gamma_i\tau)).$$

We will prove that this expression is a polynomial in X and $j(\tau)$. To see this, consider the coefficients of $\Phi_m(X,\tau)$. Being symmetric polynomials in the $j(m\gamma_i\tau)$'s, they are certainly holomorphic. To check invariance under $\mathrm{SL}(2,\mathbf{Z})$, pick $\gamma \in \mathrm{SL}(2,\mathbf{Z})$. Then the cosets $\Gamma_0(m)\gamma_i\gamma$ are a permutation of the $\Gamma_0(m)\gamma_i$'s, and since $j(m\tau)$ is invariant under $\Gamma_0(m)$, the $j(m\gamma_i\gamma\tau)$'s are a permutation of the $j(m\gamma_i\tau)$'s. This shows that the coefficients of $\Phi_m(X,\tau)$ are invariant under $\mathrm{SL}(2,\mathbf{Z})$.

We next have to show that the coefficients are meromorphic at infinity. Rather than expand in powers of q, it suffices to expand in terms of $q^{1/m} = e^{2\pi i\tau/m}$ and show that only finitely negative exponents appear.

By (11.12), we know that $j(m\gamma_i\tau) = j(\sigma\tau)$ for some $\sigma \in C(m)$, and then (11.13) shows that the q-expansion for $j(m\gamma_i\tau)$ has only finitely many exponents. Since the coefficients are polynomials in the $j(m\gamma_i\tau)$'s, they clearly are meromorphic at the cusps.

This proves that the coefficients of $\Phi_m(X,\tau)$ are holomorphic modular functions, and thus, by Lemma 11.10, they are polynomials in $j(\tau)$. This means that there is a polynomial

$$\Phi_m(X,Y) \in \mathbf{C}[X,Y]$$

such that

$$\Phi_m(X,j(\tau)) = \prod_{i=1}^{|C(m)|} (X - j(m\gamma_i\tau)). \tag{11.14}$$

The equation $\Phi_m(X,Y) = 0$ is called the *modular equation*, and by abuse of terminology we will call $\Phi_m(X,Y)$ the modular equation. Using some simple field theory, it can be proved that $\Phi_m(X,Y)$ is irreducible as a polynomial in X (see Exercise 11.10).

By (11.12), each $j(m\gamma_i\tau)$ can be written $j(\sigma\tau)$ for a unique $\sigma \in C(m)$. Thus we can also express the modular equation in the form

$$\Phi_m(X,j(\tau)) = \prod_{\sigma\in C(m)} (X - j(\sigma\tau)). \tag{11.15}$$

Note that $j(m\tau)$ is always one of the $j(\sigma\tau)$'s since $\left(\begin{smallmatrix} m & 0 \\ 0 & 1 \end{smallmatrix}\right) \in C(m)$. Hence

$$\Phi_m(j(m\tau), j(\tau)) = 0,$$

which is one of the important properties of the modular equation. Note that the degree of $\Phi_m(X,Y)$ in X is $|C(m)|$, which we know equals $m\prod_{p|m}(1 + 1/p)$.

Now let $f(\tau)$ be an arbitrary modular function for $\Gamma_0(m)$. To prove that $f(\tau)$ is a rational function in $j(\tau)$ and $j(m\tau)$, consider the function

$$\begin{aligned} G(X,\tau) &= \Phi_m(X,j(\tau)) \sum_{i=1}^{|C(m)|} \frac{f(\gamma_i\tau)}{X - j(m\gamma_i\tau)} \\ &= \sum_{i=1}^{|C(m)|} f(\gamma_i\tau) \prod_{j\neq i} (X - j(m\gamma_j\tau)). \end{aligned} \tag{11.16}$$

This is a polynomial in X, and we claim that its coefficients are modular functions for $SL(2,\mathbf{Z})$. The proof is similar to what we did for the modular equation, and the details are left to the reader (see Exercise 11.11). But once the coefficients are modular functions for $SL(2,\mathbf{Z})$, they are rational functions of $j(\tau)$ by what we proved above. Hence $G(X,\tau)$ is a polynomial $G(X,j(\tau)) \in \mathbb{C}(j(\tau))[X]$.

We can assume that γ_1 is the identity matrix. By the product rule, we obtain

$$\frac{\partial\Phi_m}{\partial X}(j(m\tau),j(\tau)) = \prod_{j\neq 1}(j(m\tau) - j(m\gamma_j\tau)).$$

Thus, substituting $X = j(m\tau)$ in (11.16) gives

$$G(j(m\tau),j(\tau)) = f(\tau)\frac{\partial\Phi_m}{\partial X}(j(m\tau),j(\tau)).$$

Now $\Phi_m(X,j(\tau))$ is irreducible (see Exercise 11.10) and hence separable, so that $(\partial/\partial X)\Phi_m(j(m\tau),j(\tau)) \neq 0$. Thus we can write

$$f(\tau) = \frac{G(j(m\tau),j(\tau))}{\dfrac{\partial\Phi_m}{\partial X}(j(m\tau),j(\tau))}, \tag{11.17}$$

which proves that $f(\tau)$ is a rational function in $j(\tau)$ and $j(m\tau)$. This completes the proof of Theorem 11.9. Q.E.D.

There is a large literature on modular functions, and the reader may wish to consult Apostol [1], Koblitz [67], Lang [73] or Shimura [90] to learn more about these remarkable functions.

C. The Modular Equation $\Phi_m(X,Y)$

The modular equation, as defined by equations (11.14) or (11.15), will play a crucial role in what follows. In particular, we will make heavy use of the arithmetic properties of $\Phi_m(X,Y)$, which are given in the following theorem:

Theorem 11.18. *Let m be a positive integer.*

(i) $\Phi_m(X,Y) \in \mathbb{Z}[X,Y]$.

(ii) *$\Phi_m(X,Y)$ is irreducible when regarded as a polynomial in X.*

(iii) $\Phi_m(X,Y) = \Phi_m(Y,X)$.

(iv) *If m is not a perfect square, then $\Phi_m(X,X)$ is a polynomial of degree > 1 whose leading coefficient is ± 1.*

(v) *If m is a prime p, then $\Phi_p(X,Y) \equiv (X^p - Y)(X - Y^p) \bmod p\mathbb{Z}[X,Y]$.*

Proof. To prove (i), it suffices to show that an elementary symmetric function $f(\tau)$ in the $j(\sigma\tau)$'s, $\sigma \in C(m)$, is a polynomial in $j(\tau)$ with integer coefficients. We begin by studying the q-expansion of $f(\tau)$ in more detail.

Let $\zeta_m = e^{2\pi i/m}$. By (11.13), each $j(\sigma\tau)$ lies in the field of formal meromorphic Laurent series $\mathbb{Q}(\zeta_m)((q^{1/m}))$, and since $f(\tau)$ is an integer polynomial in the $j(\sigma\tau)$'s, $f(\tau)$ also lies in $\mathbb{Q}(\zeta_m)((q^{1/m}))$.

We claim that $f(\tau)$ is contained in the smaller field $\mathbb{Q}((q^{1/m}))$. To see this, we will use Galois theory. An automorphism $\psi \in \mathrm{Gal}(\mathbb{Q}(\zeta_m)/\mathbb{Q})$ determines an automorphism of $\mathbb{Q}(\zeta_m)((q^{1/m}))$ by acting on the coefficients. Given $\sigma = \left(\begin{smallmatrix} a & b \\ 0 & d \end{smallmatrix}\right) \in C(m)$, let's see how ψ affects $j(\sigma\tau)$. We know that $\psi(\zeta_m) = \zeta_m^k$ for some integer k relatively prime to m, and from (11.13), it follows that

$$\psi(j(\sigma\tau)) = \frac{\zeta_m^{-abk}}{(q^{1/m})^{a^2}} + \sum_{n=1}^{\infty} c_n \zeta_m^{abkn} (q^{1/m})^{a^2 n}$$

since all of the c_n's are integers. Let b' be the unique integer $0 \le b' < d$ such that $b' \equiv bk \bmod d$. Since $ad = m$, we have $\zeta_m^{abk} = \zeta_m^{ab'}$, and consequently the above formula can be written

$$\psi(j(\sigma\tau)) = \frac{\zeta_m^{-ab'}}{(q^{1/m})^{a^2}} + \sum_{n=1}^{\infty} c_n \zeta_m^{ab'n} (q^{1/m})^{a^2 n}.$$

If we let $\sigma' = \left(\begin{smallmatrix} a & b' \\ 0 & d \end{smallmatrix}\right)$, then $\sigma' \in C(m)$, and (11.13) implies that

$$\psi(j(\sigma\tau)) = j(\sigma'\tau).$$

Thus the elements of $\mathrm{Gal}(\mathbb{Q}(\zeta_m)/\mathbb{Q})$ permute the $j(\sigma\tau)$'s. Since $f(\tau)$ is symmetric in the $j(\sigma\tau)$'s, it follows that $f(\tau) \in \mathbb{Q}((q^{1/m}))$.

We conclude that $f(\tau) \in \mathbb{Z}((q))$ since the q-expansion of $f(\tau)$ involves only integral powers of q and the coefficients of the q-expansion are algebraic integers. It remains to show that $f(\tau)$ is an integer polynomial in $j(\tau)$. By Lemma 11.10, we can find $A(X) \in \mathbb{C}[X]$ such that $f(\tau) = A(j(\tau))$. Recall from the proof of Lemma 11.10 that $A(X)$ was chosen so that the q-expansion of $f(\tau) - A(j(\tau))$ has only terms of degree > 0. Since the expansions of $f(\tau)$ and $j(\tau)$ have integer coefficients and $j(\tau) = 1/q + \cdots$, it follows that $A(X) \in \mathbb{Z}[X]$. Thus $f(\tau) = A(j(\tau))$ is an integer polynomial in $j(\tau)$, and (i) is proved.

We should mention that the passage from the coefficients of the q-expansion to the coefficients of the polynomial $A(X)$ is a special case of Hasse's q-expansion principle—see Exercise 11.12 for a precise formulation.

A proof of (ii) is given in Exercise 11.10, and a proof of (iii) may be found in Lang [73, §5.2, Theorem 3].

Turning to (iv), assume that m is not a square. We want to study the leading term of the integer polynomial $\Phi_m(X,X)$. Replacing X with $j(\tau)$, it suffices to study the coefficient of the most negative power of q in the q-expansion of $\Phi_m(j(\tau), j(\tau))$. However, given $\sigma = \left(\begin{smallmatrix} a & b \\ 0 & d \end{smallmatrix}\right) \in C(m)$, (11.13) tells us that

$$j(\tau)-j(\sigma\tau)=\frac{1}{q}-\frac{\zeta_m^{-ab}}{q^{a/d}}+\sum_{n=0}^{\infty}d_n(q^{1/m})^n \tag{11.19}$$

for some coefficients d_n. Since m is not a perfect square, we know that $a \neq d$, i.e., $a/d \neq 1$. Thus the coefficient of the most negative term in (11.19) is a root of unity. By (11.15), $\Phi_m(j(\tau),j(\tau))$ is the product of the factors (11.19), so that the coefficient of the most negative power of q in $\Phi_m(j(\tau),j(\tau))$ is also a root of unity. But this coefficient is an integer, and thus it must be ± 1, as claimed.

Finally, we turn to (v). Here, we are assuming that $m = p$, where p is prime. Let $\zeta_p = e^{2\pi i/p}$. We will use the following notation: given $f(\tau)$ and $g(\tau)$ in $\mathbf{Z}[\zeta_p]((q^{1/p}))$ and $\alpha \in \mathbf{Z}[\zeta_p]$, we will write

$$f(\tau) \equiv g(\tau) \bmod \alpha$$

to indicate that $f(\tau)-g(\tau) \in \alpha\mathbf{Z}[\zeta_p]((q^{1/p}))$.

Since p is prime, the elements of $C(p)$ are easy to write down:

$$\sigma_i = \begin{pmatrix} 1 & i \\ 0 & p \end{pmatrix}, \qquad i = 0,\ldots,p-1$$

$$\sigma_p = \begin{pmatrix} p & 0 \\ 0 & 1 \end{pmatrix}.$$

If $0 \le i \le p-1$, then (11.13) tells us that

$$j(\sigma_i\tau) = \frac{\zeta_p^{-i}}{q^{1/p}}+\sum_{n=1}^{\infty}c_n\zeta_p^{in}(q^{1/p})^n \equiv \frac{1}{q^{1/p}}+\sum_{n=1}^{\infty}c_n(q^{1/p})^n \bmod 1-\zeta_p,$$

which implies that

$$j(\sigma_i\tau) \equiv j(\sigma_0\tau) \bmod 1-\zeta_p \tag{11.20}$$

for $0 \le i \le p-1$. Turning to $j(\sigma_p\tau)$, here (11.13) tells us that

$$j(\sigma_p\tau) = \frac{1}{q^p}+\sum_{n=1}^{\infty}c_n q^{pn},$$

and since $c_n^p \equiv c_n \bmod p$, it follows easily that

$$j(\sigma_p\tau) \equiv j(\tau)^p \bmod p.$$

Since $1-\zeta_p$ divides p in $\mathbf{Z}[\zeta_p]$ (see Exercise 11.13), the above congruence can be written

$$j(\sigma_p\tau) \equiv j(\tau)^p \bmod 1-\zeta_p. \tag{11.21}$$

Then (11.20) and (11.21) imply that

$$\begin{aligned}\Phi_p(X, j(\tau)) &= \prod_{i=0}^{p}(X - j(\sigma_i\tau))\\ &\equiv (X - j(\sigma_0\tau))^p(X - j(\tau)^p) \bmod 1-\zeta_p\\ &\equiv (X^p - j(\sigma_0\tau)^p)(X - j(\tau)^p) \bmod 1-\zeta_p,\end{aligned}$$

where we are now working in the ring $\mathbf{Z}[\zeta_p]((q^{1/p}))[X]$. However, the argument used to prove (11.21) is easily adapted to prove that

$$j(\tau) \equiv j(\sigma_0\tau)^p \bmod 1-\zeta_p$$

(see Exercise 11.14), and then we obtain

$$\Phi_p(X, j(\tau)) \equiv (X^p - j(\tau))(X - j(\tau)^p) \bmod 1-\zeta_p.$$

The two sides of this congruence lie in $\mathbf{Z}((q))$, so that the coefficients of the difference are ordinary integers divisible by $1-\zeta_p$ in the ring $\mathbf{Z}[\zeta_p]$. This implies that all of the coefficients are divisible by p (see Exercise 11.13), and thus

$$\Phi_p(X, j(\tau)) \equiv (X^p - j(\tau))(X - j(\tau)^p) \bmod p\mathbf{Z}((q))[X].$$

Then the Hasse q-expansion principle (used in the proof of (i)) shows that

$$\Phi_p(X, Y) \equiv (X^p - Y)(X - Y^p) \bmod p\mathbf{Z}[X, Y],$$

as desired (see Exercise 11.15). The above congruence was first discovered by Kronecker (in a slightly different context) and is sometimes called *Kronecker's congruence*. This completes the proof of Theorem 11.18. Q.E.D.

The properties of the modular equation are straightforward consequences of the properties of the j-function, which makes the modular equation seem like a reasonable object to deal with. This is true as long as one works at the abstract level, but as soon as one asks for concrete examples, the situation gets surprisingly complicated. For example, when $m = 3$, Smith [94] showed that $\Phi_3(X, Y)$ is the polynomial

(11.22)

$$\begin{aligned}&X(X + 2^{15}\cdot 3\cdot 5^3)^3 + Y(Y + 2^{15}\cdot 3\cdot 5^3)^3\\ &\quad - X^3Y^3 + 2^3\cdot 3^2\cdot 31X^2Y^2(X+Y)\\ &\quad - 2^2\cdot 3^3\cdot 9907XY(X^2+Y^2) + 2\cdot 3^4\cdot 13\cdot 193\cdot 6367X^2Y^2\\ &\quad + 2^{16}\cdot 3^5\cdot 5^3\cdot 17\cdot 263XY(X+Y) - 2^{31}\cdot 5^6\cdot 22973XY.\end{aligned}$$

The modular equation $\Phi_m(X,Y)$ has been computed for $m = 5$, 7 and 11 (see Hermann [53] and Kaltofen and Yui [66]), and in §13 we will discuss the problem of computing $\Phi_m(X,Y)$ for general m.

Before we can apply the modular equation to complex multiplication, one task remains: we need to understand the modular equation in terms of j-invariants of lattices. The basic idea is that if L is a lattice, then the roots of $\Phi_m(X,j(L)) = 0$ are given the j-invariants of those sublattices $L' \subset L$ which satisfy:

(i) L' is a sublattice of index m in L, i.e., $[L:L'] = m$.

(ii) The quotient L/L' is a cyclic group.

In this situation, we say that L' is a *cyclic sublattice* of L of index m. Here is the precise statement of what we want to prove:

Theorem 11.23. *Let m be a positive integer. If $u,v \in \mathbb{C}$, then $\Phi_m(u,v) = 0$ if and only if there is a lattice L and a cyclic sublattice $L' \subset L$ of index m such that $u = j(L')$ and $v = j(L)$.*

Proof. We will first study the cyclic sublattices of the lattice $[1,\tau]$, $\tau \in \mathfrak{h}$:

Lemma 11.24. *Let $\tau \in \mathfrak{h}$, and consider the lattice $[1,\tau]$.*

(i) *Given a cyclic sublattice $L' \subset [1,\tau]$ of index m, there is a unique $\sigma = \left(\begin{smallmatrix} a & b \\ 0 & d \end{smallmatrix}\right) \in C(m)$ such that $L' = d[1,\sigma\tau]$.*

(ii) *Conversely, if $\sigma = \left(\begin{smallmatrix} a & b \\ 0 & d \end{smallmatrix}\right) \in C(m)$, then $d[1,\sigma\tau]$ is a cyclic sublattice of $[1,\tau]$ of index m.*

Proof. First recall that $C(m)$ is the set of matrices

$$C(m) = \left\{ \begin{pmatrix} a & b \\ 0 & d \end{pmatrix} : ad = m,\ a > 0,\ 0 \leq b < d,\ \gcd(a,b,d) = 1 \right\}.$$

A sublattice $L' \subset L = [1,\tau]$ can be written $L' = [a\tau + b, c\tau + d]$, and in Exercise 7.15 we proved that $[L:L'] = |ad - bc| = m$. Furthermore, a standard argument using elementary divisors shows that

$$(11.25) \qquad L/L' \text{ is cyclic} \iff \gcd(a,b,c,d) = 1$$

(see, for example, Lang [73, pp. 51–52]). Another proof of (11.25) is given in Exercise 11.16.

Now suppose that $L' \subset [1,\tau]$ is cyclic of index m. If d is the smallest positive integer contained in L', then it follows easily that L' is of the form $L' = [d, a\tau + b]$ (see Exercise 11.17). We may assume that $a > 0$, and then $ad = m$. However, if k is any integer, then

$$L' = [d, (a\tau + b) + kd] = [d, a\tau + (b + kd)],$$

so that by choosing k appropriately, we can assume $0 \leq b < d$. We also have $\gcd(a,b,d) = 1$ by (11.25), and thus the matrix $\sigma = \left(\begin{smallmatrix} a & b \\ 0 & d \end{smallmatrix}\right)$ lies in $C(m)$. Then

$$L' = [d, a\tau + b] = d[1, (a\tau + b)/d] = d[1, \sigma\tau]$$

shows that L' has the desired form. It is straightforward to prove that $\sigma \in C(m)$ is uniquely determined by L' (see Exercise 11.17), and (i) is proved.

The proof of (ii) follows immediately from (11.25), and we are done. Q.E.D.

By this lemma, the j-invariants of the cyclic sublattices L' of index m of $[1,\tau]$ are given by

$$j(L') = j(d[1,\sigma\tau]) = j([1,\sigma\tau]) = j(\sigma\tau).$$

By (11.15), it follows that the roots of $\Phi_m(X, j(\tau)) = 0$ are exactly the j-invariants of the cyclic sublattices of index m of $[1,\tau]$. It is now easy to complete the proof of Theorem 11.23 (see Exercise 11.18 for the details). Q.E.D.

D. Complex Multiplication and Ring Class Fields

To prove Theorem 11.1, we will apply the modular equation to lattices with complex multiplication. The key point is that such lattices have some especially interesting cyclic sublattices. To construct these sublattices, we will use the notion of a *primitive* ideal. Given an order $\mathcal{O}$, we say that a proper $\mathcal{O}$-ideal is *primitive* if it is not of the form $d\mathfrak{a}$ where $d > 1$ is an integer and $\mathfrak{a}$ is a proper $\mathcal{O}$-ideal. Then primitive ideals give us cyclic sublattices as follows:

Lemma 11.26. *Let $\mathcal{O}$ be an order in an imaginary quadratic field, and let $\mathfrak{b}$ be a proper fractional $\mathcal{O}$-ideal. Then, given a proper $\mathcal{O}$-ideal $\mathfrak{a}$, $\mathfrak{a}\mathfrak{b}$ is a sublattice of $\mathfrak{b}$ of index $N(\mathfrak{a})$, and $\mathfrak{a}\mathfrak{b}$ is a cyclic sublattice if and only if $\mathfrak{a}$ is a primitive ideal.*

Proof. Replacing $\mathfrak{b}$ by a multiple, we can assume that $\mathfrak{b} \subset \mathcal{O}$. Then the exact sequence

$$0 \longrightarrow \mathfrak{b}/\mathfrak{a}\mathfrak{b} \longrightarrow \mathcal{O}/\mathfrak{a}\mathfrak{b} \longrightarrow \mathcal{O}/\mathfrak{b} \longrightarrow 0$$

implies that $[\mathfrak{b} : \mathfrak{a}\mathfrak{b}]N(\mathfrak{b}) = N(\mathfrak{a}\mathfrak{b}) = N(\mathfrak{a})N(\mathfrak{b})$, and $[\mathfrak{b} : \mathfrak{a}\mathfrak{b}] = N(\mathfrak{a})$ follows.

Now assume that $\mathfrak{b}/\mathfrak{a}\mathfrak{b}$ is not cyclic. By part (a) of Exercise 11.16, it follows that $\mathfrak{b}/\mathfrak{a}\mathfrak{b}$ contains a subgroup isomorphic to $(\mathbf{Z}/d\mathbf{Z})^2$ for some $d > 1$, so that there is a sublattice $\mathfrak{a}\mathfrak{b} \subset \mathfrak{b}' \subset \mathfrak{b}$ such that $\mathfrak{b}'/\mathfrak{a}\mathfrak{b} \simeq (\mathbf{Z}/d\mathbf{Z})^2$. Since

$\mathfrak{b}'$ is rank 2, this implies that $\mathfrak{a}\mathfrak{b} = d\mathfrak{b}'$, and then $\mathfrak{a} = d\mathfrak{b}'\mathfrak{b}^{-1}$. But $\mathfrak{b}'\mathfrak{b}^{-1} \subset \mathcal{O}$ since $\mathfrak{b}' \subset \mathfrak{b}$, which shows that $\mathfrak{a}$ is not primitive.

The converse, that $\mathfrak{a}$ not primitive implies $\mathfrak{b}/\mathfrak{a}\mathfrak{b}$ not cyclic, is even easier to prove, and is left to the reader (see Exercise 11.19). This completes the proof of the lemma. Q.E.D.

When we apply this lemma, $\mathfrak{a}$ will often be a principal ideal $\mathfrak{a} = \alpha\mathcal{O}$, $\alpha \in \mathcal{O}$. In this case, $\alpha\mathcal{O}$ is primitive as an ideal if and only if α is primitive as an element of $\mathcal{O}$ (which means that α is not of the form $d\beta$ where $d > 1$ and $\beta \in \mathcal{O}$). Since $N(\alpha) = N(\alpha\mathcal{O})$ by Lemma 7.14, we get the following corollary of Lemma 11.26:

Corollary 11.27. *Let $\mathcal{O}$ and $\mathfrak{b}$ be as above. Then, given $\alpha \in \mathcal{O}$, $\alpha\mathfrak{b}$ is a sublattice of $\mathfrak{b}$ of index $N(\alpha)$, and $\alpha\mathfrak{b}$ is a cyclic sublattice if and only if α is primitive.* Q.E.D.

We are now ready to prove Theorem 11.1, the "First Main Theorem" of complex multiplication.

Proof of Theorem 11.1 Let $\mathfrak{a}$ be a proper fractional $\mathcal{O}$-ideal, where $\mathcal{O}$ is an order in an imaginary quadratic field K. We must prove that $j(\mathfrak{a})$ is an algebraic integer and that $K(j(\mathfrak{a}))$ is the ring class field of $\mathcal{O}$. We will follow the proof given by Deuring in [24, §10].

Let's first use the modular equation to prove that $j(\mathfrak{a})$ is an algebraic integer. The basic idea is quite simple: let $\alpha \in \mathcal{O}$ be primitive so that by the above corollary, $\alpha\mathfrak{a}$ is a cyclic sublattice of $\mathfrak{a}$ of index $m = N(\alpha)$. Then, by Theorem 11.23, we know that

$$0 = \Phi_m(j(\alpha\mathfrak{a}), j(\mathfrak{a})) = \Phi_m(j(\mathfrak{a}), j(\mathfrak{a})) = 0$$

since $j(\alpha\mathfrak{a}) = j(\mathfrak{a})$. Thus $j(\mathfrak{a})$ is a root of the polynomial $\Phi_m(X,X)$. Since $\Phi_m(X,Y)$ has integer coefficients (part (i) of Theorem 11.18), this shows that $j(\mathfrak{a})$ is an algebraic number. Furthermore, if we can pick α so that $m = N(\alpha)$ is not a perfect square, then the leading coefficient of $\Phi_m(X,X)$ is ± 1 (part (iv) of Theorem 11.18), and thus $j(\mathfrak{a})$ will be an algebraic integer. So can we find a primitive $\alpha \in \mathcal{O}$ such that $N(\alpha)$ is not a perfect square? We will see below in (11.28) that $\mathcal{O}$ has lots of α's such that $N(\alpha)$ is prime. Such an α is certainly primitive of nonsquare norm. For a more elementary proof, let f be the conductor of $\mathcal{O}$. By Lemma 7.2, $\mathcal{O} = [1, fw_K]$, $w_K = (d_K + \sqrt{d_K})/2$. Then $\alpha = fw_K$ is primitive in $\mathcal{O}$, and one easily sees that its norm $N(\alpha)$ is not a perfect square (see Exercise 11.20).

Let L denote the ring class field of $\mathcal{O}$. In order to prove $L = K(j(\mathfrak{a}))$, we will study how integer primes decompose in L and $K(j(\mathfrak{a}))$. We will make extensive use of the results of §8, especially Proposition 8.20. As usual, f and D will denote the conductor and discriminant of $\mathcal{O}$.

Let's first study how integer primes behave in the ring class field L. Let $\mathcal{S}_{L/\mathbf{Q}}$ be the set of primes that split completely in L. We claim that

$$\text{(11.28)} \qquad \mathcal{S}_{L/\mathbf{Q}} \doteq \{p \text{ prime} : p = N(\alpha) \text{ for some } \alpha \in \mathcal{O}\}.$$

(As noted above, this shows that there are α's in $\mathcal{O}$ with $N(\alpha)$ prime.) When $D \equiv 0 \bmod 4$, then $\mathcal{O} = \mathbf{Z}[\sqrt{-n}]$ for some positive integer n. Thus $N(\alpha) = N(x + y\sqrt{-n}) = x^2 + ny^2$, so that (11.28) says, with finitely many exceptions, that the primes splitting completely in L are those represented by $x^2 + ny^2$. This was proved in Theorem 9.4. The case when $D \equiv 1 \bmod 4$ is similar and was covered in Exercise 9.3. This proves (11.28).

Let $M = K(j(\mathfrak{a}))$. Since L is Galois over $\mathbf{Q}$ by Lemma 9.3, part (i) of Proposition 8.20 shows that $M \subset L$ is equivalent to

$$\text{(11.29)} \qquad \mathcal{S}_{L/\mathbf{Q}} \mathrel{\dot{\subset}} \mathcal{S}_{M/\mathbf{Q}},$$

Take $p \in \mathcal{S}_{L/\mathbf{Q}}$, and assume that p is unramified in M (this excludes only finitely many p's). By (11.28), $p = N(\alpha)$ for some $\alpha \in \mathcal{O}$. Then $\alpha\mathfrak{a} \subset \mathfrak{a}$ is a sublattice of index $N(\alpha) = p$, and is cyclic since p is prime. Thus

$$0 = \Phi_p(j(\alpha\mathfrak{a}), j(\mathfrak{a})) = \Phi_p(j(\mathfrak{a}), j(\mathfrak{a})).$$

Using the Kronecker's congruence from part (v) of Theorem 11.18, this implies that

$$0 = \Phi_p(j(\mathfrak{a}), j(\mathfrak{a})) = -(j(\mathfrak{a})^p - j(\mathfrak{a}))^2 + p\beta$$

for some $\beta \in \mathcal{O}_M$. Now let $\mathfrak{P}$ be any prime of M containing p. The above equation then implies that

$$\text{(11.30)} \qquad j(\mathfrak{a})^p \equiv j(\mathfrak{a}) \bmod \mathfrak{P}.$$

We claim the following:

(i) $\mathcal{O}_K[j(\mathfrak{a})] \subset \mathcal{O}_M$ has finite index.

(ii) If $p \nmid [\mathcal{O}_M : \mathcal{O}_K[j(\mathfrak{a})]]$, then (11.30) implies that $\alpha^p \equiv \alpha \bmod \mathfrak{P}$ for all $\alpha \in \mathcal{O}_M$.

The proof of (i) is a direct consequence of $M = K(j(\mathfrak{a}))$ and is left to the reader (see Exercise 11.21). As for (ii), note that p splits completely in L, so that it splits completely in K, and hence $p \in \mathfrak{p} \subset \mathfrak{P}$ for some ideal $\mathfrak{p}$ of norm p. This implies that $\alpha^p \equiv \alpha \bmod \mathfrak{P}$ holds for all $\alpha \in \mathcal{O}_K$, and consequently the congruence holds for all $\alpha \in \mathcal{O}_K[j(\mathfrak{a})]$ by (11.30). Then (ii) follows easily (see Exercise 11.21).

From (ii) it follows that $f_{\mathfrak{P}|p} = 1$, and since this holds for any $\mathfrak{P}$ containing p, we see that p splits completely in M. This proves (11.29), and $M \subset L$ follows.

The inclusion $M = K(j(\mathfrak{a})) \subset L$ shows that the ring class field L contains the j-invariants of *all* proper fractional $\mathcal{O}$-ideals. Let $h = h(\mathcal{O})$, and let $\mathfrak{a}_i$, $i = 1,\ldots,h$ be class representatives for $C(\mathcal{O})$. It follows that any $j(\mathfrak{a})$ equals one of $j(\mathfrak{a}_1),\ldots,j(\mathfrak{a}_h)$, and furthermore $j(\mathfrak{a}_1),\ldots,j(\mathfrak{a}_h)$ are distinct. Thus

$$(11.31) \qquad \Delta = \prod_{i<j}(j(\mathfrak{a}_i) - j(\mathfrak{a}_j))$$

is a nonzero element of $\mathcal{O}_L$.

To prove the opposite inclusion $L \subset M$, we will use the criterion $\tilde{\mathcal{S}}_{M/\mathbb{Q}} \dot{\subset} \mathcal{S}_{L/K}$ from part (ii) of Proposition 8.20. So let $p \in \tilde{\mathcal{S}}_{M/\mathbb{Q}}$, which means that p is unramified in M and $f_{\mathfrak{P}|p} = 1$ for some prime $\mathfrak{P}$ of M containing p. In particular, this implies that p splits completely in K, and thus $p = N(\mathfrak{p})$ for some prime ideal of $\mathcal{O}$. Then Proposition 7.20 tells us that $p = N(\mathfrak{p} \cap \mathcal{O})$ (we can assume that p doesn't divide f—this excludes finitely many primes). If we can show that $\mathfrak{a} \cap \mathcal{O}$ is a principal ideal $\alpha\mathcal{O}$, then $p = N(\alpha)$ implies that $p \in \mathcal{S}_{L/\mathbb{Q}}$ by (11.28). We may assume that p is relatively prime to the element Δ of (11.31).

Let $\mathfrak{a}' = (\mathfrak{p} \cap \mathcal{O})\mathfrak{a}$. Since $\mathfrak{p} \cap \mathcal{O}$ has norm p, $\mathfrak{a}' \subset \mathfrak{a}$ is a sublattice of index p by Lemma 11.26, and it is cyclic since p is prime. Thus $\Phi_p(j(\mathfrak{a}'), j(\mathfrak{a})) = 0$. Using Kronecker's congruence again, we can write this as

$$0 = \Phi_p(j(\mathfrak{a}'), j(\mathfrak{a})) = (j(\mathfrak{a}')^p - j(\mathfrak{a}))(j(\mathfrak{a}') - j(\mathfrak{a})^p) + pQ(j(\mathfrak{a}'), j(\mathfrak{a}))$$

for some polynomial $Q(X,Y) \in \mathbb{Z}[X,Y]$. Let $\tilde{\mathfrak{P}}$ be a prime of L containing $\mathfrak{P}$. Since $j(\mathfrak{a}')$ and $j(\mathfrak{a})$ are algebraic integers lying in L, the above equation implies that $pQ(j(\mathfrak{a}'), j(\mathfrak{a})) \in \tilde{\mathfrak{P}}$. Thus

$$(11.32) \qquad j(\mathfrak{a}')^p \equiv j(\mathfrak{a}) \bmod \tilde{\mathfrak{P}} \qquad \text{or} \qquad j(\mathfrak{a}') \equiv j(\mathfrak{a})^p \bmod \tilde{\mathfrak{P}}.$$

However, we also know $f_{\mathfrak{P}|p} = 1$, which tells us that $j(\mathfrak{a})^p \equiv j(\mathfrak{a}) \bmod \mathfrak{P}$, and since $\mathfrak{P} \subset \tilde{\mathfrak{P}}$, we obtain

$$(11.33) \qquad j(\mathfrak{a})^p \equiv j(\mathfrak{a}) \bmod \tilde{\mathfrak{P}}.$$

It is straightforward to show that (11.32) and (11.33) imply

$$j(\mathfrak{a}) \equiv j(\mathfrak{a}') \bmod \tilde{\mathfrak{P}}.$$

If $\mathfrak{a}$ and $\mathfrak{a}'$ lay in distinct ideal classes in $C(\mathcal{O})$, then $j(\mathfrak{a}) - j(\mathfrak{a}')$ would be one of the factors of Δ from (11.31), and p and Δ would not be relatively prime. This contradicts our choice of p, so that $\mathfrak{a}$ and $\mathfrak{a}' = (\mathfrak{p} \cap \mathcal{O})\mathfrak{a}$ must lie in the same ideal class in $C(\mathcal{O})$. This forces $\mathfrak{p} \cap \mathcal{O}$ to be a principal ideal, which as we showed above, implies that $p \in \mathcal{S}_{L/\mathbb{Q}}$. Thus $\tilde{\mathcal{S}}_{M/\mathbb{Q}} \dot{\subset} \mathcal{S}_{L/\mathbb{Q}}$, which completes the proof that $L = M$. Theorem 11.1 is proved. Q.E.D.

As an application of Theorem 11.1, let's see what it tells us about the Abelian extensions of an imaginary quadratic field K. First, we know that the Hilbert class field of K is the ring class field of the maximal order $\mathcal{O}_K$. Thus we get the following corollary of Theorem 11.1:

Corollary 11.34. *If K is an imaginary quadratic field, then $K(j(\mathcal{O}_K))$ is the Hilbert class field of K.* Q.E.D.

Besides the Hilbert class field, Theorem 11.1 also allows us to describe other Abelian extensions of K. Recall that in Theorem 9.18 we proved that an Abelian extension of K is generalized dihedral over $\mathbb{Q}$ if and only if it lies in some ring class field of K. Combining this with Theorem 11.1, we get the following theorem:

Corollary 11.35. *Let K be an imaginary quadratic field, and let $K \subset L$ be a finite extension. Then L is an Abelian extension of K which is generalized dihedral over $\mathbb{Q}$ if and only if there is an order $\mathcal{O}$ in K such that $L \subset K(j(\mathcal{O}))$.* Q.E.D.

To complete our discussion of ring class fields and complex multiplication, we need to compute the Artin map of a ring class field using j-invariants. The answer is given by the following theorem:

Theorem 11.36. *Let $\mathcal{O}$ be an order in an imaginary quadratic field K, and let L be the ring class field of $\mathcal{O}$. If $\mathfrak{a}$ is a proper fractional $\mathcal{O}$-ideal and $\mathfrak{p}$ is a prime ideal of $\mathcal{O}_K$, then*

$$\left(\frac{L/K}{\mathfrak{p}}\right)(j(\mathfrak{a})) = j(\overline{\mathfrak{p} \cap \mathcal{O}}\mathfrak{a}).$$

Proof. For analytic proofs, see Deuring [24, §15], Lang [73, Chapter 12, §3] or Cohn [21, §11.2], while algebraic proofs (which use the reduction theory of elliptic curves) may be found in Lang [73, Chapter 10, §3] or Shimura [90, §5.4]. We will use this theorem (in the guise of Corollary 11.37 below) in §12 when we compute some j-invariants, though our discussion of the class equation in §13 will use only Theorem 11.1. Q.E.D.

In terms of the ideal class group, Theorem 11.36 can be stated as follows:

Corollary 11.37. *Let $\mathcal{O}$ be an order in an imaginary quadratic field K, and let L be the ring class field of $\mathcal{O}$. Given proper fractional $\mathcal{O}$-ideals $\mathfrak{a}$ and $\mathfrak{b}$, define $\sigma_{\mathfrak{a}}(j(\mathfrak{b}))$ by*

$$\sigma_{\mathfrak{a}}(j(\mathfrak{b})) = j(\overline{\mathfrak{a}}\mathfrak{b}).$$

Then $\sigma_{\mathfrak{a}}$ *is a well-defined element of* $\mathrm{Gal}(L/K)$, *and* $\mathfrak{a} \mapsto \sigma_{\mathfrak{a}}$ *induces an isomorphism*

$$C(\mathcal{O}) \xrightarrow{\sim} \mathrm{Gal}(L/K).$$

Proof. This is a straightforward consequence of Theorem 11.36 and the isomorphisms

$$C(\mathcal{O}) \simeq I(\mathcal{O},f)/P(\mathcal{O},f) \simeq I_K(f)/P_{K,\mathbf{Z}}(f),$$

where f is the conductor of $\mathcal{O}$. See Exercise 11.22 for the details. Q.E.D.

The "First Main Theorem" of complex multiplication allowed us to describe some of the Abelian extensions of K, namely those which are generalized dihedral over $\mathbf{Q}$. The "Second Main Theorem" of complex multiplication answers the question of how to describe *all* Abelian extensions of K. By class field theory, every Abelian extension lies in a ray class field for some modulus $\mathfrak{m}$ of K, so that we need only find generators for the ray class fields of K. Rather than work with an arbitrary modulus $\mathfrak{m}$, we will describe the ray class fields only for moduli of the form $N\mathcal{O}_K$, where N is a positive integer. It is easy to see that any Abelian extension of K lies in such a ray class field (see Exercise 11.23).

The basic idea is that the ray class field of $N\mathcal{O}_K$ is obtained by adjoining, first, the j-invariant $j(L)$ of some lattice L, and second, some values of the Weierstrass $\wp$-function evaluated at N-division points of the lattice L, i.e., if $L = [\alpha, \beta]$, then we use

$$\wp\left(\frac{m\alpha + n\beta}{N}; L\right) \tag{11.38}$$

for suitable m and n. The observation that (11.38) generates Abelian extensions of K goes back to Abel. The problem is that these values aren't invariant enough: if we multiply the lattice by a constant, the j-invariant remains the same, but the values (11.38) change. To remedy this problem, we introduce a variant of the Weierstrass $\wp$-function called the *Weber function*. Given the lattice L, the Weber function $\tau(z; L)$ is defined by

$$\tau(z; L) = \begin{cases} \dfrac{g_2(L)^2}{\Delta(L)}\wp(z; L)^2 & \text{if } g_3(L) = 0 \\[2ex] \dfrac{g_3(L)}{\Delta(L)}\wp(z; L)^3 & \text{if } g_2(L) = 0 \\[2ex] \dfrac{g_2(L)g_3(L)}{\Delta(L)}\wp(z; L) & \text{otherwise,} \end{cases}$$

where $\Delta(L) = g_2(L)^3 - 27g_3(L)^2$. It is easy to check that $\tau(\lambda z; \lambda L) = \tau(z; L)$ for all $\lambda \in \mathbf{C}^*$ (see Exercise 11.24).

We can now state the "Second Main Theorem" of class field theory, which uses singular j-invariants and the Weber function to generate ray class fields:

Theorem 11.39. *Let K be an imaginary quadratic field of discriminant d_K, and let N be a positive integer.*

(i) *$K(j(\mathcal{O}_K), \tau(1/N; \mathcal{O}_K))$ is the ray class field for the modulus $N\mathcal{O}_K$.*

(ii) *Let $\mathcal{O}$ be the order of conductor N in K. Then $K(j(\mathcal{O}), \tau(w_K; \mathcal{O}))$, where $w_K = (d_K + \sqrt{d_K})/2$, is the ray class field for the modulus $N\mathcal{O}_K$.*

Proof. Notice that in each case we obtain the ray class field by adjoining the j-invariant of a lattice and the Weber function of one N-division point. The proof of (i) may be found in Deuring [24, §26] or Lang [73, §10.3, Corollary to Theorem 7], and the proof of (ii) follows from Satz 1 of Franz [37]. These references also explain how to generate the ray class field of an arbitrary modulus $\mathfrak{m}$ of K. Q.E.D.

The theory of complex multiplication, even in the one variable case described here, is still an active area of research. See, for example, the books *Elliptic Functions and Rings of Integers* [15] by Cassou–Noguès and Taylor and *Arithmetic on Elliptic Curves with Complex Multiplication* [45] by Gross.

E. Exercises

11.1. This exercise will study j-invariants and complex conjugation.

(a) Let L be a lattice, and let $\overline{L}$ denote the lattice obtained by complex conjugation. Prove that $g_2(\overline{L}) = \overline{g_2(L)}$, $g_3(\overline{L}) = \overline{g_3(L)}$ and $j(\overline{L}) = \overline{j(L)}$.

(b) Let $\mathfrak{a}$ be a proper fractional $\mathcal{O}$-ideal, where $\mathcal{O}$ is an order in an imaginary quadratic field. Show that $j(\mathfrak{a})$ is a real number if and only if the class of $\mathfrak{a}$ has order ≤ 2 in the ideal class group $C(\mathcal{O})$. Hint: use (a) and Theorem 10.9.

One consequence of (b) is that $j(\mathcal{O})$ is real for any order $\mathcal{O}$.

11.2. If $\tau \in \mathfrak{h}$ and $\gamma = \left(\begin{smallmatrix} a & b \\ c & d \end{smallmatrix}\right) \in \mathrm{SL}(2, \mathbf{Z})$, then show that

$$\gamma\tau = \frac{a\tau + b}{c\tau + d}$$

also lies in $\mathfrak{h}$. This shows that $\mathrm{SL}(2, \mathbf{Z})$ acts on $\mathfrak{h}$. Hint: use (7.9).

11.3. Let τ satisfy $|\mathrm{Re}(\tau)| \leq 1/2$ and $|\mathrm{Im}(\tau)| \geq \epsilon$, where $\epsilon < 1$ is fixed. Our goal is to show that for $x, y \in \mathbf{R}$,

$$|x + y\tau| \geq \frac{\epsilon}{2}\sqrt{x^2 + y^2}.$$

If we let $\tau = a + bi$, then the above is equivalent to

$$(x+ay)^2 + b^2y^2 \geq \frac{\epsilon^2}{4}(x^2+y^2).$$

(a) Show that the inequality is true when $|x+ay| \geq (\epsilon/2)|x|$.

(b) When $|x+ay| < (\epsilon/2)|x|$, use $|a| \leq 1/2$ and $\epsilon < 1$ to show that $|x| < |y|$.

(c) Using (b), show that the inequality is true when $|x+ay| < (\epsilon/2)|x|$.

11.4. In Lemma 11.4 we showed that every point of $\mathfrak{h}$ is $\mathrm{SL}(2,\mathbf{Z})$-equivalent to a point in the region $\{\tau \in \mathfrak{h} : |\mathrm{Re}(\tau)| \leq 1/2, |\mathrm{Im}(\tau)| \geq 1/2\}$. In this exercise we will study the smaller region

$$\begin{aligned} F = \{\tau \in \mathfrak{h} : |\mathrm{Re}(\tau)| \leq 1/2,\ |\tau| \geq 1, \text{ and} \\ \mathrm{Re}(\tau) \geq 0 \text{ if } |\mathrm{Re}(\tau)| = 1/2 \text{ or } |\tau| = 1\}, \end{aligned}$$

and we will show that every point of $\mathfrak{h}$ is $\mathrm{SL}(2,\mathbf{Z})$-equivalent to a *unique* point of F. This is usually expressed by saying that F is a *fundamental domain* for the action of $\mathrm{SL}(2,\mathbf{Z})$ on $\mathfrak{h}$. Our basic tool will be positive definite quadratic forms $f(x,y) = ax^2 + bxy + cy^2$, where we allow a, b and c to be real numbers. We say that two such forms $f(x,y)$ and $g(x,y)$ are $\mathbf{R}^+$-equivalent if there is $\left(\begin{smallmatrix} p & q \\ r & s \end{smallmatrix}\right) \in \mathrm{SL}(2,\mathbf{Z})$ such that

$$f(x,y) = \lambda g(px+qy, rx+sy)$$

for some $\lambda > 0$ in $\mathbf{R}$. Finally, we say that $f(x,y) = ax^2 + bxy + cy^2$ is reduced if

$$a \leq |b| \leq c, \text{ and } b \geq 0 \text{ if } a = |b| \text{ or } |b| = c.$$

This is consistent with the definition given in §2.

(a) Show that $\mathbf{R}^+$-equivalence of positive definite forms is an equivalence relation.

(b) Show that every positive definite form is $\mathbf{R}^+$-equivalent to a reduced form, and that two reduced forms are $\mathbf{R}^+$-equivalent if and only if one is a constant multiple of the other. Hint: see the proof of Theorem 2.8.

(c) Show that every positive definite form $f(x,y) = ax^2 + bxy + cy^2$ can be written uniquely as $f(x,y) = a|x - \tau y|^2$, where $\tau \in \mathfrak{h}$. In this case we say that τ is the *root* of $f(x,y)$ (this is consistent with the terminology used in §7). Furthermore, show that $b = 2a\mathrm{Re}(\tau)$ and $c = a|\tau|$.

(d) Show that two positive definite forms are $\mathbf{R}^+$-equivalent if and only if their roots are $SL(2,\mathbf{Z})$-equivalent. Hint: see the proof of (7.8).

(e) Show that a positive definite form is reduced if and only if its root lies in the fundamental domain F.

(f) Conclude that every $\tau \in \mathfrak{h}$ is $SL(2,\mathbf{Z})$-equivalent to a unique point of F.

This exercise shows that there is a remarkable relation between reduced forms and fundamental domains. Similar considerations led Gauss (unpublished, of course) to discover the idea of a fundamental domain in the early 1800s. See Cox [23] for more details.

11.5. In this exercise we will prove Lemma 11.5 and Corollary 11.6.

(a) Let M and ϵ be positive constants, and define $K \subset \mathfrak{h}$ by

$$K = \{\tau \in \mathfrak{h} : |\mathrm{Re}(\tau)| \le M,\ \epsilon \le |\mathrm{Im}(\tau)| \le 1/\epsilon\}.$$

We want to show that the set $\Delta(K) = \{\gamma \in SL(2,\mathbf{Z}) : \gamma(K) \cap K \neq \emptyset\}$ is finite. So take $\gamma = \left(\begin{smallmatrix} a & b \\ c & d \end{smallmatrix}\right) \in \Delta(K)$, which means that there is $\tau \in K$ such that $\gamma\tau \in K$. If we can bound $|a|$, $|b|$, $|c|$ and $|d|$ in terms of M and ϵ, then finiteness will follow.

(i) Use (7.9) to show that $|c\tau + d| \le 1/\epsilon$.

(ii) Since $|c\tau + d|^2 = (c\mathrm{Re}(\tau) + d)^2 + c^2\mathrm{Im}(\tau)^2$, conclude that $|c| \le 1/\epsilon^2$ and $|d| \le (\epsilon + M)/\epsilon^2$.

(iii) Show that $\gamma^{-1} \in \Delta(K)$. By (ii), this implies that $|a| \le (\epsilon + M)/\epsilon^2$.

(iv) Show that $|b| \le |c\tau + d||\gamma\tau| + |a||\tau|$, and conclude that $|b|$ is bounded in terms of M and ϵ.

(b) Use (a) to show that if U is a neighborhood of $\tau \in \mathfrak{h}$ such that $\overline{U} \subset \mathfrak{h}$ is compact, then $\{\gamma \in SL(2,\mathbf{Z}) : \gamma(U) \cap U \neq \emptyset\}$ is finite. This will prove Lemma 11.5.

(c) Prove Corollary 11.6.

11.6. This exercise is concerned with the proof of part (iv) of Theorem 11.2.

(a) Suppose that $\gamma = \left(\begin{smallmatrix} a & b \\ c & d \end{smallmatrix}\right) \in SL(2,\mathbf{Z})$ and that $\gamma\tau = \tau$ for some $\tau \in \mathfrak{h}$. We saw in the text that this implies $[1,\tau] = (c\tau + d)[1,\tau]$. Prove that $c \neq 0$. Hint: show that $c = 0$ implies $\gamma = \pm\left(\begin{smallmatrix} 1 & m \\ 0 & 1 \end{smallmatrix}\right)$. But such a γ has no fixed points on $\mathfrak{h}$.

(b) Let $\mathcal{O}$ be an order in an imaginary quadratic field such that $\mathcal{O}^* \neq \{\pm 1\}$. Prove that $\mathcal{O} = \mathcal{O}_K$ for $K = \mathbf{Q}(i)$ or $\mathbf{Q}(\omega)$, $\omega = e^{2\pi i/3}$. Hint: when $\mathcal{O} = \mathcal{O}_K$, see Exercise 5.9. See also Lemma 7.2.

(c) Show that the only elements of $\mathrm{SL}(2,\mathbf{Z})$ fixing i are $\pm\left(\begin{smallmatrix}1&0\\0&1\end{smallmatrix}\right)$ and $\pm\left(\begin{smallmatrix}0&1\\-1&0\end{smallmatrix}\right)$. Hint: use (a).

(d) If $\omega = e^{2\pi i/3}$, show that $j'(\omega) = j''(\omega) = 0$ but $j'''(\omega) \neq 0$.

11.7. Let $f(\tau)$ be a modular function for $\mathrm{SL}(2,\mathbf{Z})$, and assume that $f(\tau)$ has a pole of order m at $\tau = \omega$, $\omega = e^{2\pi i/3}$.

(a) Prove that m is divisible by 3. Hint: argue as in the case when $f(\tau)$ has a pole at $\tau = i$. Note that ω is fixed by $\left(\begin{smallmatrix}1&1\\-1&0\end{smallmatrix}\right)$.

(b) Prove that $j(\tau)^{m/3}f(\tau)$ is holomorphic at ω. Hint: use part (iv) of Theorem 11.2.

11.8. As in the proof of Theorem 11.9, let

$$\Gamma_0(m) = \left\{\begin{pmatrix} a & b \\ c & d \end{pmatrix} \in \mathrm{SL}(2,\mathbf{Z}) : c \equiv 0 \bmod m\right\}$$

$$C(m) = \left\{\begin{pmatrix} a & b \\ c & d \end{pmatrix} : ad = m,\ a > 0,\ 0 \le b < d,\ \gcd(a,b,d) = 1\right\},$$

and let $\sigma_0 = \left(\begin{smallmatrix}m&0\\0&1\end{smallmatrix}\right) \in C(m)$.

(a) Show that $\Gamma_0(m) = (\sigma_0^{-1}\mathrm{SL}(2,\mathbf{Z})\sigma_0) \cap \mathrm{SL}(2,\mathbf{Z})$.

(b) If $\sigma \in C(m)$, then show that $(\sigma_0^{-1}\mathrm{SL}(2,\mathbf{Z})\sigma) \cap \mathrm{SL}(2,\mathbf{Z})$ is a coset of $\Gamma_0(m)$ in $\mathrm{SL}(2,\mathbf{Z})$.

(c) In the construction of part (b), show that different σ's give different cosets, and that *all* cosets of $\Gamma_0(m)$ in $\mathrm{SL}(2,\mathbf{Z})$ arise in this way.

11.9. Let m be a positive integer, and let $f(m)$ denote the number of triples (a,b,d) of integers which satisfy $ad = m$, $a > 0$, $0 \le b < d$ and $\gcd(a,b,d) = 1$. Thus $f(m) = |C(m)|$, where $C(m)$ is the set of matrices defined in the the previous exercise. The goal of this exercise is to prove that

$$f(m) = m\prod_{p \mid m}\left(1 + \frac{1}{p}\right).$$

(a) If we fix a positive divisor d of m, then $a = m/d$ is determined. Show that the number of possible b's for this d is given by

$$\frac{d}{\gcd(d, m/d)}\phi(\gcd(d, m/d)),$$

where ϕ denotes the Euler ϕ-function.

(b) Use the formula of (a) to prove that $f(m)$ is multiplicative, i.e., that if m_1 and m_2 are relatively prime, then $f(m_1m_2) = f(m_1)f(m_2)$.

(c) Use the formula of (a) to prove that if p is a prime, then

$$f(p^r) = p^r + p^{r-1}.$$

(d) Use (b) and (c) to prove the desired formula for $f(m)$.

11.10. In this exercise we will show that $\Phi_m(X,Y)$ is irreducible as a polynomial in X (which will prove part (ii) of Theorem 11.18). Let γ_i be coset representatives for $\Gamma_0(m)$ in $\mathrm{SL}(2,\mathbf{Z})$. As we saw in (11.14), we can write

$$\Phi_m(X,j(\tau)) = \prod_{i=1}^{\Psi(m)} (X - j(m\gamma_i\tau)).$$

Let $\mathcal{F}_m$ be the field $\mathbf{C}(j(\tau), j(m\tau))$. Since $\Phi_m(X,j(\tau))$ has coefficients in $\mathbf{C}(j(\tau))$ and $j(m\tau)$ is a root, it follows that $[\mathcal{F}_m : \mathbf{C}(j(\tau))] \leq \Psi(m)$. If we can prove equality, then $\Phi_m(X,j(\tau))$ will be the minimal polynomial of $j(m\tau)$ over $\mathbf{C}(j(\tau))$, and irreducibility will follow.

(a) Let $\mathcal{F}$ be the field of all meromorphic functions on $\mathfrak{h}$, which contains $\mathcal{F}_m$ as a subfield. For $\gamma \in \mathrm{SL}(2,\mathbf{Z})$, show that $f(\tau) \mapsto f(\gamma\tau)$ is an embedding of $\mathcal{F}_m$ into $\mathcal{F}$ which is the identity on $\mathbf{C}(j(\tau))$.

(b) Use (11.13) to show that $j(m\gamma_i\tau) \neq j(m\gamma_j\tau)$ for $i \neq j$. The embeddings constructed in (a) are thus distinct, which shows that $[\mathcal{F}_m : \mathbf{C}(j(\tau))] \geq \Psi(m)$. This proves the desired equality.

11.11. Show that the coefficients of $G(X,\tau)$ (as defined in (11.16)) are modular functions for $\mathrm{SL}(2,\mathbf{Z})$. Hint: argue as in the case of the modular function. You will use the fact that $f(\gamma_i\tau)$ has a meromorphic q-expansion.

11.12. Let $A \subset \mathbf{C}$ be an additive subgroup, and let $f(\tau)$ be a holomorphic modular function. Suppose that its q-expansion is

$$f(\tau) = \sum_{n=-M}^{\infty} a_n q^n,$$

and that $a_n \in A$ for all $n \leq 0$. Then prove the Hasse q-expansion principle, which states that $f(\tau)$ is a polynomial in $j(\tau)$ with coefficients in A. Hint: since the q-expansion of $j(\tau)$ has integer coefficients and begins with $1/q$, the polynomial $A(x)$ used in part (ii) of Lemma 11.10 must have coefficients in A.

11.13. Let p be a prime, and let $\zeta_p = e^{2\pi i/p}$.

(a) Prove that $p = (1-\zeta_p)(1-\zeta_p^2)\cdots(1-\zeta_p^{p-1})$. Hint: factor $x^{p-1} + \cdots + x + 1$.

(b) Given $\alpha \in \mathbf{Z}[\zeta_p]$, define the norm $N_{\mathbf{Q}(\zeta_p)/\mathbf{Q}}(\alpha)$ to be the number

$$N_{\mathbf{Q}(\zeta_p)/\mathbf{Q}}(\alpha) = \prod_{\sigma \in \mathrm{Gal}(\mathbf{Q}(\zeta_p)/\mathbf{Q})} \sigma(\alpha).$$

For simplicity, we will write $N(\alpha)$ instead of $N_{\mathbf{Q}(\zeta_p)/\mathbf{Q}}(\alpha)$. Then prove that $N(\alpha)$ is an integer, and show that $N(\alpha\beta) = N(\alpha)N(\beta)$ and $N(1-\zeta_p) = p$.

(c) If an integer a can be written $a = (1-\zeta_p)\alpha$ where $\alpha \in \mathbf{Z}[\zeta_p]$, then use (b) to prove that a is divisible by p.

11.14. Adapt the proof of (11.21) to show that $j(\tau) \equiv j(\sigma_0\tau)^p \bmod p$.

11.15. Let $f(X,Y) \in \mathbf{Z}[X,Y]$ be a polynomial such that $f(X, j(\tau)) \in p\mathbf{Z}((q))[X]$. Prove that $f(X,Y) \in p\mathbf{Z}[X,Y]$. Hint: apply the q-expansion principle (Exercise 11.12) to the coefficients of X.

11.16. Let $M = \mathbf{Z}^2$, and let A be a 2×2 integer matrix with $\det(A) \neq 0$. We know by Exercise 7.15 that M/AM is a finite group of order $|\det(A)|$. The object of this exercise is to prove that M/AM is cyclic if and only if the entries of A are relatively prime.

(a) Let G be a finite Abelian group. Prove that G is not cyclic if and only if G contains a subgroup isomorphic to $(\mathbf{Z}/d\mathbf{Z})^2$ for some integer $d > 1$. Hint: use the structure theorem for finite Abelian groups.

(b) Assume that the entries of A have a common divisor $d > 1$, and prove that M/AM is not cyclic. Hint: write $A = dA'$, where A' is an integer matrix, and note that $A'M/dA'M \subset M/AM$. Then use (a).

(c) Finally, assume that M/AM is not cyclic, and prove that that the entries of A have a common divisor $d > 1$. Hint: by (a), there is $AM \subset M' \subset M$ such that $M'/AM \simeq (\mathbf{Z}/d\mathbf{Z})^2$ for some $d > 1$. Prove that $AM = dM'$, and conclude that d divides the entries of A.

11.17. This exercise is concerned with the proof of Lemma 11.24.

(a) Let L' be a sublattice of $[1,\tau]$ of finite index, and let d be the smallest positive integer in L'. Then prove that $L' = [d, a\tau + b]$ for some integers a and b.

(b) Let $\tau \in \mathfrak{h}$, and let $C(m)$ be the set of matrices defined in the text. If $\sigma, \sigma' \in C(m)$ and $d[1,\sigma\tau] = d'[1,\sigma'\tau]$, then prove that $\sigma = \sigma'$.

11.18. In the text, we proved that for $\tau \in \mathfrak{h}$, the roots of $\Phi_m(X, j(\tau)) = 0$ are the j-invariants of the cyclic sublattices of index m of $[1, \tau]$. Use this fact and the surjectivity of the j-function to prove Theorem 11.23.

11.19. Let $\mathcal{O}$ be an order, and let $\mathfrak{b}$ be a proper fractional $\mathcal{O}$-ideal. If $\mathfrak{a}$ is a proper $\mathcal{O}$-ideal which is not primitive, then prove that $\mathfrak{b}/\mathfrak{a}\mathfrak{b}$ is not cyclic. Hint: use part (a) of Exercise 11.16.

11.20. Let $\mathcal{O}$ be an order in an imaginary quadratic field K of conductor f. Letting $w_K = (d_K + \sqrt{d_K})/2$, we proved in Lemma 7.2 that $\mathcal{O} = [1, f w_K]$. Prove that $\alpha = f w_K$ is a primitive element of $\mathcal{O}$ whose norm is not a perfect square.

11.21. Let $K \subset L$ be an extension of number fields, and let $\alpha \in \mathcal{O}_L$ satisfy $L = K(\alpha)$.

(a) Prove that $\mathcal{O}_K[\alpha]$ has finite index in $\mathcal{O}_L$. Hint: By Theorem 5.3, we know that $\mathcal{O}_L$ is a free $\mathbb{Z}$-module of rank $[L : \mathbb{Q}]$. Then show that $\mathcal{O}_K[\alpha]$ has the same rank.

(b) Let $\mathfrak{P}$ be a prime ideal of $\mathcal{O}_L$, and suppose that $N(\mathfrak{P}) = p^f$, where p is relatively prime to $[\mathcal{O}_L : \mathcal{O}_K[\alpha]]$. If $\beta^p \equiv \beta \bmod \mathfrak{P}$ holds for all $\beta \in \mathcal{O}_K[\alpha]$, then show that the same congruence holds for all $\beta \in \mathcal{O}_L$. Hint: if $N = [\mathcal{O}_L : \mathcal{O}_K[\alpha]]$, then multiplication by N induces an isomorphism of $\mathcal{O}_L/\mathfrak{P}$.

11.22. Complete the proof of Corollary 11.37.

11.23. Let K be an imaginary quadratic field, and let L be an Abelian extension of K. Prove that there is a positive integer N such that L is contained in the ray class field for the modulus $N\mathcal{O}_K$.

11.24. If L is a lattice and $\tau(z; L)$ is the Weber function defined in the text, then prove that $\tau(\lambda z; \lambda L) = \tau(z; L)$ for any $\lambda \in \mathbb{C}^*$.

§12. MODULAR FUNCTIONS AND SINGULAR j-INVARIANTS

The j-invariant $j(L)$ of a lattice with complex multiplication is often called a *singular j-invariant* or a *singular modulus*. In §11 we learned about the fields generated by singular moduli, and in this section we will compute some of these remarkable numbers. One of our main tools will be the function $\gamma_2(\tau)$, which is defined by

$$\gamma_2(\tau) = \sqrt[3]{j(\tau)}.$$

We will show that $\gamma_2(3\tau)$ is a modular function for $\Gamma_0(9)$, and we will use $\gamma_2(\tau)$ to generate ring class fields for orders of discriminant not divisible by 3. This will explain why the j-invariants computed in §10 were perfect cubes.

We will then give a modern treatment of some of the results contained in Volume III of Weber's monumental *Lehrbuch der Algebra* [102]. There is a wealth of material in this book, far more than we could ever cover here. We will concentrate on some applications of the Dedekind η-function $\eta(\tau)$ and the three Weber functions $\mathfrak{f}(\tau)$, $\mathfrak{f}_1(\tau)$ and $\mathfrak{f}_2(\tau)$. These functions are closely related to $\gamma_2(\tau)$ and $j(\tau)$ and make it easy to compute the j-invariants of most orders of class number 1. The Weber functions also give some interesting modular functions, which will enable us to compute that

(12.1)
$$j(\sqrt{-14}) = 2^3\left(323 + 228\sqrt{2} + (231 + 161\sqrt{2})\sqrt{2\sqrt{2}-1}\right)^3.$$

At the end of the section, we will present Heegner's proof of the Baker–Heegner–Stark theorem on imaginary quadratic fields of class number 1.

A. The Cube Root of the j-Function

Our first task is to study the cube root $\gamma_2(\tau)$ of the j-function. Recall from §11 that $j(\tau)$ can be written as the quotient

$$j(\tau) = 1728\frac{g_2(\tau)^3}{\Delta(\tau)}.$$

The function $\Delta(\tau)$ is nonvanishing and holomorphic on the simply connected domain $\mathfrak{h}$, and hence has a holomorphic cube root $\sqrt[3]{\Delta(\tau)}$. Since $\Delta(\tau)$ is real-valued on the imaginary axis (see Exercise 12.1), we can choose $\sqrt[3]{\Delta(\tau)}$ with the same property. Using this cube root, we define

$$\gamma_2(\tau) = 12\frac{g_2(\tau)}{\sqrt[3]{\Delta(\tau)}}.$$

Since $g_2(\tau)$ is also real on the imaginary axis (see Exercise 12.1), it follows that $\gamma_2(\tau)$ is the unique cube root of $j(\tau)$ which is real-valued on the imaginary axis.

For us, the main property of $\gamma_2(\tau)$ is that it can be used to generate all ring class fields of orders of discriminant not divisible by 3. Note that τ needs to be chosen carefully, for replacing τ by $\tau+1$ doesn't affect $j(\tau)$, but we will see below that $\gamma_2(\tau+1) = \zeta_3^{-1}\gamma_2(\tau)$, where $\zeta_3 = e^{2\pi i/3}$. The necessity to normalize τ leads to the following theorem:

Theorem 12.2. *Let $\mathcal{O}$ be an order of disciminant D in an imaginary qua-*

dratic field K. Assume that $3 \nmid D$, and write $\mathcal{O} = [1, \tau_0]$, where

$$\tau_0 = \begin{cases} \sqrt{-m}, & D = -4m \equiv 0 \bmod 4 \\ \dfrac{3+\sqrt{-m}}{2}, & D = -m \equiv 1 \bmod 4. \end{cases}$$

Then $\gamma_2(\tau_0)$ is an algebraic integer and $K(\gamma_2(\tau_0))$ is the ring class field of $\mathcal{O}$. Furthermore, $\mathbb{Q}(\gamma_2(\tau_0)) = \mathbb{Q}(j(\tau_0))$.

Let's first see how this theorem relates to the j-invariants computed in §10. When $\mathcal{O}$ has class number one, we know that $j(\mathcal{O})$ is an integer, so that by Theorem 12.2, $\gamma_2(\tau)$ is also an integer when $3 \nmid D$. This explains why

$$\begin{aligned} j(i) &= 12^3 \\ j(\sqrt{-2}) &= 20^3 \\ j\left(\frac{1+\sqrt{-7}}{2}\right) &= -15^3 \end{aligned}$$

are all perfect cubes. (In the last case, note that $j((1+\sqrt{-7})/2) = j((3+\sqrt{-7})/2)$, so that Theorem 12.2 does apply.)

Proof of Theorem 12.2. By the Theorem 11.1, we know that $K(j(\tau_0))$ is the ring class field of of $\mathcal{O} = [1, \tau_0]$. Thus, to prove Theorem 12.2, it suffices to prove that

$$\mathbb{Q}(\gamma_2(\tau_0)) = \mathbb{Q}(j(\tau_0))$$

whenever $3 \nmid D$. The first proof of this theorem was due to Weber [102, §125], and modern proofs have been given by Birch [7] and Schertz [87]. Our presentation is based on [87].

The first step of the proof is to show that $\gamma_2(3\tau)$ is a modular function.

Proposition 12.3. *$\gamma_2(3\tau)$ is a modular function for the group $\Gamma_0(9)$.*

Proof. We first study how $\gamma_2(\tau)$ transforms under elements of $\mathrm{SL}(2, \mathbb{Z})$. We claim that

$$\begin{aligned} \gamma_2(-1/\tau) &= \gamma_2(\tau) \\ \gamma_2(\tau+1) &= \zeta_3^{-1}\gamma_2(\tau), \end{aligned} \tag{12.4}$$

where $\zeta_3 = e^{2\pi i/3}$. The first line of (12.4) is easy to prove, for $\gamma_2(-1/\tau)$ is a cube root of $j(-1/\tau) = j(\tau)$. But $-1/\tau$ lies on the imaginary axis whenever τ does, so that $\gamma_2(-1/\tau)$ is a cube root of $j(\tau)$ which is real on the imaginary axis. By the definition of $\gamma_2(\tau)$, this implies $\gamma_2(-1/\tau) = \gamma_2(\tau)$.

To prove the second line of (12.4), consider the q-expansion of $\gamma_2(\tau)$. We know that

$$j(\tau) = q^{-1} + \sum_{n=1}^{\infty} c_n q^n = q^{-1}h(q),$$

where $h(q)$ is holomorphic for $|q| < 1$ and $h(0) = 1$. We can therefore write $h(q) = u(q)^3$, where $u(q)$ is holomorphic and $u(0) = 1$. Note also that $u(q)$ has rational coefficients since $h(q)$ does (see Exercise 12.2). Then $q^{-1/3}u(q)$ is a cube root of $j(\tau)$ which is real-valued on the imaginary axis, and it follows that

$$\gamma_2(\tau) = q^{-1/3}u(q) = q^{-1/3}\Big(1 + \sum_{n=1}^{\infty} b_n q^n\Big), \qquad b_n \in \mathbf{Q}. \tag{12.5}$$

It is now trivial to see that $\gamma_2(\tau+1) = \zeta_3^{-1}\gamma_2(\tau)$, and (12.4) is proved.

We next claim that if $\left(\begin{smallmatrix} a & b \\ c & d \end{smallmatrix}\right) \in \mathrm{SL}(2,\mathbf{Z})$, then

$$\gamma_2\left(\frac{a\tau + b}{c\tau + d}\right) = \zeta_3^{ac-ab+a^2cd-cd}\gamma_2(\tau). \tag{12.6}$$

To see this, first note that (12.6) holds for $S = \left(\begin{smallmatrix} 0 & -1 \\ 1 & 0 \end{smallmatrix}\right)$ and $T = \left(\begin{smallmatrix} 1 & 1 \\ 0 & 1 \end{smallmatrix}\right)$ by (12.4). It is well-known that these two matrices generate $\mathrm{SL}(2,\mathbf{Z})$ (see Serre [88, §VII.1] or Exercise 12.3). Then (12.6) follows by induction on the length of $\left(\begin{smallmatrix} a & b \\ c & d \end{smallmatrix}\right)$ as a word in S and T (see Exercise 12.5).

Given (12.6), it follows easily that $\gamma_2(\tau)$ is invariant under the group of matrices

$$\tilde{\Gamma}(3) = \left\{\begin{pmatrix} a & b \\ c & d \end{pmatrix} : b \equiv c \equiv 0 \bmod 3\right\}.$$

This group is related to $\Gamma_0(9)$ by the identity

$$\Gamma_0(9) = \begin{pmatrix} 1/3 & 0 \\ 0 & 1 \end{pmatrix} \tilde{\Gamma}(3) \begin{pmatrix} 3 & 0 \\ 0 & 1 \end{pmatrix},$$

and a simple computation then shows that $\gamma_2(3\tau)$ is invariant under $\Gamma_0(9)$ (see Exercise 12.5). The group $\tilde{\Gamma}(3)$ is not the largest subgroup of $\mathrm{SL}(2,\mathbf{Z})$ fixing $\gamma_2(\tau)$, but it's the one that relates most easily to the $\Gamma_0(m)$'s (see Exercise 12.5).

To finish the proof that $\gamma_2(3\tau)$ is a modular function for $\Gamma_0(9)$, we need to check its behavior at the cusps. Let $\gamma \in \mathrm{SL}(2,\mathbf{Z})$. By Theorem 11.9, $j(3\tau)$ is a modular function for $\Gamma_0(3)$, so that $j(3\gamma\tau)$ has a meromorphic expansion in powers of $q^{1/3}$. Taking cube roots, this implies that $\gamma_2(3\gamma\tau)$ has a meromorphic expansion in powers of $q^{1/9}$, which proves that $\gamma_2(3\tau)$ is meromorphic at the cusps. This proves the proposition. Q.E.D.

Once we know that $\gamma_2(3\tau)$ is a modular function for $\Gamma_0(9)$, Theorem 11.9 tells us that it is a rational function in $j(\tau)$ and $j(9\tau)$. The following proposition will give us information about the coefficients of this rational function:

Proposition 12.7. *Let $f(\tau)$ be a modular function for $\Gamma_0(m)$.*

(i) *If the q-expansion of $f(\tau)$ has rational coefficients, then $f(\tau) \in \mathbb{Q}(j(\tau), j(m\tau))$.*

(ii) *Assume in addition $f(\tau)$ is holomorphic on $\mathfrak{h}$, and let $\tau_0 \in \mathfrak{h}$. If*

$$\frac{\partial \Phi_m}{\partial X}(j(m\tau_0), j(\tau_0)) \neq 0,$$

then $f(\tau_0) \in \mathbb{Q}(j(\tau_0), j(m\tau_0))$.

Remark. Note that the hypothesis of (i) involves only the expansion of $f(\tau)$ in powers of $q^{1/m}$. For general $\gamma \in \mathrm{SL}(2, \mathbb{Z})$, the expansion of $f(\gamma\tau)$ need not have coefficients in $\mathbb{Q}$.

Proof. To prove (i), we will use the representation

$$f(\tau) = \frac{G(j(m\tau), j(\tau))}{\frac{\partial}{\partial X}\Phi_m(j(m\tau), j(\tau))} \tag{12.8}$$

given by (11.17). Since the denominator clearly lies in $\mathbb{Q}(j(\tau), j(m\tau))$ (part (i) of Theorem 11.18), it suffices to show that the same holds for the numerator. We know that $G(j(m\tau), j(\tau))$ lies in $\mathbb{C}(j(\tau))[j(m\tau)]$, so that

$$G(j(m\tau), j(\tau)) = \frac{P(j(m\tau), j(\tau))}{Q(j(\tau))},$$

where $P(X,Y)$ and $Q(Y) \neq 0$ are polynomials with complex coefficients. Let's write these polynomials as

$$P(X,Y) = \sum_{i=1}^{N}\sum_{k=1}^{M} a_{ik}X^iY^k$$

$$Q(Y) = = \sum_{l=1}^{L} b_l Y^l.$$

Then (12.8) implies that

$$P(j(m\tau), j(\tau)) = f(\tau)\frac{\partial \Phi_m}{\partial X}(j(m\tau), j(\tau))Q(j(\tau)),$$

which we can write as

$$\sum_{i=1}^{N}\sum_{k=1}^{M} a_{ik} j(m\tau)^i j(\tau)^k = f(\tau)\frac{\partial \Phi_m}{\partial X}(j(m\tau), j(\tau))\left(\sum_{l=1}^{L} b_l j(\tau)^l\right).$$

Substituting in the q-expansions of $f(\tau)$, $j(\tau)$ and $j(m\tau)$ and equating coefficients of powers of $q^{1/m}$, we get an infinite system of homogeneous linear equations with the a_{ik}'s and b_l's as unknowns. The q-expansions of $f(\tau)$, $j(\tau)$ and $j(m\tau)$ all have coefficients in $\mathbb{Q}$, and the coefficients of $(\partial/\partial X)\Phi_m(X,Y)$ are also rational. Thus the coefficients of our system of equations all lie in $\mathbb{Q}$. This system has a solution over $\mathbb{C}$ which is nontrivial in the b_l's (since $Q(j(\tau)) \neq 0$), and hence must have a solution over $\mathbb{Q}$ also nontrivial in the b_l's. This proves that $P(X,Y)$ and $Q(Y) \neq 0$ can be chosen to have rational coefficients, which proves part (i).

To prove (ii), let's go back to the definition of $G(X, j(\tau))$ given in (11.16). Since $f(\tau)$ is holomorphic on $\mathfrak{h}$, the coefficients of $G(X, j(\tau))$ are also holomorphic on $\mathfrak{h}$. As we saw in Lemma 11.10, this means that the coefficients are polynomials in $j(\tau)$. Thus, in the representation of $f(\tau)$ given by (12.8), the numerator $G(j(m\tau), j(\tau))$ is a polynomial in $j(m\tau)$ and $j(\tau)$. By a slight modification of the argument for part (i), we can assume that it has rational coefficients (see Exercise 12.6). Consequently, whenever the denominator doesn't vanish at τ_0, we can evaluate this expression at $\tau = \tau_0$ to conclude that $f(\tau_0)$ lies in $\mathbb{Q}(j(\tau_0), j(m\tau_0))$. Q.E.D.

We want to apply this proposition to $\gamma_2(\tau_0)$, where τ_0 is given in the statement of Theorem 12.2. By (12.5), we see that the q-expansion of $\gamma_2(3\tau)$ has rational coefficients. Since it is a modular function for $\Gamma_0(9)$, Proposition 12.7 tells us that

$$\gamma_2(3\tau) \in \mathbb{Q}(j(\tau), j(9\tau)).$$

Since we're concerned about $\gamma_2(\tau_0)$, we need to evaluate the above expression at $\tau = \tau_0/3$. We will for the moment assume that

$$\frac{\partial \Phi_9}{\partial X}(j(3\tau_0), j(\tau_0/3)) \neq 0. \tag{12.9}$$

Since $\gamma_2(3\tau)$ is holomorphic, the second part of Proposition 12.7 then implies that $\gamma_2(\tau_0) \in \mathbb{Q}(j(\tau_0/3), j(3\tau_0))$, which we can write as

$$\gamma_2(\tau_0) \in \mathbb{Q}(j([1,\tau_0/3]), j([1,3\tau_0])). \tag{12.10}$$

To see what this says about $\gamma_2(\tau_0)$, recall that $\mathcal{O} = [1,\tau_0]$. Then $\mathcal{O}' = [1,3\tau_0]$ is the order of index 3 in $\mathcal{O}$, and the special form of τ_0 implies that $[1,\tau_0/3]$ is a proper fractional $\mathcal{O}'$-ideal (this follows from Lemma 7.5 and $3 \nmid D$—see Exercise 12.7). Thus, by Theorem 11.1, both $j(\tau_0/3)$ and $j(3\tau_0)$ generate the ring class field L' of the order $\mathcal{O}'$. Consequently, (12.10) implies that $\gamma_2(\tau_0)$ lies in the ring class field L'.

Let L denote the ring class field of $\mathcal{O}$, so that $L \subset L'$. To compute the degree of this extension, recall that the class number is the degree of the ring class field over K. Since the discriminant of $\mathcal{O}$ is D, this means that $[L' : L] = h(9D)/h(D)$. Corollary 7.28 implies that

$$h(9D) = \frac{3h(D)}{[\mathcal{O}^* : \mathcal{O}'^*]}\left(1 - \left(\frac{D}{3}\right)\frac{1}{3}\right),$$

and since $3 \nmid D$, we see that $L \subset L'$ is an extension of degree 2 or 4.

Now consider the following diagram of fields:

$$\begin{array}{ccc} \mathbb{Q}(j(\tau_0)) & \subset & L \\ \cap & & \cap \\ \mathbb{Q}(\gamma_2(\tau_0)) & \subset & L' \end{array}.$$

We know that L has degree 2 over $\mathbb{Q}(j(\tau_0))$, and by the above computation, L' has degree 2 or 4 over L. It follows that the degree of $\mathbb{Q}(\gamma_2(\tau_0))$ over $\mathbb{Q}(j(\tau_0))$ is a power of 2. But recall that $\gamma_2(\tau_0)$ is the real cube root of $j(\tau_0)$, which means that the extension $\mathbb{Q}(j(\tau_0)) \subset \mathbb{Q}(\gamma_2(\tau_0))$ has degree 1 or 3. Hence this degree must be 1, which proves that $\mathbb{Q}(j(\tau_0)) = \mathbb{Q}(\gamma_2(\tau_0))$.

We are not quite done with the theorem, for we still have to verify that (12.9) is satisfied, i.e., that

$$\frac{\partial \Phi_9}{\partial X}(j(3\tau_0), j(\tau_0/3)) \neq 0.$$

For later purposes, we will prove the following general lemma:

Lemma 12.11. *Let $\mathcal{O}$ be an order in an imaginary quadratic field, and assume that $\mathcal{O}^* = \{\pm 1\}$. Write $\mathcal{O} = [1, \alpha]$, and assume that for some integer s, $s \mid T(\alpha)$ and $\gcd(s^2, N(\alpha))$ is squarefree, where $T(\alpha)$ and $N(\alpha)$ are the trace and norm of α. Then for any positive integer m,*

$$\frac{\partial \Phi_m}{\partial X}(j(m\alpha/s), j(\alpha/s)) \neq 0.$$

Proof. Since $\Phi_m(j(m\alpha/s), j(\alpha/s)) = 0$, the nonvanishing of the partial derivative means that $j(m\alpha/s)$ is not a multiple root of the polynomial

$$\Phi_m(X, j(\alpha/s)) = \prod_{\sigma \in C(m)} (X - j(\sigma\alpha/s)).$$

Thus we must show that

$$j(m\alpha/s) \neq j(\sigma\alpha/s), \qquad \sigma \in C(m), \quad \sigma \neq \sigma_0 = \begin{pmatrix} m & 0 \\ 0 & 1 \end{pmatrix}.$$

So pick $\sigma = \left(\begin{smallmatrix} a & b \\ 0 & d \end{smallmatrix}\right) \in C(m)$, $\sigma \neq \sigma_0$, and assume that $j(m\alpha/s) = j(\sigma\alpha/s)$. In terms of lattices, this means that there is a complex number λ such that

$$\lambda[1, m\alpha/s] = [d, a\alpha/s + b]. \tag{12.12}$$

We will show that this leads to a contradiction when $\mathcal{O}^* = \{\pm 1\}$.

The idea is to prove that λ is a unit of $\mathcal{O}$. To see this, note that by Lemma 11.24, both $[1, m\alpha/s]$ and $[d, a\alpha/s + b]$ have index m in $[1, \alpha/s]$, so that λ must have norm 1. Furthermore, we have

$$s\lambda \in s[d, a\alpha/s + b] = [sd, a\alpha + sb] \subset [s, \alpha].$$

Writing $s\lambda = us + v\alpha$, $u, v \in \mathbf{Z}$, and taking norms, we obtain

$$s^2 = s^2 N(\lambda) = N(us + v\alpha) = u^2 s^2 + usvT(\alpha) + v^2 N(\alpha).$$

Since $s \mid T(\alpha)$, it follows that $s^2 | v^2 N(\alpha)$, and since $\gcd(s^2, N(\alpha))$ is squarefree, we must have $s \mid v$. This shows that $\lambda \in [1, \alpha] = \mathcal{O}$, so that λ is a unit since it has norm 1. Then $\mathcal{O}^* = \{\pm 1\}$ implies that $\lambda = \pm 1$, and hence $[1, m\alpha/s] = [d, a\alpha/s + b]$, which contradicts $\sigma \neq \sigma_0$ by the uniqueness part of Lemma 11.24. The lemma is proved. Q.E.D.

We want to apply this lemma to the case $s = 3$, $m = 9$ and $\alpha = \tau_0$. Using the special form of τ_0, it is easy to see that the norm and trace conditions are satisfied (note that the discriminant of $\mathcal{O} = [1, \tau_0]$ is $D = T(\tau_0)^2 - 4N(\tau_0)$). Thus (12.9) holds except possibly when $\mathcal{O}$ is $\mathbf{Z}[i]$ or $\mathbf{Z}[\zeta_3]$. The latter can't occur since 3 doesn't divide the discriminant, and when $\mathcal{O} = \mathbf{Z}[i]$, a simple argument shows that (12.12) is impossible (see Exercise 12.8). This completes the proof of Theorem 12.2. Q.E.D.

This theorem tells us about the behavior of $\gamma_2(\tau_0)$ when 3 doesn't divide the discriminant D. For completeness, let's record what happens when D is a multiple of 3 (see Schertz [87] for a proof):

Theorem 12.13. *Let $\mathcal{O}$ be an order of discriminant D in an imaginary quadratic field K. Assume $3 \mid D$ and $D < -3$, and write $\mathcal{O} = [1, \tau_0]$, where*

$$\tau_0 = \begin{cases} \sqrt{-m}, & D = -4m \equiv 0 \bmod 4 \\ \dfrac{3 + \sqrt{-m}}{2}, & D = -m \equiv 1 \bmod 4. \end{cases}$$

Then $K(\gamma_2(\tau_0))$ is the ring class field of the order $\mathcal{O}' = [1, 3\tau_0]$ and is an extension of degree 3 of the ring class field of $\mathcal{O}$. Furthermore, $\mathbf{Q}(\gamma_2(\tau_0)) = \mathbf{Q}(j(3\tau_0))$. Q.E.D.

B. The Weber Functions

To work effectively with $\gamma_2(\tau)$, we need good formulas for computing it. This leads us to our next topic, the Dedekind η-function $\eta(\tau)$ and the three Weber functions $\mathfrak{f}(\tau)$, $\mathfrak{f}_1(\tau)$ and $\mathfrak{f}_2(\tau)$. If $\tau \in \mathfrak{h}$, we let $q = e^{2\pi i\tau}$ as usual, and then the Dedekind η-function is defined by the formula

$$\eta(\tau) = q^{1/24}\prod_{n=1}^{\infty}(1-q^n).$$

Note that this product converges (and is nonzero) for $\tau \in \mathfrak{h}$ since $0 < |q| < 1$.

We then define the Weber functions $\mathfrak{f}(\tau)$, $\mathfrak{f}_1(\tau)$ and $\mathfrak{f}_2(\tau)$ in terms of the η-function as follows:

$$\begin{aligned} \mathfrak{f}(\tau) &= \zeta_{48}^{-1}\frac{\eta((\tau+1)/2)}{\eta(\tau)} \\ \mathfrak{f}_1(\tau) &= \frac{\eta(\tau/2)}{\eta(\tau)} \\ \mathfrak{f}_2(\tau) &= \sqrt{2}\frac{\eta(2\tau)}{\eta(\tau)}, \end{aligned} \tag{12.14}$$

where $\zeta_{48} = e^{2\pi i/48}$. From these definitions, one gets the following product expansions for the Weber functions:

$$\begin{aligned} \mathfrak{f}(\tau) &= q^{-1/48}\prod_{n=1}^{\infty}(1+q^{n-1/2}) \\ \mathfrak{f}_1(\tau) &= q^{-1/48}\prod_{n=1}^{\infty}(1-q^{n-1/2}) \\ \mathfrak{f}_2(\tau) &= \sqrt{2}q^{1/24}\prod_{n=1}^{\infty}(1+q^n) \end{aligned} \tag{12.15}$$

(see Exercise 12.9), and we also get the following useful identities connecting the Weber functions:

$$\begin{aligned} \mathfrak{f}(\tau)\mathfrak{f}_1(\tau)\mathfrak{f}_2(\tau) &= \sqrt{2} \\ \mathfrak{f}_1(2\tau)\mathfrak{f}_2(\tau) &= \sqrt{2} \end{aligned} \tag{12.16}$$

(see Exercise 12.9).

Much deeper lie the following relations between $\eta(\tau)$, $\mathfrak{f}(\tau)$, $\mathfrak{f}_1(\tau)$ and $\mathfrak{f}_2(\tau)$ and the previously defined functions $j(\tau)$, $\gamma_2(\tau)$ and $\Delta(\tau)$:

Theorem 12.17. *If $\tau \in \mathfrak{h}$, then $\Delta(\tau) = (2\pi)^{12}\eta(\tau)^{24}$ and*

$$\gamma_2(\tau) = \frac{\mathfrak{f}(\tau)^{24} - 16}{\mathfrak{f}(\tau)^8} = \frac{\mathfrak{f}_1(\tau)^{24} + 16}{\mathfrak{f}_1(\tau)^8} = \frac{\mathfrak{f}_2(\tau)^{24} + 16}{\mathfrak{f}_2(\tau)^8}.$$

Remark. Since $j(\tau) = \gamma_2(\tau)^3$, this theorem gives us some remarkable formulas for computing the j-function.

Proof. We need to relate $\eta(\tau)$ and the Weber functions to the Weierstrass $\wp$-function. Let $\wp(z) = \wp(z;\tau)$ denote the $\wp$-function for the lattice $[1,\tau]$, and set

$$e_1 = \wp(\tau/2), \qquad e_2 = \wp(1/2), \qquad e_3 = \wp((\tau+1)/2).$$

We will prove the following formulas for the differences $e_i - e_j$:

$$\begin{aligned} e_2 - e_1 &= \pi^2\eta(\tau)^4\mathfrak{f}(\tau)^8 \\ e_2 - e_3 &= \pi^2\eta(\tau)^4\mathfrak{f}_1(\tau)^8 \\ e_3 - e_1 &= \pi^2\eta(\tau)^4\mathfrak{f}_2(\tau)^8. \end{aligned} \tag{12.18}$$

The basic strategy of the proof is to express $e_i - e_j$ in terms of the Weierstrass σ-function, and then use the product expansion of the σ-function to get product expansions for $e_i - e_j$. Proofs will appear in the exercises.

The Weierstrass σ-function is defined as follows. Let $\tau \in \mathfrak{h}$, and let L be the lattice $[1,\tau]$. Then the Weierstrass σ-function is the product

$$\sigma(z;\tau) = z \prod_{\omega \in L - \{0\}} \left(1 - \frac{z}{\omega}\right) e^{z/\omega + (1/2)(z/\omega)^2}.$$

Note that $\sigma(z;\tau)$ is an odd function in z. We will usually write $\sigma(z;\tau)$ more simply as $\sigma(z)$. The σ-function is not periodic, but there are complex numbers η_1 and η_2, depending only on τ, such that

$$\begin{aligned} \sigma(z+\tau) &= -e^{\eta_1(z+\tau/2)}\sigma(z) \\ \sigma(z+1) &= -e^{\eta_2(z+1/2)}\sigma(z), \end{aligned}$$

and the numbers η_1 and η_2 satisfy the Legendre relation $\eta_2\tau - \eta_1 = 2\pi i$ (see Exercise 12.10). The σ-function is related to the $\wp$-function by the formula

$$\wp(z) - \wp(w) = -\frac{\sigma(z+w)\sigma(z-w)}{\sigma^2(z)\sigma^2(w)}$$

whenever z and w do not lie in L (see Exercise 12.11). Since $e_1 = \wp(\tau/2)$, $e_2 = \wp(1/2)$ and $e_3 = \wp((\tau+1)/2)$, it follows easily that

$$e_2 - e_1 = e^{-\eta_2\tau/2}\frac{\sigma^2\left(\dfrac{\tau+1}{2}\right)}{\sigma^2\left(\dfrac{1}{2}\right)\sigma^2\left(\dfrac{\tau}{2}\right)}$$

$$e_2 - e_3 = e^{\eta_2(\tau+1)/2}\frac{\sigma^2\left(\dfrac{\tau}{2}\right)}{\sigma^2\left(\dfrac{1}{2}\right)\sigma^2\left(\dfrac{\tau+1}{2}\right)}$$

$$e_3 - e_1 = e^{\eta_1(\tau+1)/2}\frac{\sigma^2\left(\dfrac{1}{2}\right)}{\sigma^2\left(\dfrac{\tau+1}{2}\right)\sigma^2\left(\dfrac{\tau}{2}\right)}$$

(see Exercise 12.12).

There is also the following q-product expansion for the σ-function:

$$\sigma(z;\tau) = \frac{1}{2\pi i}e^{\eta_2 z^2/2}(q_z^{1/2} - q_z^{-1/2})\prod_{n=1}^{\infty}\frac{(1-q_\tau^n q_z)(1-q_\tau^n/q_z)}{(1-q_\tau^n)^2},$$

where $q_\tau = e^{2\pi i\tau}$ and $q_z = e^{2\pi i z}$ (see Exercise 12.13). Using this product expansion, we obtain

$$\sigma\left(\frac{1}{2}\right) = \frac{1}{2\pi}e^{\eta_2/8}\frac{\mathfrak{f}_2(\tau)^2}{\eta(\tau)^2}$$

$$\sigma\left(\frac{\tau}{2}\right) = \frac{i}{2\pi}e^{\eta_2\tau^2/8}q^{-1/8}\frac{\mathfrak{f}_1(\tau)^2}{\eta(\tau)^2}$$

$$\sigma\left(\frac{\tau+1}{2}\right) = \frac{1}{2\pi}e^{\eta_2(\tau+1)^2/8}q^{-1/8}\frac{\mathfrak{f}(\tau)^2}{\eta(\tau)^2}$$

(see Exercise 12.14). It is now straightforward to derive the desired formulas (12.18) for $e_i - e_j$ (see Exercise 12.14).

To relate this to $\Delta(\tau)$, recall from (10.6) that $\Delta(\tau) = 16(e_2-e_1)^2(e_2-e_3)^2(e_3-e_1)^2$. By (12.18), it is now easy to express $\Delta(\tau)$ in terms of the η-function:

$$\begin{aligned}\Delta(\tau) &= 16(e_2-e_1)^2(e_2-e_3)^2(e_3-e_1)^2\\ &= 16\pi^{12}\eta(\tau)^{24}\mathfrak{f}(\tau)^{16}\mathfrak{f}_1(\tau)^{16}\mathfrak{f}_2(\tau)^{16}\\ &= (2\pi)^{12}\eta(\tau)^{24},\end{aligned}$$

where the last line follows by (12.16).

Turning to $\gamma_2(\tau)$, we know that

$$\gamma_2(\tau) = \sqrt[3]{j(\tau)} = \frac{12g_2(\tau)}{\sqrt[3]{\Delta(\tau)}},$$

where the cube root is chosen to be real-valued on the imaginary axis. Using what we just proved about $\Delta(\tau)$, this formula can be written

$$\gamma_2(\tau) = \frac{3g_2(\tau)}{4\pi^4\eta(\tau)^8}$$

since $\eta(\tau)$ is real valued on the imaginary axis. Thus, to express $\gamma_2(\tau)$ in terms of Weber functions, we need to express $g_2(\tau)$ in terms of $\eta(\tau)$, $\mathfrak{f}(\tau)$, $\mathfrak{f}_1(\tau)$ and $\mathfrak{f}_2(\tau)$.

The idea is to write $g_2(\tau)$ in terms of the $e_i - e_j$'s. Recall from the proof of Proposition 10.7 that the e_i's are the roots of $4x^3 - g_2(\tau)x - g_3(\tau)$, which implies that $g_2(\tau) = -4(e_1e_2 + e_1e_3 + e_2e_3)$ (see Exercise 10.8). Then, using $e_1 + e_2 + e_3 = 0$, one obtains

$$3g_2(\tau) = 4((e_2 - e_1)^2 - (e_2 - e_3)(e_3 - e_1))$$

(see Exercise 12.15). Substituting in the formulas from (12.18) yields

$$3g_2(\tau) = 4\pi^4\eta(\tau)^8(\mathfrak{f}(\tau)^{16} - \mathfrak{f}_1(\tau)^8\mathfrak{f}_2(\tau)^8),$$

so that

$$\begin{aligned}\gamma_2(\tau) &= \mathfrak{f}(\tau)^{16} - \mathfrak{f}_1(\tau)^8\mathfrak{f}_2(\tau)^8\\ &= \mathfrak{f}(\tau)^{16} - \frac{16}{\mathfrak{f}(\tau)^8}\\ &= \frac{\mathfrak{f}(\tau)^{24} - 16}{\mathfrak{f}(\tau)^8},\end{aligned}$$

where we have again used the basic identity (12.16). The other two formulas for $\gamma_2(\tau)$ are proved similarly and are left to the reader (see Exercise 12.15). This completes the proof of the theorem. Q.E.D.

Using these formulas it is easy to show that the q-expansions of $\gamma_2(\tau)$ and $j(\tau)$ have integer coefficients (see Exercise 12.16), and this proves Theorem 11.8. We can also use Theorem 12.17 to study the transformation properties of $\eta(\tau)$, $\mathfrak{f}(\tau)$, $\mathfrak{f}_1(\tau)$ and $\mathfrak{f}_2(\tau)$:

Corollary 12.19. *For a positive integer n, let $\zeta_n = e^{2\pi i/n}$. Then*

$$\begin{aligned}\eta(\tau + 1) &= \zeta_{24}\eta(\tau)\\ \eta(-1/\tau) &= \sqrt{-i\tau}\,\eta(\tau),\end{aligned}$$

where the square root is chosen to be positive on the imaginary axis. Furthermore,

$$\mathfrak{f}(\tau+1) = \zeta_{48}^{-1}\mathfrak{f}_1(\tau)$$
$$\mathfrak{f}_1(\tau+1) = \zeta_{48}^{-1}\mathfrak{f}(\tau)$$
$$\mathfrak{f}_2(\tau+1) = \zeta_{24}\mathfrak{f}_2(\tau),$$

and

$$\mathfrak{f}(-1/\tau) = \mathfrak{f}(\tau)$$
$$\mathfrak{f}_1(-1/\tau) = \mathfrak{f}_2(\tau)$$
$$\mathfrak{f}_2(-1/\tau) = \mathfrak{f}_1(\tau).$$

Proof. The definition of $\eta(\tau)$ makes the formula for $\eta(\tau+1)$ obvious. Turning to $\eta(-1/\tau)$, first consider $\Delta(\tau) = (2\pi)^{12}\eta(\tau)^{24}$. For a lattice L, we know

$$\Delta(L) = g_2(L)^3 - 27g_3(L)^2.$$

In (10.10) we showed that $g_2(\lambda L) = \lambda^{-4}g_2(L)$ and $g_3(\lambda L) = \lambda^{-6}g_3(L)$, which implies that

$$\Delta(\lambda L) = \lambda^{-12}\Delta(L).$$

This gives us the formula

$$\Delta(-1/\tau) = \Delta([1,-1/\tau]) = \Delta(\tau^{-1}[1,\tau]) = \tau^{12}\Delta([1,\tau]) = \tau^{12}\Delta(\tau),$$

and taking 24th roots, we obtain

$$\eta(-1/\tau) = \epsilon\sqrt{-i\tau}\,\eta(\tau)$$

for some root of unity ϵ. Both sides take positive real values on the imaginary axis, which forces ϵ to be 1. This proves that $\eta(\tau)$ transforms as desired.

Turning to the Weber functions, their behavior under $\tau \mapsto \tau+1$ and $\tau \mapsto -1/\tau$ are simple consequences of their definitions and the transformation properties of $\eta(\tau)$ (see Exercise 12.17). Q.E.D.

We will make extensive use of these transformation properties in the latter part of this section.

C. j-Invariants of Orders of Class Number 1

Using the properties of the Weber functions, we can now compute the j-invariants for orders of class number 1. In §7 we saw that there are exactly 13 such orders, with discriminants

$$-3,\ -4,\ -7,\ -8,\ -11,\ -16,\ -19,\ -27,\ -28,\ -43,\ -67,\ -163$$

(we will prove this in Theorem 12.34 below). The j-invariants of these orders are integers, and if we restrict ourselves to those where 3 doesn't divide the discriminant (10 of the above 13), then Theorem 12.2 tells us that the j-invariant is a cube. So in these cases we need only compute $\gamma_2(\tau_0)$, where τ_0 is an appropriately chosen element of the order. Rather than compute $\gamma_2(\tau_0)$ directly, we will use the Weber functions to approximate its value to within $\pm .5$. Since $\gamma(\tau_0)$ is an integer, this will determine its value uniquely. This scheme for computing these j-invariants is due to Weber [102, §125].

The ten j-invariants we want to compute are given in the following table:

(12.20)

d_K	τ_0	$\gamma_2(\tau_0)$	$j(\mathcal{O}) = j(\tau_0)$
-4	i	$12 = 2^2 \cdot 3$	12^3
-7	$(3+\sqrt{-7})/2$	$-15 = -3 \cdot 5$	-15^3
-8	$\sqrt{-2}$	$20 = 2^2 \cdot 5$	20^3
-11	$(3+\sqrt{-11})/2$	$-32 = -2^5$	-32^3
-16	$2i$	$66 = 2 \cdot 3 \cdot 11$	66^3
-19	$(3+\sqrt{-19})/2$	$-96 = -2^5 \cdot 3$	-96^3
-28	$\sqrt{-7}$	$255 = 3 \cdot 5 \cdot 17$	255^3
-43	$(3+\sqrt{-43})/2$	$-960 = -2^6 \cdot 3 \cdot 5$	-960^3
-67	$(3+\sqrt{-67})/2$	$-5280 =$ $-2^5 \cdot 3 \cdot 5 \cdot 11$	-5280^3
-163	$(3+\sqrt{-163})/2$	$-640320 =$ $-2^6 \cdot 3 \cdot 5 \cdot 23 \cdot 29$	-640320^3

For completeness, here are the j-invariants of the orders of discriminant divisible by 3:

d_K	τ_0	$j(\mathcal{O}) = j(\tau_0)$
-3	$(1+\sqrt{-3})/2$	0
-12	$\sqrt{-3}$	$54000 = 2^4 \cdot 3^3 \cdot 5^3$
-27	$(1+3\sqrt{-3})/2$	$-12288000 = -2^{15} \cdot 3 \cdot 5^3$

We computed $j((1+\sqrt{-3})/2) = 0$ in §10, and we will prove $j(\sqrt{-3}) = 54000$ in §13. As predicted by Theorem 12.13, the last two entries are not perfect cubes.

To start the computation, first consider the case of even discriminant. Here, $\tau_0 = \sqrt{-m}$, where $m = 1, 2, 4$ or 7. Setting $q = e^{2\pi i\sqrt{-m}} = e^{-2\pi\sqrt{m}}$, we claim that

$$\gamma_2(\sqrt{-m}) = [\![256q^{2/3} + q^{-1/3}]\!], \tag{12.21}$$

where $[\![\ \ ]\!]$ is the nearest integer function (i.e., for a real number $x \notin \mathbf{Z} + \frac{1}{2}$, $[\![x]\!]$ is the integer nearest to x).

To prove this, we will write $\gamma_2(\tau)$ in terms of the Weber function $\mathfrak{f}_2(\tau)$:

$$\gamma_2(\sqrt{-m}) = \mathfrak{f}_2(\sqrt{-m})^{16} + \frac{16}{\mathfrak{f}_2(\sqrt{-m})^8}. \tag{12.22}$$

Using $q = e^{-2\pi\sqrt{m}}$ as above, (12.15) gives us

$$\mathfrak{f}_2(\sqrt{-m}) = \sqrt{2}q^{1/24}\prod_{n=1}^{\infty}(1+q^n),$$

and to estimate the infinite product, we use the inequality $1+x < e^x$ for $x > 0$. This yields

$$1 < \prod_{n=1}^{\infty}(1+q^n) < \prod_{n=1}^{\infty} e^{q^n} = e^{q/(1-q)},$$

and we can simplify the exponent by noting that $q/(1-q) \le q/(1-e^{-2\pi}) < 1.002q$ since $q \le e^{-2\pi}$. Thus we have the following inequalities for $\mathfrak{f}_2(\sqrt{-m})$:

$$\sqrt{2}q^{1/24} < f_2(\sqrt{-m}) < \sqrt{2}q^{1/24}e^{1.002q},$$

and applying this to (12.22), we get upper and lower bounds for $\gamma_2(\sqrt{-m})$:

$$256q^{2/3} + q^{-1/3}e^{-8.016q} < \gamma_2(\sqrt{-m}) < 256q^{2/3}e^{16.032q} + q^{-1/3}. \tag{12.23}$$

To see how sharp these bounds are, consider their difference

$$E = 256q^{2/3}(e^{16.032q} - 1) + q^{-1/3}(1 - e^{-8.016q}).$$

Using the inequality

$$1 - e^{-x} < \frac{x}{1-x}, \qquad 0 < x < 1,$$

one sees that

$$\begin{aligned} E &< 256q^{2/3}(e^{16.032q} - 1) + q^{-1/3}8.016q/(1-8.016q) \\ &= 256q^{2/3}(e^{16.032q} - 1) + 8.016q^{2/3}/(1-8.016q). \end{aligned}$$

The last quantity is an increasing function in q, and then $q < e^{-2\pi}$ easily implies that $E < .25$. Since $\gamma_2(\sqrt{-m})$ is an integer, this means that $[\![x]\!] = \gamma_2(\sqrt{-m})$ for any x lying between the upper and lower limits of (12.23). In particular, $256q^{2/3} + q^{-1/3}$ lies between these limits, which proves (12.21). Using a hand calculator, it is now trivial to compute the corresponding entries in table (12.20) (see Exercise 12.18).

Turning to the case of odd discriminant, let $\tau_0 = (3+\sqrt{-m})/2$, $m = 7$, 11, 19, 43, 67 or 163, and we again want to compute

$$\gamma_2(\tau_0) = \mathfrak{f}_2(\tau_0)^{16} + \frac{16}{\mathfrak{f}_2(\tau_0)^8}.$$

Our previous techniques won't work, for $q = e^{2\pi i(3+\sqrt{-m})/2} = -e^{-\pi\sqrt{m}}$ is negative in this case. But Weber uses the following clever trick: from (12.16), we know that

$$\mathfrak{f}_2(\tau_0) = \frac{\sqrt{2}}{\mathfrak{f}_1(2\tau_0)},$$

and then the transformation properties from Corollary 12.19 imply

$$\begin{aligned}\mathfrak{f}_1(2\tau_0) = \mathfrak{f}_1(3+\sqrt{-m}) &= \zeta_{48}^{-1}\mathfrak{f}(2+\sqrt{-m}) \\ &= \zeta_{48}^{-2}\mathfrak{f}_1(1+\sqrt{-m}) = \zeta_{48}^{-3}\mathfrak{f}(\sqrt{-m}).\end{aligned}$$

Combining the above equations implies that

$$\mathfrak{f}_2(\tau_0) = \frac{\sqrt{2}\zeta_{16}}{\mathfrak{f}(\sqrt{-m})},$$

and thus

$$\gamma_2(\tau_0) = \frac{256}{\mathfrak{f}(\sqrt{-m})^{16}} - \mathfrak{f}(\sqrt{-m})^8.$$

From here, our previous methods easily imply that if $m = 7, 11, 19, 43, 67$ or 163, and $q = e^{-2\pi\sqrt{m}}$, then

$$\gamma_2((3+\sqrt{-m})/2) = [\![-q^{-1/6} + 256q^{1/3}]\!],$$

where $[\![\ \ ]\!]$ is again the nearest integer function. Using a hand calculator, we can now complete our table (12.20) of singular j-invariants (see Exercise 12.18).

D. Weber's Computation of $j(\sqrt{-14})$

We next want to compute some singular j-invariants when the class number is greater than 1. There are several ways one can proceed. For example, when the class number is 2, the Kronecker Limit Formula gives an elegant method to determine the j-invariant, and this method generalizes to the case of orders with only one class per genus. (Recall from §3 that for discriminants $-4n$, this condition means that n is one of Euler's convenient numbers.) For example, when $n = 105$, Weber [102, §143] shows that

$$\mathfrak{f}(\sqrt{-105})^6 = \sqrt{2}^{-13}(1+\sqrt{3})^3(1+\sqrt{5})^3(\sqrt{3}+\sqrt{7})^3(\sqrt{5}+\sqrt{7}),$$

which would then allow us to compute $\gamma_2(\sqrt{-105})$ and hence $j(\sqrt{-105})$. (The radicals appearing in the above formula are not surprising, since in this case the Hilbert class field equals the genus field, which we know by Theorem 6.1—see Exercise 12.19.) Other examples may be found in Weber [102, pp. 721–726] or [103], and a modern treatment of the Kronecker Limit Formula is in Lang [73, Chapter 20].

We will instead take a different route and compute $j(\sqrt{-14})$, an example particularly relevant to earlier sections. Namely, $K(j(\sqrt{-14}))$ is the Hilbert class field of $K = \mathbf{Q}(\sqrt{-14})$ since $\mathcal{O}_K = [1, \sqrt{-14}]$. We determined this field in §5, so that finding $j(\sqrt{-14})$ will give us a second and quite different way of finding the Hilbert class field of $\mathbf{Q}(\sqrt{-14})$. Our exposition will again follow Weber [102, §144], using ideas from Schertz [87] to give a modern proof.

A key fact we will use is that in many cases, one can generate ring class fields using small powers of the Weber functions. Weber gives a long list of such theorems in [102, §§126–127], and modern proofs have been given by Birch [7] and Schertz [87]. We will discuss two cases which will be useful to our purposes:

Theorem 12.24. *Given a positive integer m not divisible by 3, let $\mathcal{O} = [1, \sqrt{-m}]$, which is an order in $K = \mathbf{Q}(\sqrt{-m})$. Then:*

(i) *For $m \equiv 6 \bmod 8$, $\mathfrak{f}_1(\sqrt{-m})^2$ is an algebraic integer and $K(\mathfrak{f}_1(\sqrt{-m})^2)$ is the ring class field of $\mathcal{O}$.*

(ii) *For $m \equiv 3 \bmod 4$, $\mathfrak{f}(\sqrt{-m})^2$ is an algebraic integer and $K(\mathfrak{f}(\sqrt{-m})^2)$ is the ring class field of $\mathcal{O}$.*

Proof. We begin with (i). Multiplying out the identity

$$j(\sqrt{-m}) = \gamma_2(\sqrt{-m})^3 = \left(\frac{\mathfrak{f}_1(\sqrt{-m})^{24} + 16}{\mathfrak{f}_1(\sqrt{-m})^8}\right)^3,$$

it follows that $\mathfrak{f}_1(\sqrt{-m})^2$ is a root of a monic polynomial with coefficients in $\mathbf{Z}[j(\sqrt{-m})]$. But $j(\sqrt{-m})$ is an algebraic integer, which implies that the same is true for $\mathfrak{f}_1(\sqrt{-m})^2$.

We know that $L = K(j(\sqrt{-m}))$ is the ring class field of $[1, \sqrt{-m}]$, and since $j(\sqrt{-m})$ is a polynomial in $\mathfrak{f}_1(\sqrt{-m})^2$, we need only show that $\mathfrak{f}_1(\sqrt{-m})^2$ lies in L. Actually, it suffices to show that $\mathfrak{f}_1(\sqrt{-m})^6$ lies in L. This is a consequence of Theorems 12.2 and 12.17, for since $3 \nmid m$, we have $\gamma_2(\sqrt{-m}) \in L$, and we also know that

$$\gamma_2(\sqrt{-m}) = \frac{\mathfrak{f}_1(\sqrt{-m})^{24} + 16}{\mathfrak{f}_1(\sqrt{-m})^8}.$$

When $\mathfrak{f}_1(\sqrt{-m})^6$ lies in L, so does $\mathfrak{f}_1(\sqrt{-m})^{24}$. The above equation implies $\mathfrak{f}_1(\sqrt{-m})^8 \in L$, and then $\mathfrak{f}_1(\sqrt{-m})^2 \in L$ follows immediately.

The next step in the proof is to show that $\mathfrak{f}_1(8\tau)^6$ is a modular function:

Proposition 12.25. *$\mathfrak{f}_1(8\tau)^6$ is a modular function for the group $\Gamma_0(32)$.*

Proof. We first study the transformation properties of $f_1(\tau)^6$. Consider the group

$$\Gamma_0(2)^t = \left\{ \begin{pmatrix} a & b \\ c & d \end{pmatrix} : b \equiv 0 \bmod 2 \right\}.$$

In Exercise 12.4, we will show that the matrices

$$-I = \begin{pmatrix} -1 & 0 \\ 0 & -1 \end{pmatrix}, \quad U = \begin{pmatrix} 1 & 0 \\ 1 & 1 \end{pmatrix}, \quad V = \begin{pmatrix} 1 & 2 \\ 0 & 1 \end{pmatrix}$$

generate $\Gamma_0(2)^t$. Using Corollary 12.19, $\mathfrak{f}_1(\tau)^6$ transforms under U and V as follows:

$$\mathfrak{f}_1(U\tau)^6 = -i\mathfrak{f}_1(\tau)^6$$
$$\mathfrak{f}_1(V\tau)^6 = -i\mathfrak{f}_1(\tau)^6$$

(see Exercise 12.20.). Then we get the general transformation law for $\mathfrak{f}_1(\tau)^6$:

$$(12.26) \quad \mathfrak{f}_1(\gamma\tau)^6 = i^{-ac-(1/2)bd+(1/2)b^2c}\mathfrak{f}_1(\tau)^6, \qquad \gamma = \begin{pmatrix} a & b \\ c & d \end{pmatrix} \in \Gamma_0(2)^t.$$

This can be proved by induction on the length of γ as a word in $-I$, U and V. A more enlightening way to prove (12.26) is sketched in Exercise 12.21.

Now consider the function $\mathfrak{f}_1(8\tau)^6$, and let $\gamma \in \Gamma_0(32)$. Then

$$8\gamma\tau = 8\begin{pmatrix} a & b \\ 32c & d \end{pmatrix}\tau = \begin{pmatrix} a & 8b \\ 4c & d \end{pmatrix} 8\tau = \tilde{\gamma}8\tau.$$

Since $\tilde{\gamma} \in \Gamma_0(2)^t$, it follows easily from (12.26) that $\mathfrak{f}_1(8\gamma\tau)^6 = \mathfrak{f}_1(\tilde{\gamma}8\tau)^6 = \mathfrak{f}_1(8\tau)^6$, which proves that $\mathfrak{f}_1(8\tau)^6$ is invariant under $\Gamma_0(32)$. To check the cusps, take $\gamma \in \mathrm{SL}(2,\mathbf{Z})$. Under the correspondence between cosets of $\Gamma_0(8)$ and matrices in $C(8)$ given by Lemma 11.11, there is $\sigma \in C(8)$ and $\tilde{\gamma} \in \mathrm{SL}(2,\mathbf{Z})$ such that

$$8\gamma\tau = \tilde{\gamma}\sigma\tau.$$

Writing $\tilde{\gamma}$ as a product of various powers of $\left(\begin{smallmatrix} 1 & 1 \\ 0 & 1 \end{smallmatrix}\right)$ and $\left(\begin{smallmatrix} 0 & -1 \\ 1 & 0 \end{smallmatrix}\right)$, the transformation properties of Corollary 12.19 imply that

$$\mathfrak{f}_1(8\gamma\tau)^6 = \mathfrak{f}_1(\tilde{\gamma}\sigma\tau)^6 = \epsilon\mathfrak{f}(\sigma\tau)^6, \epsilon\mathfrak{f}_1(\sigma\tau)^6, \text{ or } \epsilon\mathfrak{f}_2(\sigma\tau)^6$$

for some root of unity ϵ. Since $\sigma = \left(\begin{smallmatrix} a & b \\ 0 & d \end{smallmatrix}\right)$, where $ad = 8$, we have

$$e^{2\pi i\sigma\tau} = \zeta_d^a q^{a/d} = \zeta_d^a (q^{1/8})^{a^2},$$

and consequently, the product expansions for the Weber functions imply that $\mathfrak{f}_1(8\gamma\tau)^6$ is meromorphic in $q^{1/8}$. This proves that $\mathfrak{f}_1(8\tau)^6$ is a modular function for $\Gamma_0(32)$. Q.E.D.

The next step in proving Theorem 12.24 is to determine some field (not necessarily the smallest) containing $\mathfrak{f}_1(\sqrt{-m})^6$. The key point is that $\mathfrak{f}_1(8\tau)^6$ is not only a modular function for $\Gamma_0(32)$, it's also holomorphic and its q-expansion is integral. Thus Proposition 12.7 tells us that $\mathfrak{f}_1(8\tau)^6 = R(j(\tau), j(32\tau))$ for some rational function $R(X,Y) \in \mathbb{Q}(X,Y)$. We will write this in the form

$$\mathfrak{f}_1(\tau)^6 = R(j(\tau/8), j(4\tau)). \tag{12.27}$$

Using Lemma 12.11 with $m = 32$ and $s = 8$, we see that

$$\frac{\partial \Phi_{32}}{\partial X}(j(4\sqrt{-m}), j(\sqrt{-m}/8)) \neq 0,$$

and thus, by Proposition 12.7, we conclude that

(12.28)

$$\mathfrak{f}_1(\sqrt{-m})^6 = R(j(\sqrt{-m}/8), j(4\sqrt{-m})) = R(j([8,\sqrt{-m}]), j([1,4\sqrt{-m}])).$$

To identify what field this lies in, let L' denote the ring class field of the order $\mathcal{O}' = [1, 4\sqrt{-m}]$. Since $[8, \sqrt{-m}]$ is a fractional proper ideal for $\mathcal{O}'$ (this uses Lemma 7.5 and $m \equiv 6 \bmod 8$—see Exercise 12.22), it follows that $\mathfrak{f}_1(\sqrt{-m})^6 \in L'$.

We want to prove that $\mathfrak{f}_1(\sqrt{-m})^6$ lies in the smaller field L. This is the situation that occurred in the proof of Theorem 12.2, but here we will need more than just a degree calculation. The crucial new idea will be to relate Galois theory and modular functions.

Let's first study the Galois theory of $L \subset L'$. The orders $\mathcal{O}'$ and $\mathcal{O}$ have discriminants $-64m$ and $-4m$ respectively, so that Corollary 7.28 implies that $h(-64m)) = 4h(-4m)$. Thus L' has degree 4 over L. Furthermore, the isomorphisms $C(\mathcal{O}') \simeq \mathrm{Gal}(L'/K)$ and $C(\mathcal{O}) \simeq \mathrm{Gal}(L/K)$ imply that

$$\mathrm{Gal}(L'/L) \simeq \ker(C(\mathcal{O}') \to C(\mathcal{O})).$$

In Exercise 12.22 we show that $[4, 1 + \sqrt{-m}]$ is a proper $\mathcal{O}'$-ideal which lies in the above kernel and has order 4. It follows that $L \subset L'$ is a cyclic extension of degree 4.

The goal of the remainder of the proof will be to compute $\sigma(\mathfrak{f}_1(\sqrt{-m})^6)$ for some generator σ of $\mathrm{Gal}(L'/L)$. At the end of §11 we described an isomorphism

$$C(\mathcal{O}') \simeq \mathrm{Gal}(L'/K)$$

as follows. Given a class $[\mathfrak{a}] \in C(\mathcal{O}')$, let the corresponding automorphism be $\sigma_{\mathfrak{a}} \in \mathrm{Gal}(L'/K)$. If we write $L' = K(j(\mathfrak{b}))$ for some proper fractional $\mathcal{O}'$-ideal $\mathfrak{b}$, then Corollary 11.37 states that

$$\sigma_{\mathfrak{a}}(j(\mathfrak{b})) = j(\overline{\mathfrak{a}}\mathfrak{b}).$$

To exploit this, let $\mathfrak{b} = [8, \sqrt{-m}] = 8[1, \sqrt{-m}/8]$, so that (12.28) can be written

$$\mathfrak{f}_1(\sqrt{-m})^6 = R(j(\mathfrak{b}), j(\mathcal{O}')).$$

Now let $\mathfrak{a} = [4, 1+\sqrt{-m}]$, and let the corresponding automorphism be $\sigma = \sigma_{\mathfrak{a}} \in \mathrm{Gal}(L'/L)$. Note that σ is a generator of $\mathrm{Gal}(L'/L)$, and hence to prove that $\mathfrak{f}_1(\sqrt{-m})^6$ lies in L, we need only prove that it is fixed by σ. Using the above formula for $\mathfrak{f}_1(\sqrt{-m})^6$, we compute

$$\sigma(\mathfrak{f}_1(\sqrt{-m})^6) = R(\sigma(j(\mathfrak{b})), \sigma(j(\mathcal{O}'))) = R(j(\overline{\mathfrak{a}}\mathfrak{b}), j(\overline{\mathfrak{a}})).$$

Since $m \equiv 6 \bmod 8$, one easily sees that

$$\overline{\mathfrak{a}}\mathfrak{b} = [8, -2+\sqrt{-m}], \qquad \overline{\mathfrak{a}} = [4, -1+\sqrt{-m}]$$

(see Exercise 12.22), and hence $\sigma(\mathfrak{f}_1(\sqrt{-m})^6)$ can be written

(12.29)
$$\sigma(\mathfrak{f}_1(\sqrt{-m})^6) = R(j([8, -2+\sqrt{-m}]), j([4, -1+\sqrt{-m}])).$$

Now let $\gamma = \begin{pmatrix} 1 & -2 \\ 1 & -1 \end{pmatrix} \in \Gamma_0(2)^t$. If we substitute $\gamma\tau$ for τ in (12.27), we get

$$\mathfrak{f}_1(\gamma\tau)^6 = R(j(\gamma\tau/8), j(4\gamma\tau)).$$

Since $\gamma\tau = (\tau-2)/(\tau-1)$, one sees that

$$\begin{aligned} &[1, \gamma\tau/8] \quad \text{is homothetic to} \quad [8(\tau-1), \tau-2] = [8, -2+\tau] \\ &[1, 4\gamma\tau] \quad \text{is homothetic to} \quad [\tau-1, 4(\tau-2)] = [4, -1+\tau], \end{aligned}$$

and thus

$$\mathfrak{f}_1(\gamma\tau)^6 = R(j([8, -2+\tau]), j([4, -1+\tau])).$$

Evaluating this at $\tau = \sqrt{-m}$ and using (12.29), we see that

$$\sigma(\mathfrak{f}_1(\sqrt{-m})^6) = \mathfrak{f}_1(\gamma\sqrt{-m})^6.$$

However, (12.26) shows that $\mathfrak{f}_1(\gamma\tau)^6 = \mathfrak{f}_1(\tau)^6$ for all τ, which proves that $\mathfrak{f}_1(\sqrt{-m})^6$ is fixed by σ and hence lies in the ring class field L. This completes the proof of (i).

The equation $\sigma(\mathfrak{f}_1(\sqrt{-m})^6) = \mathfrak{f}_1(\gamma\sqrt{-m})^6$ used above is significant, for it allows us to compute the action of $\sigma \in \mathrm{Gal}(L'/K)$ using the matrix $\gamma \in \mathrm{SL}(2, \mathbb{Z})$. This correspondence between Galois automorphisms and linear fractional transformations is not unexpected, for the $\mathfrak{f}_1(\gamma\tau)^6$'s are the conjugates of $\mathfrak{f}_1(\tau)^6$ over $\mathbb{Q}(j(\tau))$, hence when we specialize to $\tau = \sqrt{-m}$, the conjugates of $\mathfrak{f}_1(\sqrt{-m})^6$ should lie among the $\mathfrak{f}_1(\gamma\sqrt{-m})^6$'s. What's surprising is that there's a systematic way of finding γ. This is the basic content of the *Shimura Reciprocity Law*. A complete statement of the theorem requires the adeles, so that we refer the reader to Lang [73, Chapter 11] or Shimura [90, §6.8] for further details.

The proof of (ii) is similar to what we did for (i), though this case is a little more difficult. We will sketch the main steps of the proof in Exercise 12.23. This completes the proof of Theorem 12.24. Q.E.D.

We can now begin Weber's computation of $j(\sqrt{-14})$ from [102, §144]. Let $K = \mathbb{Q}(\sqrt{-14})$. Since $\mathcal{O}_K = [1, \sqrt{-14}]$, $L = K(j(\sqrt{-14})$ is the Hilbert class field of K. As we saw in §5, $\mathrm{Gal}(L/K) \simeq C(\mathcal{O}_K)$ is cyclic of order 4. Furthermore, we can use the results of §6 to determine part of this extension. Recall that the *genus field* M of K is the intermediate field $K \subset M \subset L$ corresponding to the subgroup of squares. When $K = \mathbb{Q}(\sqrt{-14})$, Theorem 6.1 tells us that $M = K(\sqrt{-7}) = K(\sqrt{2})$. Thus

$$K \subset K(\sqrt{2}) \subset L.$$

We will compute $\mathfrak{f}_1(\sqrt{-14})^2$, which lies in the Hilbert class field L since $m = 14$ satisfies the hypothesis of the first part of Theorem 12.24. Let σ be the unique element of $\mathrm{Gal}(L/K)$ of order 2, so that the fixed field of σ is the genus field $K(\sqrt{2})$. The key step in the computation is to show that

$$\sigma(\mathfrak{f}_1(\sqrt{-14})^2) = \mathfrak{f}_2(\sqrt{-14}/2)^2. \tag{12.30}$$

We start with the equation

$$\mathfrak{f}_1(\sqrt{-m})^6 = R(j(\mathfrak{b}), j(\mathcal{O}'))$$

from Theorem 12.24, where $\mathcal{O}' = [1, 4\sqrt{-14}]$ and $\mathfrak{b} = [8, \sqrt{-14}]$. If $\mathcal{O}'$ and L' are as in the proof of Theorem 12.24, then $\mathfrak{b}$ determines a class in $C(\mathcal{O}')$ and hence an automorphism $\sigma_{\mathfrak{b}} \in \mathrm{Gal}(L'/K)$. It is easy to check that $\mathfrak{b}$ maps to the unique element of order 2 in $C(\mathcal{O}_K)$ (see Exercise 12.24), and consequently, the restriction of $\sigma_{\mathfrak{b}}$ to L is the above automorphism σ. By abuse of notation, we will write $\sigma = \sigma_{\mathfrak{b}}$. Then, using Corollary 11.37, we obtain

$$\sigma(\mathfrak{f}_1(\sqrt{-14})^6) = R(j(\overline{\mathfrak{b}}\mathfrak{b}), j(\overline{\mathfrak{b}})) = R(j(\mathcal{O}'), j(\mathfrak{b}))$$

since $\overline{\mathfrak{b}} = \mathfrak{b}$ and $\overline{\mathfrak{b}}\mathfrak{b} = [2, 8\sqrt{-14}] = 2\mathcal{O}'$. Thus

$$\sigma(\mathfrak{f}_1(\sqrt{-14})^6) = R(j([1, 4\sqrt{-14}]), j([8, \sqrt{-14}])). \tag{12.31}$$

Let $\gamma = \left(\begin{smallmatrix} 0 & -1 \\ 1 & 0 \end{smallmatrix}\right)$, and note that $\mathfrak{f}_2(\tau) = \mathfrak{f}_1(\gamma\tau)$ by Corollary 12.19. Combining this with (12.27), we get

$$\begin{aligned} \mathfrak{f}_2(\tau)^6 = \mathfrak{f}_1(\gamma\tau)^6 &= R(j(\gamma\tau/8), j(4\gamma\tau)) \\ &= R(j([1, 8\tau]), j([4, \tau])). \end{aligned}$$

Evaluating this at $\tau = \sqrt{-14}/2$ and using (12.31), we obtain

$$\sigma(\mathfrak{f}_1(\sqrt{-14})^6) = \mathfrak{f}_2(\sqrt{-14}/2)^6.$$

If we take the cube root of each side, we see that

$$\sigma(\mathfrak{f}_1(\sqrt{-14})^2) = \zeta_3^i \mathfrak{f}_2(\sqrt{-14}/2)^2$$

for some cube root of unity ζ_3^i. It remains to prove that the cube root is 1. From (12.16) we know that $\mathfrak{f}_1(\tau)\mathfrak{f}_2(\tau/2) = \sqrt{2}$, so that

$$\mathfrak{f}_1(\sqrt{-14})^2\sigma(\mathfrak{f}_1(\sqrt{-14}/2)^2) = \zeta_3^i\mathfrak{f}_1(\sqrt{-14})^2\mathfrak{f}_2(\sqrt{-14}/2)^2 = 2\zeta_3^i. \tag{12.32}$$

Since $\mathfrak{f}_1(\sqrt{-14})^2\sigma(\mathfrak{f}_1(\sqrt{-14})^2)$ is fixed by σ, it lies in $K(\sqrt{2})$, and hence $\zeta_3^i \in K(\sqrt{2}) = \mathbb{Q}(\sqrt{2}, \sqrt{-7})$. This forces the root of unity to be 1, and (12.30) is proved.

Now let $\alpha = f_1(\sqrt{-14})^2$. From (12.32) we see that $\alpha\sigma(\alpha) = 2$, so that $\alpha + \sigma(\alpha) = \alpha + 2/\alpha$ lies in $K(\sqrt{2})$. But α is clearly real, so that $\alpha + 2/\alpha \in \mathbb{Q}(\sqrt{2})$, and furthermore, α and $2/\alpha = \sigma(\alpha)$ are algebraic integers by Theorem 12.24. It follows that

$$\alpha + \frac{2}{\alpha} = a + b\sqrt{2}, \qquad a, b \in \mathbb{Z}. \tag{12.33}$$

We will use a wonderful argument of Weber to show that a and b are both positive. Namely, (12.33) gives a quadratic equation for α, and since α is real and positive (see the product formula for $\mathfrak{f}_1(\tau)$), the discriminant must be nonnegative, i.e.,

$$(a + b\sqrt{2})^2 \geq 8.$$

Let σ_1 be a generator of $\mathrm{Gal}(L/K)$ (so $\sigma = \sigma_1^2$). Then $\sigma_1(\sqrt{2}) = -\sqrt{2}$, and hence

$$\sigma_1(\alpha) + \frac{2}{\sigma_1(\alpha)} = a - b\sqrt{2}.$$

But $\sigma_1(\alpha)$ cannot be real, for then $L \cap \mathbb{R} = \mathbb{Q}(\alpha)$ would be Galois over $\mathbb{Q}$, which contradicts $\mathrm{Gal}(L/\mathbb{Q}) \simeq D_8$ (see Lemma 9.3). Thus the discriminant of the resulting quadratic equation must be negative, i.e.,

$$(a - b\sqrt{2})^2 < 8.$$

Subtracting these two inequalities gives

$$4ab\sqrt{2} > 0,$$

so that a and b are positive since $\alpha > 0$.

As a and b range over all positive integers, the resulting numbers $a + b\sqrt{2}$ form a discrete subset of $\mathbb{R}$ (by contrast, $\mathbb{Z}[\sqrt{2}]$ is dense in $\mathbb{R}$). Thus we can compute a and b by approximating $\alpha + 2/\alpha$ sufficiently closely. Setting $q = e^{-\pi\sqrt{14}}$, (12.15) implies

$$\frac{2}{\alpha} = \mathfrak{f}_2(\sqrt{-14}/2)^2 = 2q^{1/12}\prod_{n=1}^{\infty}(1 + q^n)^2.$$

Applying the methods used in our class number 1 calculations, we see that

$$2q^{1/12} < \frac{2}{\alpha} < 2q^{1/12}e^{2.002q},$$

and thus

$$q^{-1/12}e^{-2.002q} < \alpha < q^{-1/12}.$$

These inequalities imply that

$$\alpha + \frac{2}{\alpha} \approx q^{-1/12} + 2q^{1/12} \approx 2.6633 + .7509 = 3.4142,$$

with an error of at most 10^{-4}. Compare this to the smallest values of $a + b\sqrt{2}$, $a, b > 0$:

$$1 + \sqrt{2} \approx 2.4142 < 2 + \sqrt{2} \approx 3.4142 < 1 + 2\sqrt{2} \approx 3.8284.$$

It follows that $\alpha + 2/\alpha = 2 + \sqrt{2}$, and then the quadratic formula implies

$$\alpha = \frac{2 + \sqrt{2} \pm \sqrt{4\sqrt{2} - 2}}{2} = \frac{\sqrt{2} + 1 \pm \sqrt{2\sqrt{2} - 1}}{\sqrt{2}}.$$

Since $\alpha \approx 2.6633$ is the larger root, we have

$$\alpha = \mathfrak{f}_1(\sqrt{-14})^2 = \frac{\sqrt{2} + 1 + \sqrt{2\sqrt{2} - 1}}{\sqrt{2}},$$

and we can now compute $\gamma_2(\sqrt{-14})$:

$$\begin{aligned}
\gamma_2(\sqrt{-14}) &= \mathfrak{f}_1(\sqrt{-14})^{16} + \frac{16}{\mathfrak{f}_1(\sqrt{-14})^8} \\
&= \alpha^8 + \frac{16}{\alpha^4} = \alpha^8 + \left(\frac{2}{\alpha}\right)^4 \\
&= \left(\frac{\sqrt{2} + 1 + \sqrt{2\sqrt{2} - 1}}{\sqrt{2}}\right)^8 + \left(\frac{\sqrt{2} + 1 - \sqrt{2\sqrt{2} - 1}}{\sqrt{2}}\right)^4 \\
&= 2\left(323 + 228\sqrt{2} + (231 + 161\sqrt{2})\sqrt{2\sqrt{2} - 1}\right),
\end{aligned}$$

where the last step was done using REDUCE. Cubing this, we get the formula for $j(\sqrt{-14})$ given in (12.1).

One corollary is that $L = K(\sqrt{2\sqrt{2} - 1})$ is the Hilbert class field of $K = \mathbb{Q}(\sqrt{-14})$. This method of determining L is more difficult than what we did in §5, but it's worth the effort—the formulas are simply wonderful! These same techniques can be used to determine $j(\sqrt{-46})$ and $j(\sqrt{-142})$ (see

Exercise 12.25), and in [102, §144] Weber does 7 other cases by similar methods.

The examples done so far represent only a small fraction of the singular j-invariants computed by Weber in [102]. He uses a wide variety of methods and devotes many sections to computations—the interested reader should consult §§125, 128, 129, 130, 131, 135, 139, 143, and 144 for more examples. We should also mention that in 1927, Berwick [4] published the j-invariants (in factored form) of all known orders of class number ≤ 3. For a modern discussion of how to compute singular moduli, see Herz [55].

E. Imaginary Quadratic Fields of Class Number 1

We will end this section with another application of the Weber functions: the determination of all imaginary quadratic fields of class number 1.

Theorem 12.34. *Let K be an imaginary quadratic field of discriminant d_K. Then*

$$h(d_K) = 1 \Longleftrightarrow d_K = -3, -4, -7, -8, -11, -19, -43, -67, -163.$$

Remark. As we saw in Theorem 7.30, this theorem enables us to determine all discriminants D with $h(D) = 1$.

Proof. This theorem was first proved by Heegner [52] in 1952, but his proof was not accepted at first, partly because of his heavy reliance on Weber. In 1966 complete proofs were found independently by Baker [3] and Stark [96], which led people to look back at Heegner's work and realize that he did have a complete proof after all (see Birch [6] and Stark [98]). We will follow Stark's presentation [98] of Heegner's argument.

The first part of the proof is quite elementary. Let d_K be a discriminant such that $h(d_K) = 1$. Recall from Theorem 2.18 that $h(-4n) = 1$ if and only if $-4n = -4, -8, -12, -16$ or -28. Thus, if $d_K \equiv 0 \bmod 4$, then $d_K = -4$ or -8 since d_K is a field discriminant. So we may assume $d_K \equiv 1 \bmod 4$, and then Theorem 3.15 implies that there are $2^{\mu-1}$ genera of forms of discriminant d_K, where μ is the number of primes dividing d_K. Since $h(d_K) = 1$, it follows that $\mu = 1$, so that $d_K = -p$, where $p \equiv 3 \bmod 4$ is prime.

If $p \equiv 7 \bmod 8$, then Theorem 7.24 implies that

$$h(-4p) = 2h(-p)\left(1 - \left(\frac{-p}{2}\right)\frac{1}{2}\right) = h(-p) = 1,$$

and using Theorem 2.18 again, we see that $p = 7$.

We are thus reduced to the case $p \equiv 3 \bmod 8$, and of course we may assume that $p \neq 3$. Then Theorem 7.24 tells us that

$$h(-4p) = 2h(-p)\left(1 - \left(\frac{-p}{2}\right)\frac{1}{2}\right) = 3h(-p) = 3.$$

This implies that $\mathbb{Q}(j(\sqrt{-p}))$ has degree 3 over $\mathbb{Q}$. By the second part of Theorem 12.24, we know that $\mathfrak{f}(\sqrt{-p})^2 \in K(j(\sqrt{-p}))$, and since $\mathfrak{f}(\sqrt{-p})^2$ is real, we see that $\mathfrak{f}(\sqrt{-p})^2$ generates a cubic extension of $\mathbb{Q}$.

Let $\tau_0 = (3+\sqrt{-p})/2$, and set $\alpha = \zeta_8\, \mathfrak{f}_2(\tau_0)^2$. We can relate this to $\mathfrak{f}(\sqrt{-p})^2$ as follows. We know from (12.16) that

$$\mathfrak{f}_1(2\tau_0)\mathfrak{f}_2(\tau_0) = \sqrt{2},$$

and Corollary 12.19 tells us that

$$\mathfrak{f}_1(2\tau_0) = \mathfrak{f}_1(3+\sqrt{-p}) = \zeta_{48}^{-3}\mathfrak{f}(\sqrt{-p}) = \zeta_{16}^{-1}\mathfrak{f}(\sqrt{-p}).$$

These formulas imply that $\alpha = 2/\mathfrak{f}(\sqrt{-p})^2$, and hence α generates the cubic extension $\mathbb{Q}(\mathfrak{f}(\sqrt{-p})^2)$. Note also that α^4 generates the same cubic extension.

Let's study the minimal polynomial of α^4. Since $\mathcal{O} = [1, \tau_0]$ and $h(-p) = 1$, we know that $j(\tau_0)$ is an integer, and then $\gamma_2(\tau_0)$ is also an integer by Theorem 12.2. Since

$$\gamma_2(\tau_0) = \frac{\mathfrak{f}_2(\tau_0)^{24} + 16}{\mathfrak{f}_2(\tau_0)^8},$$

it follows that $\alpha^4 = -\mathfrak{f}_2(\tau_0)^8$ is a root of the cubic equation

$$x^3 - \gamma_2(\tau_0)x - 16 = 0. \tag{12.35}$$

This is the minimal polynomial of α^4 over $\mathbb{Q}$.

However, α is also cubic over $\mathbb{Q}$, and thus satisfies an equation of the form

$$x^3 + ax^2 + bx + c = 0,$$

where a, b and c lie in $\mathbb{Z}$ since α is an algebraic integer. Heegner's insight was that this equation put some very strong constraints on the equation satisfied by α^4. In fact, moving the even degree terms to the right and squaring, we get

$$(x^3 + bx)^2 = (-ax^2 - c)^2,$$

so that α satisfies

$$x^6 + (2b - a^2)x^4 + (b^2 - 2ac)x^2 - c^2 = 0.$$

Hence α^2 satisfies the cubic equation

$$x^3 + ex^2 + fx + g = 0, \qquad e = 2b - a^2, \quad f = b^2 - 2ac, \quad g = -c^2,$$

and repeating this process, we see that α^4 satisfies the cubic equation

$$x^3 + (2f - e^2)x^2 + (f^2 - 2eg)x - g^2.$$

By the uniqueness of the minimal polynomial, this equation must equal (12.35). Comparing coefficients, we obtain

$$\begin{aligned} 2f - e^2 &= 0 \\ f^2 - 2eg &= -\gamma_2(\tau_0) \\ g^2 &= 16. \end{aligned} \tag{12.36}$$

The third equation of (12.36) implies $g = \pm 4$, and since $g = -c^2$, we have $g = -4$ and $c = \pm 2$. However, changing α to $-\alpha$ leaves α^4 fixed but takes a, b, c to $-a, b, -c$. Thus we may assume $c = 2$, and it follows that

$$\gamma_2(\tau_0) = -f^2 - 8e = -(b^2 - 4a)^2 - 8(2b - a^2).$$

It remains to determine the possible a's and b's.

The first equation $2f = e^2$ of (12.36) may be written

$$2(b^2 - 4a) = (2b - a^2)^2,$$

which implies that a and b are even. If we set $X = -a/2$ and $Y = (b - a^2)/2$, then a little algebra shows that X and Y are integer solutions of the Diophantine equation

$$2X(X^3 + 1) = Y^2$$

(see Exercise 12.26). This equation has the following integer solutions:

Proposition 12.37. *The only integer solutions of the Diophantine equation* $2X(X^3 + 1) = Y^2$ *are* $(X, Y) = (0,0)$, $(-1,0)$, $(1,\pm 2)$, *and* $(2,\pm 6)$.

Proof. Let (X, Y) be an integer solution. Since X and $X^3 + 1$ are relatively prime, the equation $2X(X^3 + 1) = Y^2$ implies that $\pm(X^3 + 1)$ is a square or twice a square. Thus (X, Y) gives an integer solution of one of four Diophantine equations. These equations, together with some of their obvious solutions, may be written as follows:

(i) $X^3 + 1 = Z^2$, $(X, Z) = (-1,0)$, $(0,\pm 1)$, $(2,\pm 3)$.
(ii) $X^3 + 1 = -Z^2$, $(X, Z) = (-1,0)$.
(iii) $W^6 + 1 = 2Z^2$, $(W, Z) = (1,\pm 1)$.
(iv) $X^3 + 1 = -2Z^2$, $(X, Z) = (-1,0)$.

To explain (iii), note that if $X^3 + 1 = 2Z^2$, then $2X(X^3 + 1) = Y^2$ implies that $X = W^2$ for some W, which by substitution gives us $W^6 + 1 = 2Z^2$. In Exercises 12.27–12.29, we will show that the solutions listed above are *all* integer solutions of these four equations. Once this is done, the proposition follows easily.

The integer solutions of equations (ii)–(iv) are relatively easy to find. We need nothing more than the techniques used when we considered the equation $Y^2 = X^3 - 2$ in Exercises 5.21 and 5.22. See Exercise 12.27 for the details of these three cases.

The integer solutions of equation (i) are more difficult to find, and the elementary methods used in (ii)–(iv) don't suffice. Fortunately, we can turn to Euler for help, for in 1738 he used Fermat's technique of infinite descent to determine all integer (and rational) solutions of (iv) (see [33, Vol. II, pp. 56–58]). A version of Euler's argument may be found in Exercises 12.28 and 12.29. This completes the proof of the proposition. Q.E.D.

Once we know the solutions of $2X(X^3+1) = Y^2$, we can compute a, b and hence $\gamma_2(\tau_0)$. This gives us the following table:

X	Y	$a = -2X$	$b = 4X^2 + 2Y$	$\gamma_2(\tau_0) = -(b^2-4a)^2 - 8(2b-a^2)$
0	0	0	0	0
−1	0	2	4	−96
1	2	−2	8	−5280
1	−2	−2	0	−32
2	6	−4	28	−640320
2	−6	−4	4	−960.

Note that these $\gamma_2(\tau_0)$'s are among those computed earlier in table (12.20). Since $j(\mathcal{O}_K)$ determines K uniquely (see Exercise 12.30), it follows that we now know all imaginary quadratic fields of class number 1. This proves the theorem. Q.E.D.

Note that Heegner's argument is clever but elementary—the hard part is proving that $\mathfrak{f}(\sqrt{-p})^2$ lies in the appropriate ring class field. Thus Weber could have solved the class number 1 problem in 1908! We should also mention that there is a more elementary version of the above argument which makes no use of the Weber functions (see Stark [98]).

F. Exercises

12.1. Show that $g_2(\tau)$, $g_3(\tau)$ and $\Delta(\tau)$ are real-valued when τ is purely imaginary. Hint: use Exercise 11.1.

12.2. Let $F(q) = 1 + \sum_{n=1}^{\infty} a_n q^n$ be a power series which converges in a neighborhood of the origin.

(a) Show that for any positive integer m, there is a unique power series $G(q)$, converging in a possibly smaller neighborhood of 0, such that $F(q) = G(q)^m$ and $G(0) = 1$.

(b) If in addition the coefficients of $F(q)$ are rational numbers, show that the power series $G(q)$ from part (a) also has rational coefficients.

12.3. In this exercise we will prove that $S = \left(\begin{smallmatrix} 0 & -1 \\ 1 & 0 \end{smallmatrix}\right)$ and $T = \left(\begin{smallmatrix} 1 & 1 \\ 0 & 1 \end{smallmatrix}\right)$ generate $\mathrm{SL}(2, \mathbf{Z})$. To start, let Γ be the subgroup of $\mathrm{SL}(2, \mathbf{Z})$ generated by S and T.

(a) Show that every element of $\mathrm{SL}(2, \mathbf{Z})$ of the form $\left(\begin{smallmatrix} a & b \\ 0 & d \end{smallmatrix}\right)$ or $\left(\begin{smallmatrix} 0 & b \\ c & d \end{smallmatrix}\right)$ lies in Γ.

(b) Fix $\gamma_0 \in \mathrm{SL}(2, \mathbf{Z})$, and choose $\gamma \in \Gamma$ so that $\gamma\gamma_0 = \left(\begin{smallmatrix} a & b \\ c & d \end{smallmatrix}\right)$ has the minimal $|c|$.

(i) If $a = 0$ or $c = 0$, then use (a) to show that $\gamma_0 \in \Gamma$.

(ii) If $c \neq 0$, then, of the $\gamma's$ that give the minimal $|c|$, choose one that has the minimal $|a|$. Use

$$T^{\pm 1}\begin{pmatrix} a & b \\ c & d \end{pmatrix} = \begin{pmatrix} a \pm c & * \\ c & * \end{pmatrix}$$

to show that $|a| \geq |c|$, and then use

$$S\begin{pmatrix} a & b \\ c & d \end{pmatrix} = \begin{pmatrix} -c & * \\ a & * \end{pmatrix}$$

to show that $a = 0$. Conclude that $\gamma_0 \in \Gamma$.

(c) Use (a) and (b) to show that S and T generate $\mathrm{SL}(2, \mathbf{Z})$.

12.4. In this exercise we will give generators for the following subgroups of $\mathrm{SL}(2, \mathbf{Z})$:

$$\Gamma_0(2) = \left\{\begin{pmatrix} a & b \\ c & d \end{pmatrix} \in \mathrm{SL}(2, \mathbf{Z}) : c \equiv 0 \bmod 2\right\}$$

$$\Gamma_0(2)^t = \left\{\begin{pmatrix} a & b \\ c & d \end{pmatrix} \in \mathrm{SL}(2, \mathbf{Z}) : b \equiv 0 \bmod 2\right\}$$

$$\Gamma(2) = \left\{\begin{pmatrix} a & b \\ c & d \end{pmatrix} \in \mathrm{SL}(2, \mathbf{Z}) : b \equiv c \equiv 0 \bmod 2\right\}.$$

Let $I = \left(\begin{smallmatrix} 1 & 0 \\ 0 & 1 \end{smallmatrix}\right)$, $A = \left(\begin{smallmatrix} 1 & 0 \\ 1 & 1 \end{smallmatrix}\right)$ and $B = \left(\begin{smallmatrix} 1 & 1 \\ 0 & 1 \end{smallmatrix}\right)$.

(a) Modify the argument of Exercise 12.3 to show that $-I$, A^2 and B generate $\Gamma_0(2)$. Hint: let Γ be generated by $-I$, A^2 and B. Given $\gamma_0 \in \Gamma_0(2)$, choose $\gamma \in \Gamma$ so that $\gamma\gamma_0 = \left(\begin{smallmatrix} a & b \\ c & d \end{smallmatrix}\right)$ is minimal in the sense of Exercise 12.3. If $c \neq 0$, show that $|a| < |c|$, and then use

$$A^{\pm 2}\begin{pmatrix} a & b \\ c & d \end{pmatrix} = \begin{pmatrix} a & * \\ c \pm 2a & * \end{pmatrix}$$

to prove that $a = 0$, which is impossible in this case.

(b) Show that $-I$, A and B^2 generate $\Gamma_0(2)'$. In the text, these generators are denoted $-I$, U and V respectively.

(c) Adapt the argument of (a) to show that $-I$, A^2 and B^2 generate $\Gamma(2)$.

12.5. This exercise is concerned with the properties of $\gamma_2(\tau)$.

(a) Prove (12.6) by induction on the length of $\left(\begin{smallmatrix} a & b \\ c & d \end{smallmatrix}\right)$ as a word in the matrices S and T of Exercise 12.3.

(b) Use (12.6) to show that $\gamma_2(\tau)$ is invariant under the group

$$\tilde{\Gamma}(3) = \left\{ \begin{pmatrix} a & b \\ c & d \end{pmatrix} : b \equiv c \equiv 0 \bmod 3 \right\}.$$

(c) Show that

$$\Gamma_0(9) = \begin{pmatrix} 1/3 & 0 \\ 0 & 1 \end{pmatrix} \tilde{\Gamma}(3) \begin{pmatrix} 3 & 0 \\ 0 & 1 \end{pmatrix},$$

and conclude that $\gamma_2(3\tau)$ is invariant under $\Gamma_0(9)$.

(d) Use (12.6) to show that the exact subgroup of $SL(2, \mathbf{Z})$ under which $\gamma_2(\tau)$ is invariant is

$$\left\{ \begin{pmatrix} a & b \\ c & d \end{pmatrix} \in SL(2, \mathbf{Z}) : a \equiv d \equiv 0 \bmod 3 \text{ or } b \equiv c \bmod 3 \right\}.$$

12.6. Complete the proof of part (ii) of Proposition 12.7 using the hints given in the text.

12.7. Let $\mathcal{O} = [1, \tau_0]$ be an order of discriminant D in an imaginary quadratic field, and assume that $\tau_0 = \sqrt{-m}$ or $(3 + \sqrt{-m})/2$, depending on whether $D \equiv 0$ or $1 \bmod 4$. Let $\mathcal{O}' = [1, 3\tau_0]$ be the order of index 3 in $\mathcal{O}$. If $3 \nmid D$, then prove that $[1, \tau_0/3]$ is a proper fractional $\mathcal{O}'$-ideal. Hint: use Lemma 7.5.

12.8. Adapt the argument of Lemma 12.11 to show that

$$\frac{\partial \Phi_9}{\partial X}(j(3i), j(i/3)) \neq 0.$$

Hint: it suffices to show that (12.12) cannot hold.

12.9. This exercise is concerned with the elementary properties of the Weber functions.

(a) Prove the product expansions (12.15).

(b) Prove the top line of (12.16). Hint: use the product exansions to show that

$$\eta(\tau)\mathfrak{f}(\tau)\mathfrak{f}_1(\tau)\mathfrak{f}_2(\tau) = \sqrt{2}\eta(\tau).$$

(c) Prove the bottom line of (12.16). Hint: use the definitions.

12.10. Exercises 12.10, 12.11 and 12.13 are concerned with the Weierstrass σ-function. The basic properties of $\sigma(z;\tau)$ will be covered, though we will neglect the details of convergence. For careful treatment of this material, see Chandrasekharan [16, Chapter IV], Lang [73, Chapter 18], and Whittaker and Watson [109, Chapter XX]. As in the text, the σ-function is defined by

$$\sigma(z;\tau) = z \prod_{\omega \in L-\{0\}} \left(1 - \frac{z}{\omega}\right) e^{z/\omega + (1/2)(z/\omega)^2},$$

where $L = [1,\tau]$. Note that $\sigma(z;\tau)$ is an odd function in z. We will write $\sigma(z)$ instead of $\sigma(z;\tau)$.

(a) Define the Weierstrass ζ-function $\zeta(z)$ (not to be confused with Riemann's) by

$$\zeta(z) = \frac{\sigma'(z)}{\sigma(z)}.$$

Using the definition of $\sigma(z)$, show that

$$\zeta(z) = \frac{1}{z} + \sum_{\omega \in L-\{0\}} \left(\frac{1}{z-\omega} + \frac{1}{\omega} + \frac{z}{\omega^2}\right).$$

(b) Show that the ζ-function is related to the $\wp$-function by the formula

$$\wp(z) = -\zeta'(z).$$

(c) By (b), it follows that if $\omega \in L$, then $\zeta(z+\omega) - \zeta(z)$ is a constant depending only on ω. Since $L = [1,\tau]$, we define η_1 and η_2 by the formulas

$$\begin{aligned} \eta_1 &= \zeta(z+\tau) - \zeta(z) \\ \eta_2 &= \zeta(z+1) - \zeta(z). \end{aligned}$$

Then prove Legendre's relation

$$\eta_2\tau - \eta_1 = 2\pi i.$$

Hint: consider $\int_\Gamma \zeta(z)dz$, where Γ is the boundary, oriented counterclockwise, of the parallelogram **P** used in the proof of Lemma 10.4. By standard residue theory, the integral equals $2\pi i$ by (a). But the defining relations for η_1 and η_2 allow one to compute the integral directly.

(d) We can now show that

$$\begin{aligned} \sigma(z+\tau) &= -e^{\eta_1(z+\frac{\tau}{2})}\sigma(z) \\ \sigma(z+1) &= -e^{\eta_2(z+\frac{1}{2})}\sigma(z). \end{aligned}$$

(i) Show that

$$\frac{d}{dz}\frac{\sigma(z+\tau)}{\sigma(z)} = \eta_1 \frac{\sigma(z+\tau)}{\sigma(z)},$$

and conclude that for some constant C,

$$\sigma(z+\tau) = Ce^{\eta_1 z}\sigma(z).$$

(ii) Determine the constant C in (i) by evaluating the above identity at $z = -\tau/2$. This will prove the desired formula for $\sigma(z+\tau)$. Hint: recall that $\sigma(z)$ is an odd function.

(iii) In a similar way, prove the formula for $\sigma(z+1)$.

12.11. The goal of this exercise is to prove the formula

$$\wp(z) - \wp(w) = -\frac{\sigma(z+w)\sigma(z-w)}{\sigma^2(z)\sigma^2(w)}.$$

Fix $w \notin L = [1,\tau]$, and consider the function

$$f(z) = -\frac{\sigma(z+w)\sigma(z-w)}{\sigma^2(z)\sigma^2(w)}.$$

(a) Show that $f(z)$ is an even elliptic function for L. By Lemma 10.17, this implies that $f(z)$ is a rational function in $\wp(z)$.

(b) Show that $f(z)$ is holomorphic on $\mathbf{C} - L$ and that its Laurent expansion at $z = 0$ begins with $1/z^2$.

(c) Conclude from (b) that $f(z) = \wp(z) + C$ for some constant C, and evaluate the constant by setting $z = w$. This proves the desired formula.

12.12. Use the previous exercise to show that

$$e_2 - e_1 = e^{-\eta_2\tau/2}\frac{\sigma^2\left(\dfrac{\tau+1}{2}\right)}{\sigma^2\left(\dfrac{1}{2}\right)\sigma^2\left(\dfrac{\tau}{2}\right)}$$

$$e_2 - e_3 = e^{\eta_2(\tau+1)/2}\frac{\sigma^2\left(\dfrac{\tau}{2}\right)}{\sigma^2\left(\dfrac{1}{2}\right)\sigma^2\left(\dfrac{\tau+1}{2}\right)}$$

$$e_3 - e_1 = e^{\eta_1(\tau+1)/2}\frac{\sigma^2\left(\dfrac{1}{2}\right)}{\sigma^2\left(\dfrac{\tau+1}{2}\right)\sigma^2\left(\dfrac{\tau}{2}\right)}.$$

Hint: for $e_2 - e_1$, use the fact that

$$\sigma\left(\frac{1-\tau}{2}\right) = \sigma\left(-\frac{1+\tau}{2}+1\right) = -e^{\eta_2(-(\tau+1)/2+1/2)}\sigma\left(-\frac{1+\tau}{2}\right)$$
$$= e^{-\eta_2\tau/2}\sigma\left(\frac{\tau+1}{2}\right).$$

12.13. The final fact we need to know about the σ-function is its q-product expansion

$$\sigma(z;\tau) = \frac{1}{2\pi i}e^{\eta_2 z^2/2}(q_z^{1/2} - q_z^{-1/2})\prod_{n=1}^{\infty}\frac{(1-q_\tau^n q_z)(1-q_\tau^n/q_z)}{(1-q_\tau^n)^2},$$

where $q_\tau = e^{2\pi i\tau}$ and $q_z = e^{2\pi i z}$. To prove this, let $f(z)$ denote the right-hand side of the above equation.

(a) Show that the zeros of $f(z)$ and $\sigma(z)$ are exactly the points of L. Thus $\sigma(z)/f(z)$ is holomorphic on $\mathbb{C} - L$.

(b) Show that $\sigma(z)/f(z)$ has periods $L = [1,\tau]$.

(c) Show that $\sigma(z)/f(z)$ is holmorphic at $z = 0$ and takes the value 1 there.

(d) Conclude that $\sigma(z) = f(z)$. Hint: use Exercise 10.5.

12.14. This exercise will complete the proof of the formulas (12.18) expressing $e_i - e_j$ in terms of $\eta(\tau)$ and the Weber functions.

(a) Use the product expansion from Exercise 12.13 to show that

$$\sigma\left(\frac{1}{2}\right) = \frac{1}{2\pi}e^{\eta_2/8}\frac{\mathfrak{f}_2(\tau)^2}{\eta(\tau)^2}$$
$$\sigma\left(\frac{\tau}{2}\right) = \frac{i}{2\pi}e^{\eta_2\tau^2/8}q^{-1/8}\frac{\mathfrak{f}_1(\tau)^2}{\eta(\tau)^2}$$
$$\sigma\left(\frac{\tau+1}{2}\right) = \frac{1}{2\pi}e^{\eta_2(\tau+1)^2/8}q^{-1/8}\frac{\mathfrak{f}(\tau)^2}{\eta(\tau)^2}.$$

(b) Use (a) and the formulas from Exercise 12.12 to prove

$$e_2 - e_1 = \pi^2\eta(\tau)^4\mathfrak{f}(\tau)^8$$
$$e_2 - e_3 = \pi^2\eta(\tau)^4\mathfrak{f}_1(\tau)^8$$
$$e_3 - e_1 = \pi^2\eta(\tau)^4\mathfrak{f}_2(\tau)^8.$$

This proves (12.18). Hint: use (12.16).

12.15. In this exercise we will complete the proof of Theorem 12.17. Recall from Exercise 10.8 that $g_2(\tau) = -4(e_1e_2 + e_1e_3 + e_2e_3)$ and $e_1 + e_2 + e_3 = 0$.

(a) Show that

$$3g_2(\tau) = 4((e_2 - e_1)^2 - (e_2 - e_3)(e_3 - e_1)).$$

(b) The identity of part (a), together with the formulas for $e_i - e_j$, were used in the text to derive a formula for $\gamma_2(\tau)$ in terms of $\mathfrak{f}(\tau)$. Find two other identities for $3g_2(\tau)$ similar to the one given in part (a), and use them to derive formulas for $\gamma_2(\tau)$ in terms of $\mathfrak{f}_1(\tau)$ and $\mathfrak{f}_2(\tau)$.

12.16. Use the formulas for $\gamma_2(\tau)$ given in Theorem 12.17 to show that the q-expansion of the j-function has integral coefficients. This proves Theorem 11.8.

12.17. Complete the proof of Corollary 12.19.

12.18. Verify the calculations made in table 12.20.

12.19. Use Theorem 6.1 to determine the Hilbert class field of $K = \mathbf{Q}(\sqrt{-105})$, and show that its maximal real subfield is $\mathbf{Q}(\sqrt{2}, \sqrt{3}, \sqrt{5}, \sqrt{7})$. Hint: use Theorem 3.22 to show that the genus field equals the Hilbert class field in this case.

12.20. This exercise is concerned with the properties of the Weber function $\mathfrak{f}_1(\tau)$. Let $I = \left(\begin{smallmatrix} 1 & 0 \\ 0 & 1 \end{smallmatrix}\right)$, $U = \left(\begin{smallmatrix} 1 & 0 \\ 1 & 1 \end{smallmatrix}\right)$ and $V = \left(\begin{smallmatrix} 1 & 2 \\ 0 & 1 \end{smallmatrix}\right)$.

(a) Use Corollary 12.19 to show $\mathfrak{f}_1(U\tau)^6 = \mathfrak{f}_1(V\tau)^6 = -i\mathfrak{f}_1(\tau)^6$.

(b) In Exercise 12.4 we proved that $-I$, U and V generate $\Gamma_0(2)^t$. Use induction on the length of $\gamma = \left(\begin{smallmatrix} a & b \\ c & d \end{smallmatrix}\right) \in \Gamma_0(2)^t$ as a word in $-I$, U and V to show that

$$\mathfrak{f}_1(\gamma\tau)^6 = i^{-ac-(1/2)bd+(1/2)b^2c}\mathfrak{f}_1(\tau)^6.$$

12.21. In this exercise we will show how to discover the transformation law for $\mathfrak{f}_1(\tau)$ proved in part (b) of Exercise 12.20. Let $-I$, U and V be as in Exercise 12.20. We will be using the groups

$$\Gamma(2) = \left\{ \begin{pmatrix} a & b \\ c & d \end{pmatrix} \in SL(2, \mathbf{Z}) : b \equiv c \equiv 0 \bmod 2 \right\}$$

$$\tilde{\Gamma}(8) = \left\{ \begin{pmatrix} a & b \\ c & d \end{pmatrix} \in SL(2, \mathbf{Z}) : b \equiv c \equiv 0 \bmod 8 \right\}.$$

Note that $\tilde{\Gamma}(8) \subset \Gamma(2) \subset \Gamma_0(2)^t$, and recall from Exercise 12.4 that $-I$, U^2 and V generate $\Gamma(2)$.

(a) Show that $\Gamma(2)$ has index 2 in $\Gamma_0(2)^t$ with I and U as coset representatives.

(b) Show that $\tilde{\Gamma}(8)$ is normal in $\mathrm{SL}(2,\mathbb{Z})$ and that the quotient $\Gamma(2)/\tilde{\Gamma}(8)$ is Abelian. Hint: compute $[U^2,V]$.

(c) We can now discover how $\mathfrak{f}_1(\tau)^6$ transforms under $\gamma = \left(\begin{smallmatrix} a & b \\ c & d \end{smallmatrix}\right) \in \Gamma(2)$. Write

$$\gamma = \pm \prod_{i=1}^{s} U^{2a_i} V^{b_i},$$

and set $A = \sum_{i=1}^{s} a_i$ and $B = \sum_{i=1}^{s} b_i$.

(i) Show that $\mathfrak{f}_1(\gamma\tau)^6 = i^{-2A-B}\mathfrak{f}_1(\tau)^6$.

(ii) Use (b) to show that $\gamma \equiv U^{2A}V^B \bmod \tilde{\Gamma}(8)$, which means that

$$\begin{pmatrix} a & b \\ c & d \end{pmatrix} \equiv \begin{pmatrix} 1 & 2B \\ 2A & 1+4AB \end{pmatrix} \bmod \tilde{\Gamma}(8).$$

(iii) Use (ii) to show that $ac \equiv 2A \bmod 8$ and $bd \equiv 2B \bmod 8$.

(iv) Conclude that for all $\gamma \in \Gamma(2)$,

$$\mathfrak{f}_1(\gamma\tau)^6 = i^{-ac-(1/2)bd}\mathfrak{f}_1(\tau)^6.$$

(d) Now take $\gamma = \left(\begin{smallmatrix} a & b \\ c & d \end{smallmatrix}\right) \in \Gamma_0(2)^t$, $\gamma \notin \Gamma(2)$. By (a), we can write $\gamma = U\tilde{\gamma}$ for some $\tilde{\gamma} \in \Gamma(2)$. Then use (c) to show that

$$\mathfrak{f}_1(\gamma\tau)^6 = i^{-ac-(1/2)bd+(1/2)b^2}\mathfrak{f}_1(\tau)^6.$$

Hint: observe that $a^2 \equiv 1 \bmod 4$ in this case.

(e) To unify the formulas of (c) and (d), take $\gamma = \left(\begin{smallmatrix} a & b \\ c & d \end{smallmatrix}\right) \in \Gamma_0(2)^t$. Show that

$$\tfrac{1}{2}b^2c \equiv \begin{cases} 0 \bmod 4 & \gamma \in \Gamma(2) \\ \tfrac{1}{2}b^2 \bmod 4 & \gamma \notin \Gamma(2). \end{cases}$$

From here, it follows immediately that

$$\mathfrak{f}_1(\gamma\tau)^6 = i^{-ac-(1/2)bd+(1/2)b^2c}\mathfrak{f}_1(\tau)^6$$

for all $\gamma \in \Gamma_0(2)^t$.

12.22. Let $\mathcal{O} = [1,\sqrt{-m}]$ and $\mathcal{O}' = [1,4\sqrt{-m}]$, where $m > 0$ is an integer satsifying $m \equiv 6 \bmod 8$. Note that $\mathcal{O}'$ is the order of index 4 in $\mathcal{O}$. Let $\mathfrak{a} = [4,1+\sqrt{-m}]$ and $\mathfrak{b} = [8,\sqrt{-m}]$.

(a) Show that $\mathfrak{a}$ and $\mathfrak{b}$ are proper fractional $\mathcal{O}'$-ideals. Hint: use Lemma 7.5.

(b) Show that the class of $\mathfrak{a}$ has order 4 in $C(\mathcal{O}')$ and is in the kernel of the natural map $C(\mathcal{O}') \to C(\mathcal{O})$.

(c) Verify that $\bar{\mathfrak{a}}\mathfrak{b} = [8,-2+\sqrt{-m}]$ and $\bar{\mathfrak{a}} = [4,-1+\sqrt{-m}]$.

12.23. In this exercise we will prove part (ii) of Theorem 12.24. We are thus concerned with $\mathfrak{f}(\sqrt{-m})^2$, where $m \equiv 3 \bmod 4$ is a positive integer not divisible by 3. Let L denote the ring class field of the order $\mathcal{O} = [1, \sqrt{-m}]$.

(a) Show that $\mathfrak{f}(\sqrt{-m})^6 \in L$ implies that $L = K(\mathfrak{f}(\sqrt{-m})^2)$.

(b) By Corollary 12.19, we have $\mathfrak{f}(\tau)^6 = \zeta_8\, \mathfrak{f}_1(\tau+1)^6$. Use this to prove that $\mathfrak{f}(8\tau)^6$ is a modular function for $\Gamma_0(64)$. Hint: show that $\mathfrak{f}_1(\tau)^6$ is invariant under

$$\Gamma(8) = \left\{ \begin{pmatrix} a & b \\ c & d \end{pmatrix} \in \mathrm{SL}(2,\mathbb{Z}) : \begin{pmatrix} a & b \\ c & d \end{pmatrix} \equiv \begin{pmatrix} 1 & 0 \\ 0 & 1 \end{pmatrix} \bmod 8 \right\}.$$

Since $\Gamma(8)$ is normal in $\mathrm{SL}(2,\mathbb{Z})$, this implies that $\mathfrak{f}(\tau)^6$ is also invariant under $\Gamma(8)$.

(c) Use Proposition 12.7 and Lemma 12.11 to show that

$$f(\sqrt{-m})^6 = S(j([8,\sqrt{-m}]), j([1,8\sqrt{-m}]))$$

for some rational function $S(X,Y) \in \mathbb{Q}(X,Y)$.

(d) Let $\mathcal{O}'$ be the order $[1, 8\sqrt{-m}]$. Show that $\mathfrak{a} = [8, 2+\sqrt{-m}]$ and $\mathfrak{b} = [8, \sqrt{-m}]$ are proper fractional $\mathcal{O}'$-ideals. Then use (c) to conclude that $f(\sqrt{-m})^6$ lies in the ring class field L' of $\mathcal{O}'$.

(e) Show that the extension $L \subset L'$ has degree 8 and that under the isomorphism $C(\mathcal{O}') \simeq \mathrm{Gal}(L'/K)$, the classes of the ideals $\mathfrak{a}$ and $\mathfrak{b}$ map to generators σ_1 and σ_2 of $\mathrm{Gal}(L'/L)$. Thus we need to prove that $f(\sqrt{-m})^6$ is fixed by both σ_1 and σ_2.

(f) Using (c) and Corollary 11.37, show that

$$\sigma_1(\mathfrak{f}(\sqrt{-m})^6) = S(j([4,3+\sqrt{-m}]), j([8,6+\sqrt{-m}]))$$
$$\sigma_2(\mathfrak{f}(\sqrt{-m})^6) = S(j([1,8\sqrt{-m}]), j([8,\sqrt{-m}]))$$

(this is where $m \equiv 3 \bmod 4$ is used).

(g) Let $\gamma_1 = \left(\begin{smallmatrix} 2 & 11 \\ 1 & 6 \end{smallmatrix}\right)$ and $\gamma_2 = \left(\begin{smallmatrix} 0 & -1 \\ 1 & 0 \end{smallmatrix}\right)$. Then show that

$$\mathfrak{f}(\gamma_1\tau)^6 = S(j([4,3+\tau]), j([8,6+\tau]))$$
$$\mathfrak{f}(\gamma_2\tau)^6 = S(j([1,8\tau]), j([8,\tau])).$$

(h) Use Corollary 12.19 to show that $f(\tau)^6$ is invariant under both γ_1 and γ_2. Then (f) and (g) imply that $\mathfrak{f}(\sqrt{-m})^6$ is fixed by σ_1 and σ_2, which completes the proof.

12.24. Let $\mathcal{O} = [1, \sqrt{-14}]$ and $\mathcal{O}' = [1, 4\sqrt{-14}]$. By part (a) of Exercise 12.22, we know that $\mathfrak{b} = [8, \sqrt{-14}]$ is a proper fractional $\mathcal{O}'$ ideal.

Under the natural map $C(\mathcal{O}') \to C(\mathcal{O})$, show that $\mathfrak{b}$ maps to the unique element of order 2 of $C(\mathcal{O})$.

12.25. Compute $j(\sqrt{-46})$ and $j(\sqrt{-142})$. Hint: in each case the class number is 4. Note also that $46 \equiv 142 \equiv 6 \bmod 8$, so that part (i) of Theorem 12.24 applies.

12.26. Let (a,b) be a solution of the Diophantine equation $2(b^2-4a) = (2b-a^2)^2$.

(a) Show that a and b must be even.

(b) If we set $X = -a/2$ and $Y = (b-a^2)/2$, then show that X and Y are integer solutions of the Diophantine equation $2X(X^3+1) = Y^2$.

12.27. This exercise will discuss three of the Diophantine equations that arose in the proof of Proposition 12.37. In each case, the methods used in Exercises 5.21 and 5.22 are sufficient to determine the integer solutions.

(a) Show that the only integer solutions of $X^3+1 = -Z^2$ are $(X,Z) = (-1,0)$. Hint: work in the ring $\mathbb{Z}[i]$.

(b) Show that the only integer solutions of $W^6+1 = 2Z^2$ are $(W,Z) = (\pm 1, \pm 1)$. Hint: work in the ring $\mathbb{Z}[\omega]$, $\omega = e^{2\pi i/3}$. The fact that $3 \nmid W^2+1$ will be useful.

(c) Show that the only integer solutions of $X^3+1 = -2Z^2$ are $(X,Z) = (-1,0)$. Hint: work in the ring $\mathbb{Z}[\sqrt{-2}]$.

12.28. Exercises 12.28 and 12.29 will present Euler's proof [33, Vol. II, pp. 56–58] that the only rational solutions of $X^3+1 = Z^2$ are $(X,Y) = (-1,0)$, $(0,\pm 1)$ and $(2,\pm 3)$. In this exercise we will show that there are no relatively prime positive integers c and b such that $bc(c^2-3bc+3b^2)$ is a perfect square when $c \neq b$ and $3 \nmid c$. The proof will use infinite descent. Then Exercise 12.29 will use this result to study $X^3+1 = Z^2$.

(a) Assume that c and b are positive relatively prime integers such that $bc(c^2-3bc+3b^2)$ is a perfect square, and assume also that $c \neq b$ and $3 \nmid c$. Show that b, c and $c^2-3bc+3b^2$ are relatively prime, and conclude that each is a perfect square. Then write $c^2-3bc+3b^2 = (\frac{m}{n}b-c)^2$, where $n > 0$, $m > 0$ and $\gcd(m,n) = 1$. Show that this implies

$$\frac{b}{c} = \frac{2mn-3n^2}{m^2-3n^2}.$$

There are two cases to consider, depending on whether $3 \nmid m$ or $3 \mid m$.

(b) Preserving the notation of (a), let's consider the case $3 \nmid m$.

(i) Show that $b = 2mn - 3n^2$ and $c = m^2 - 3n^2$.

(ii) Since c is a perfect square, we can write $m^2 - 3n^2 = (\frac{p}{q}n - m)^2$, where $p > 0$, $q > 0$ and $\gcd(p,q) = 1$. Show that p and q may be chosen so that $3 \nmid p$, and show also that

$$\frac{m}{n} = \frac{p^2 + 3q^2}{2pq}.$$

(iii) Prove that

$$\frac{b}{n^2} = \frac{p^2 - 3pq + 3q^2}{pq},$$

and conclude that $pq(p^2 - 3pq + 3q^2)$ is a perfect square. Show also that $p \neq q$. Hint: use (i) and (ii) to show that $p = q$ implies $c = 3$.

(iv) By (ii) and (iii) we see that p and q satisfy the same conditions as c and b. Now prove that $q < b$, which shows that the new solution is "smaller". Hint: note that $q \mid b$, so that $q < b$ unless $q = n = b$. Use (i) and (ii) to show that $c = 3$ in this case.

(c) With the same notation as (a), we will now consider the case $3 \mid m$. Then $m = 3k$, so that by (a),

$$\frac{b}{c} = \frac{n^2 - 2nk}{n^2 - 3k^2}.$$

Since $3 \nmid n$, the argument of (b) implies that $b = n^2 - 2nk$ and $c = n^2 - 3k^2$, and since c is a perfect square, we can write $n^2 - 3k^2 = (\frac{p}{q}k - n)^2$, where $p > 0$, $q > 0$ and $\gcd(p,q) = 1$. As in (b), we may assume $3 \nmid p$, and we also have

$$\frac{n}{k} = \frac{p^2 + 3q^2}{2pq}.$$

(i) Show that

$$\frac{b}{n^2} = \frac{p^2 - 4pq + 3q^2}{p^2 + 3q^2} = \frac{(p-q)(p-3q)}{p^2 + 3q^2},$$

and conclude that $(p-q)(p-3q)(p^2 + 3q^2)$ is a perfect square.

(ii) Let $t = |p - q|$ and $u = |p - 3q|$. Show that

$$(p-q)(p-3q)(p^2 + 3q^2) = tu(u^2 - 3tu + 3t^2).$$

Show also that $3 \nmid u$ and that t and u are positive and unequal.

(iii) It follows from (i) and (ii) that u and t, divided by their greatest common divisor, satisfy the same conditions as c and b. Now prove that $t < b$, so that the new solution is "smaller". Hint: consider the cases $t = q - p$ and $t = p - q$ separately. In the latter case, note that $p \mid n \pm \sqrt{n^2 - 3k^2}$, and that $p = n + \sqrt{n^2 - 3k^2}$ implies $q = k$.

Thus, given c and b satisfying the above conditions, we can always produce a pair of integers satisfying the same conditions, but with strictly smaller b. By infinite descent, no such c and b can exist.

12.29. We can now show that the only rational solutions of $X^3 + 1 = Z^2$ are $(X,Y) = (-1,0)$, $(0,\pm 1)$ and $(2,\pm 3)$. Let (X,Y) be a rational solution, and write $X = a/b$, where $b > 0$ and $\gcd(a,b) = 1$. Assume in addition that $a/b \neq -1$, 0 or 2, and set $c = a + b$. Our goal is to derive a contradiction.

(a) Show that $b(a^3 + b^3) = bc(c^2 - 3bc + 3b^2)$ is a perfect square and that b and c are relatively prime, positive, and unequal.

(b) It follows from Exercise 12.28 that $3 \mid c$. Then $c = 3d$ and $3 \nmid b$. Show that $bd(b^2 - 3bd + 3d^2)$ is a perfect square, and use Exercise 12.28 to show that $b = d$. This implies $b = d = 1$, and hence $c = 3$. Then $a/b = 2$, which contradicts our initial assumption.

12.30. If K and K' are imaginary quadratic fields and $j(\mathcal{O}_K) = j(\mathcal{O}_{K'})$, then prove that $K = K'$. Hint: use Theorem 10.9.

§13. THE CLASS EQUATION

Now that we have discussed singular j-invariants and computed some examples, it is time to turn our attention to their minimal polynomials. Given an order $\mathcal{O}$ in an imaginary quadratic field K, $H_{\mathcal{O}}(X)$ will denote the monic minimal polynomial of $j(\mathcal{O})$ over $\mathbb{Q}$. Note that $H_{\mathcal{O}}(X)$ has integer coefficients since $j(\mathcal{O})$ is an algebraic integer. The equation $H_{\mathcal{O}}(X) = 0$ is called the *class equation*, and by abuse of terminology we will refer to $H_{\mathcal{O}}(X)$ as the class equation. Since $\mathcal{O}$ is uniquely determined by its discriminant D, we will often write $H_D(X)$ instead of $H_{\mathcal{O}}(X)$.

For an example of a class equation, consider the order $\mathbb{Z}[\sqrt{-14}]$ of discriminant -56. It's j-invariant is $j(\sqrt{-14})$, which we computed in (12.1). Thus the minimal polynomial of $j(\sqrt{-14})$ is

(13.1)
$$H_{-56}(X) = X^4 - 2^8 \cdot 19 \cdot 937 \cdot 3559 X^3 + 2^{13} \cdot 251421776987 X^2 + 2^{20} \cdot 3 \cdot 11^6 \cdot 21323 X + (2^8 \cdot 11^2 \cdot 17 \cdot 41)^3,$$

where the coefficients have been factored into primes. Note that the constant term, being the norm of $j(\sqrt{-14}) = \gamma_2(\sqrt{-14})^3$, is a cube by Theorem 12.2.

The first part of §13 will describe an algorithm for computing the class equation $H_D(X)$ for *any* discriminant D. We have a special reason to be interested in this question, for by Theorem 9.2, the polynomial $H_{-4n}(X)$ gives us the criterion for when a prime is of the form $x^2 + ny^2$. Thus our algorithm will provide a constructive version of Theorem 9.2. In the second part of §13, we will discuss some more recent work of Deuring, Gross and Zagier on the class equation. We will see that there are strong restrictions on primes dividing the discriminant and constant term of the class equation. The small size of the primes appearing in the constant term of (13.1) is thus no accident.

A. Computing the Class Equation

We will begin by giving a more precise description of the class equation:

Proposition 13.2. *Let $\mathcal{O}$ be an order in an imaginary quadratic field K, and let $\mathfrak{a}_i$, $i = 1,\ldots,h$ be ideal class representatives (so that h is the class number). Then the class equation is given by the formula*

$$H_{\mathcal{O}}(X) = \prod_{i=1}^{h}(X - j(\mathfrak{a}_i)).$$

Proof. This result is an easy consequence of Corollary 11.37 (see Exercise 13.1), but there is a more elementary argument which we will now give.

By Theorem 11.1, $K(j(\mathcal{O}))$ is the ring class field of $\mathcal{O}$. Thus $[K(j(\mathcal{O})):K] = h$, and since $j(\mathcal{O})$ is real, it follows that $[\mathbb{Q}(j(\mathcal{O})):\mathbb{Q}] = h$. This shows that $H_{\mathcal{O}}(X)$ has degree h. Now let α be a root of $H_{\mathcal{O}}(X)$, and let σ be an automorphism of $\mathbb{C}$ that takes $j(\mathcal{O})$ to α. In the proof of Theorem 10.23 we showed that $\sigma(j(\mathcal{O})) = j(\mathfrak{a})$ for some proper fractional $\mathcal{O}$-ideal $\mathfrak{a}$ (see (10.26)). Hence every root of $H_{\mathcal{O}}(X)$ is also a root of $\prod_{i=1}^{h}(X - j(\mathfrak{a}_i))$, and since both polynomials are monic of degree h, they must be equal. Q.E.D.

An important consequence of this proposition is that $H_{\mathcal{O}}(X)$ is the minimal polynomial of $j(\mathfrak{a})$, where $\mathfrak{a}$ is *any* proper fractional $\mathcal{O}$-ideal.

The algorithm we will present for computing $H_{\mathcal{O}}(X)$ uses the theory of complex multiplication, and in particular, the polynomial $\Phi_m(X,X)$ obtained by setting $X = Y$ in the modular equation plays an important role. The reason for this is the following observation:

Lemma 13.3. *Let $m > 1$. If $\mathcal{O}$ has a primitive element of norm m, then the class equation $H_{\mathcal{O}}(X)$ is an irreducible factor of $\Phi_m(X,X)$. Furthermore, every irreducible factor of $\Phi_m(X,X)$ arises in this way.*

Proof. Let $\alpha \in \mathcal{O}$ be a primitive element of norm m. Corollary 11.27 tells us that $\alpha\mathcal{O} \subset \mathcal{O}$ is a cyclic sublattice of index m, and it follows that

$$0 = \Phi_m(j(\alpha\mathcal{O}), j(\mathcal{O})) = \Phi_m(j(\mathcal{O}), j(\mathcal{O})).$$

Thus $j(\mathcal{O})$ is a root of $\Phi_m(X,X)$, which implies that its minimal polynomial $H_{\mathcal{O}}(X)$ is a factor of $\Phi_m(X,X)$.

To show that every irreducible factor of $\Phi_m(X,X)$ is a class equation, suppose that $\Phi_m(\beta,\beta) = 0$. Then Theorem 11.23 implies that $\beta = j(L) = j(L')$, where $L' \subset L$ is a cyclic sublattice of index m. By Theorem 10.9, $L' = \alpha L$ for some complex number α, and then α is primitive of norm m by Corollary 11.27. Thus $\alpha \notin \mathbf{Z}$, so that L has complex multiplication by α. By Theorem 10.14, this means that up to homothety, L is a proper fractional $\mathcal{O}$-ideal for some order $\mathcal{O}$ in an imaginary quadratic field. Then $\beta = j(L)$ has $H_{\mathcal{O}}(X)$ as its minimal polynomial, and hence $H_{\mathcal{O}}(X)$ is the corresponding irreducible factor of $\Phi_m(X,X)$. This proves the lemma. Q.E.D.

The next question is, what power of $H_{\mathcal{O}}(X)$ appears in the factorization of $\Phi_m(X,X)$? The answer involves the number $r(\mathcal{O},m)$, which is defined as follows: given an order $\mathcal{O}$ in an imaginary quadratic field and a positive integer m, set

$$r(\mathcal{O},m) = |\{\alpha \in \mathcal{O} : \alpha \text{ is primitive}, \ N(\alpha) = m\}/\mathcal{O}^*|,$$

where the units $\mathcal{O}^*$ act by sending α to $\epsilon\alpha$ for $\epsilon \in \mathcal{O}^*$. It is easy to see that $r(\mathcal{O},m)$ is finite, and for a given m, there are only finitely many orders with $r(\mathcal{O},m) > 0$ (see Exercise 13.2). Then the following theorem tells us how to factor $\Phi_m(X,X)$:

Theorem 13.4. *If $m > 1$, there is a constant $c_m \in \mathbf{C}^*$ such that*

$$\Phi_m(X,X) = c_m \prod_{\mathcal{O}} H_{\mathcal{O}}(X)^{r(\mathcal{O},m)}.$$

Proof. Fix an order $\mathcal{O}$, and pick a number τ_0 in the upper half plane such that $\mathcal{O} = [1,\tau_0]$. To prove the theorem, it suffices to show that $j(\mathcal{O}) = j(\tau_0)$ is a root of $\Phi_m(X,X)$ of multiplicity $r(\mathcal{O},m)$.

We begin by studying the multiplicity of $j(\tau_0)$ as a root of $\Phi_m(X, j(\tau_0))$. Using the standard factorization

$$\Phi_m(X, j(\tau_0)) = \prod_{\sigma \in C(m)} (X - j(\sigma\tau_0)),$$

we see that

$$\Phi_m(X, j(\tau_0)) = (X - j(\tau_0))^r \prod_{j(\sigma\tau_0) \neq j(\tau_0)} (X - j(\sigma\tau_0)),$$

where

$$r = |\{\sigma \in C(m) : j(\sigma\tau_0) = j(\tau_0)\}|. \tag{13.5}$$

Thus $j(\tau_0)$ is a root of multiplicity r of $\Phi_m(X, j(\tau_0))$.

We will next show that the number r of (13.5) is the multiplicity of $j(\tau_0)$ as a root of $\Phi_m(X,X)$. To see what's involved, suppose that we have a polynomial $F(X,Y)$ and a number X_0 such that $F(X_0,X_0) = 0$. Then X_0 is a root of both $F(X,X)$ and $F(X,X_0)$, but in general, the multiplicities of these roots are different (see Exercise 13.3 for an example). So it will take a special argument to show that $j(\tau_0)$ has the same multiplicity for both $\Phi_m(X,X)$ and $\Phi_m(X, j(\tau_0))$. The basic idea is to show that

$$\lim_{u \to j(\tau_0)} \frac{\Phi_m(u,u)}{\Phi_m(u, j(\tau_0))}$$

is nonzero, which will force the multiplicities to be equal (see Exercise 13.3). To study this limit, note that

$$\begin{aligned} \lim_{u \to j(\tau_0)} \frac{\Phi_m(u,u)}{\Phi_m(u, j(\tau_0))} &= \lim_{\tau \to \tau_0} \frac{\Phi_m(j(\tau), j(\tau))}{\Phi_m(j(\tau), j(\tau_0))} \\ &= \lim_{\tau \to \tau_0} \prod_{\sigma \in C(m)} \frac{j(\tau) - j(\sigma\tau)}{j(\tau) - j(\sigma\tau_0)}. \end{aligned}$$

It suffices to compute the limit of each factor individually, and note that if $j(\tau_0) \neq j(\sigma\tau_0)$, then the limit of the corresponding factor is 1. Thus it remains to study the limit

$$\lim_{\tau \to \tau_0} \frac{j(\tau) - j(\sigma\tau)}{j(\tau) - j(\sigma\tau_0)} \tag{13.6}$$

when $\sigma \in C(m)$ satisfies $j(\tau_0) = j(\sigma\tau_0)$.

The equality $j(\tau_0) = j(\sigma\tau_0)$ implies that there is some $\gamma \in \mathrm{SL}(2,\mathbb{Z})$ such that $\sigma\tau_0 = \gamma\tau_0$. If we set $\tilde{\sigma} = \gamma^{-1}\sigma$, then $\tilde{\sigma}$ fixes τ_0. Note also that $\det(\tilde{\sigma}) = m$ and that the entries of $\tilde{\sigma}$ are relatively prime. Using $\tilde{\sigma}$, the limit (13.6) can be written

$$\lim_{\tau \to \tau_0} \frac{j(\tau) - j(\tilde{\sigma}\tau)}{j(\tau) - j(\tau_0)}.$$

Consider the Taylor expansion of $j(\tau)$ about $\tau = \tau_0$:

$$j(\tau) = j(\tau_0) + a_k(\tau - \tau_0)^k + \cdots, \qquad a_k \neq 0.$$

Substituting $\tilde{\sigma}\tau$ for τ, we get the series

$$j(\tilde{\sigma}\tau) = j(\tau_0) + a_k(\tilde{\sigma}\tau - \tau_0)^k + \cdots,$$

and then one computes that

$$\frac{j(\tau)-j(\tilde{\sigma}\tau)}{j(\tau)-j(\tau_0)} = \frac{a_k((\tau-\tau_0)^k-(\tilde{\sigma}\tau-\tau_0)^k)+\cdots}{a_k(\tau-\tau_0)^k+\cdots}$$

$$= 1-\left(\frac{\tilde{\sigma}\tau-\tau_0}{\tau-\tau_0}\right)^k+\cdots.$$

Since

$$\lim_{\tau\to\tau_0}\frac{\tilde{\sigma}\tau-\tau_0}{\tau-\tau_0} = \lim_{\tau\to\tau_0}\frac{\tilde{\sigma}\tau-\tilde{\sigma}\tau_0}{\tau-\tau_0} = \tilde{\sigma}'(\tau_0),$$

it follows that the limit (13.6) equals $1-\tilde{\sigma}'(\tau_0)^k$, and thus we need to prove that

$$(13.7) \qquad \tilde{\sigma}'(\tau_0)^k \neq 1,$$

where k is the order of vanishing of $j(\tau)-j(\tau_0)$ at τ_0.

If we write $\tilde{\sigma} = \left(\begin{smallmatrix} a & b \\ c & d\end{smallmatrix}\right)$, then an easy computation shows that

$$\tilde{\sigma}'(\tau_0) = \frac{m}{(c\tau_0+d)^2}.$$

Note also that $c \neq 0$ since $\tilde{\sigma}$ fixes τ_0 (see Exercise 13.4). Now suppose that $j(\tau_0) \neq 0$ or 1728. Then, by part (iv) of Theorem 11.2, it follows that $k = 1$, so that (13.7) reduces to

$$\frac{m}{(c\tau_0+d)^2} \neq 1,$$

which is obvious since $c \neq 0$ and τ_0 is not a real number. When $j(\tau_0) = 1728$, we can assume that $\tau_0 = i$ (recall that $j(i) = 1728$), and then Theorem 11.2 tells us that $k = 2$. Thus if (13.7) failed to hold, we would have

$$\frac{m^2}{(ci+d)^4} = 1,$$

which implies that $c = \pm\sqrt{m}$ and $d = 0$ (see Exercise 13.4). Then $\tilde{\sigma}(i) = i$ tells us that $a = 0$ and $b = \mp\sqrt{m}$. So either $\tilde{\sigma}$ doesn't have integer entries (when m is not a perfect square), or the entries are integers with a common divisor (since $m > 1$). Both cases contradict what we know about $\tilde{\sigma}$, so that (13.7) holds when $j(\tau_0) = 1728$. The case when $j(\tau_0) = 0$ is similar and is left to the reader (see Exercise 13.4).

We should mention that the standard treatment of (13.6) in the literature (see Deuring [24, §12] or Lang [73, Appendix to §10]) seems to be incomplete.

We have thus shown that the multiplicity of $j(\tau_0)$ as a root of $\Phi_m(X,X)$ is

$$r = |\{\sigma \in C(m) : j(\sigma\tau_0) = j(\tau_0)\}|,$$

and it remains to show that $r = r(\mathcal{O}, m)$, where

$$r(\mathcal{O}, m) = |\{\alpha \in \mathcal{O} : \alpha \text{ is primitive, } N(\alpha) = m\}/\mathcal{O}^*|.$$

To prove the desired equality, we will construct a map $\alpha \mapsto \sigma$. Namely, if $\alpha \in \mathcal{O}$ is primitive of norm m, then by Corollary 11.27, $\alpha\mathcal{O}$ is a cyclic sublattice of $\mathcal{O}$ of index m, and since $\mathcal{O} = [1, \tau_0]$, Lemma 11.24 implies that there is a unique $\sigma = \left(\begin{smallmatrix} a & b \\ 0 & d \end{smallmatrix}\right) \in C(m)$ such that $\alpha\mathcal{O} = d[1, \sigma\tau_0]$. Then σ satisfies $j(\sigma\tau_0) = j(\tau_0)$, and note also that if $\epsilon \in \mathcal{O}^*$, then $\epsilon\alpha$ maps to the same σ that α does. Thus we have constructed a well-defined map

$$\begin{aligned} &\{\alpha \in \mathcal{O} : \alpha \text{ is primitive and } N(\alpha) = m\}/\mathcal{O}^* \\ &\longrightarrow \{\sigma \in C(m) : j(\sigma\tau_0) = j(\tau_0)\}. \end{aligned}$$

This map is easily seen to be bijective (see Exercise 13.5), which proves that $r = r(\mathcal{O}, m)$. This completes the proof of Theorem 13.4. Q.E.D.

Besides knowing the factorization of $\Phi_m(X, X)$, its degree is easy to compute:

Proposition 13.8. *If $m > 1$, then the degree of $\Phi_m(X, X)$ is*

$$2 \sum_{\substack{a|m \\ a > \sqrt{m}}} \frac{a}{\gcd(a, m/a)} \phi(\gcd(a, m/a)) + \phi(\sqrt{m}),$$

where ϕ is the Euler ϕ-function and $\phi(\sqrt{m}) = 0$ when m is not a perfect square.

Proof. The proof of this proposition is given in Exercise 13.6. Q.E.D.

If we write $r(\mathcal{O}, m)$ as $r(D, m)$, where D is the discriminant of $\mathcal{O}$, then Proposition 13.8 and Theorem 13.4 allow us to express the degree of $\Phi_m(X, X)$ in two ways. This gives us the following corollary, which is one of Kronecker's class number relations:

Corollary 13.9. *If $m > 1$, then*

$$\sum_D r(D, m) h(D) = 2 \sum_{\substack{a|m \\ a > \sqrt{m}}} \frac{a}{\gcd(a, m/a)} \phi(\gcd(a, m/a)) + \phi(\sqrt{m}).$$

Q.E.D.

To illustrate the above theorems, let's study the case $m = 3$. There are only four orders with primitive elements of norm 3, namely $\mathbf{Z}[\omega]$, $\mathbf{Z}[\sqrt{-3}]$,

$\mathbb{Z}[\sqrt{-2}]$ and $\mathbb{Z}[(1+\sqrt{-11})/2]$, and the corresponding $r(D,3)$'s are 1, 1, 2 and 2 respectively (see Exercise 13.7). Then Theorem 13.4 tells us that

$$\Phi_3(X,X) = \pm H_{-3}(X)H_{-12}(X)H_{-8}(X)^2 H_{-11}(X)^2, \tag{13.10}$$

and since $\Phi_3(X,X)$ has degree 6 by Proposition 13.8, we get the following class number relation:

$$6 = h(-3) + h(-12) + 2h(-8) + 2h(-11).$$

This equation implies that all four class numbers must be one.

We can work out (13.10) more explicitly, for we know $\Phi_3(X,Y)$ from (11.22). Setting $X = Y$ gives us

$$\begin{aligned}\Phi_3(X,X) = -X^6 + 4464X^5 + 2585778176X^4 + 17800519680000X^3 \\ - 769939996672000000X^2 + 37108517437440000000000,\end{aligned}$$

and factoring this over $\mathbb{Q}$, we obtain

$$\Phi_3(X,X) = -X(X-54000)(X-8000)^2(X+32768)^2.$$

However, in §§10 and 12, we computed the j-invariants $j((1+\sqrt{-3})/2) = 0$, $j(\sqrt{-2}) = 8000$ and $j((1+\sqrt{-11})/2) = -32768$. Thus we recognize three of the above four factors, and it follows that the fourth must be $H_{-12}(X)$, i.e.,

$$H_{-12}(X) = X - j(\sqrt{-3}) = X - 54000.$$

This proves that $j(\sqrt{-3}) = 54000$.

Let's now turn to the general problem of computing a given class equation $H_D(X)$. Since $\Phi_m(X,X)$ will have many factors, we need to know which one is the particular $H_D(X)$ we're interested in. The basic idea is to use *multiplicities* to distinguish the factors we seek. In particular, the factors of multiplicity one play an especially important role. Let's define the polynomial

$$\Phi_{m,1}(X,X) = \prod_{r(D,m)=1} H_D(X).$$

By Theorem 13.4, we know that $\Phi_{m,1}(X,X)$ is the product of the multiplicity one factors of $\Phi_m(X,X)$. We can describe $\Phi_{m,1}(X,X)$ as follows:

Proposition 13.11. *If $m > 1$, then $\Phi_{m,1}(X,X)$ equals*

$$\begin{array}{ll} H_{-4}(X)H_{-8}(X), & \textit{if } m = 2 \\ H_{-m}(X)H_{-4m}(X), & \textit{if } m \equiv 3 \bmod 4 \textit{ and } m \neq 3k^2,\ k > 1 \\ H_{-4m}(X), & \textit{if } m > 2,\ m \not\equiv 3 \bmod 4 \textit{ or } m = 3k^2,\ k > 1. \end{array}$$

Proof. Let's first show that the $H_D(X)$'s listed above are factors of multiplicity one of $\Phi_m(X,X)$. Since $\pm\sqrt{-m}$ are the only primitive norm m elements of $\mathbf{Z}[\sqrt{-m}]$, it follows that $H_{-4m}(X)$ is a factor of multiplicity 1. When $m = 2$, the elements of norm 2 in $\mathbf{Z}[i]$ are $\pm 1 \pm i$, which are all associate under $\mathbf{Z}[i]^*$. Thus $H_{-4}(X)$ is also a factor of multiplicity one. Finally, when $m \equiv 3 \bmod 4$ and $m \neq 3k^2$, $k > 1$, we need to consider the multiplicity of $H_{-m}(X)$. The order $\mathbf{Z}[(1+\sqrt{-m})/2]$ has at least two primitive norm m elements, namely $\pm\sqrt{-m}$. To see if there are any others, suppose that $a + b(1+\sqrt{-m})/2$ is also primitive of norm m. Then $b \neq 0$ and, taking norms,

$$4m = (2a+b)^2 + mb^2.$$

Thus $b = \pm 1$ or ± 2, and $b = \pm 2$ leads to the solutions we already know. So what happens if $b = \pm 1$? This clearly implies $3m = (2a+b)^2$, so that $m = 3k^2$, and since $k > 1$ is excluded by hypothesis, we see that $m = 3$. Here, $b = \pm 1$ leads to 4 more solutions, but since $|\mathbf{Z}[\zeta_3]^*| = 6$, we still get a multiplicity one factor.

The next step is to show that these are the *only* factors of multiplicity one. So suppose that $r(\mathcal{O}, m) = 1$ for some order $\mathcal{O}$. For simplicity, let's also assume that $\mathcal{O}^* = \{\pm 1\}$. Given $\alpha \in \mathcal{O}$ primitive of norm m, note that $\pm\alpha$ and $\pm\overline{\alpha}$ are also primitive of the same norm. Then $r(\mathcal{O}, m) = 1$ implies that $\overline{\alpha} = \pm\alpha$. But $\overline{\alpha} = \alpha$ is easily seen to be impossible (α is primitive and $m > 1$), so that $\overline{\alpha} = -\alpha$. This means that α is a rational multiple of $\sqrt{D}$, where D is the discriminant of $\mathcal{O}$. The argument now breaks up into two cases.

If $D \equiv 0 \bmod 4$, then $\mathcal{O} = [1, \sqrt{D}/2]$, so that α, being primitive, must be $\pm\sqrt{D}/2$. This implies that $m = N(\alpha) = -D/4$, hence $D = -4m$. The corresponding factor is thus $H_{-4m}(X)$, which is one of the ones we know.

If $D \equiv 1 \bmod 4$, then $\mathcal{O} = [1, (1+\sqrt{D})/2]$, so that $\alpha = a + b(1+\sqrt{D})/2$. Since α is a multiple of $\sqrt{D}$, we have $2a + b = 0$, and since a and b are relatively prime (α is primitive), we have $b = \pm 2$. This means that $\alpha = \pm\sqrt{D}$, so that $m = N(\alpha) = -D$. Thus $D = -m$, and this will be the other case we know once we prove that $m \neq 3k^2$, $k > 1$. So suppose that m has this form. Then $D = -3k^2$, which means that $\mathcal{O}$ is the order of conductor k in $\mathbf{Z}[\zeta_3]$. Note that $\mathcal{O}^* = \{\pm 1\}$ since $k > 1$. One easily computes that $\pm k\sqrt{-3}$ and $\pm k(1-\zeta_3)$ are primitive elements of $\mathcal{O}$ of norm $3k^2 = m$, which contradicts our assumption that $r(\mathcal{O}, m) = 1$.

It remains to consider the case when $\mathcal{O}^* \neq \{\pm 1\}$. We leave it to the reader to check that when $\mathcal{O} = \mathbf{Z}[\zeta_3]$ (resp. $\mathcal{O} = \mathbf{Z}[i]$), $r(\mathcal{O}, m) = 1$ implies that $m = 3$ (resp. $m = 2$) (see Exercise 13.8). This completes the proof of Proposition 13.11. Q.E.D.

It is now fairly easy to compute $H_D(X)$ using the $\Phi_m(X,X)$'s. In the discussion that follows, m will denote a positive integer, and for simplicity we will assume $m > 3$. It turns out that there are three cases to consider.

If $m \not\equiv 3 \bmod 4$ or $m = 3k^2$, then Proposition 13.11 tells us that

$$H_{-4m}(X) = \Phi_{m,1}(X,X),$$

so that once we factor $\Phi_m(X,X)$ into irreducibles, we know $H_{-4m}(X)$.

Next, if $m \equiv 3 \bmod 8$ and $m \neq 3k^2$, then Proposition 13.11 tells us that

$$H_{-m}(X)H_{-4m}(X) = \Phi_{m,1}(X,X). \tag{13.12}$$

However, since $m > 3$ and $m \equiv 3 \bmod 8$, it follows from Corollary 7.28 that $h(-4m) = 3h(-m)$, so that $H_{-4m}(X)$ has greater degree than $H_{-m}(X)$. Thus, factoring $\Phi_m(X,X)$ determines both $H_{-m}(X)$ and $H_{-4m}(X)$.

Finally, if $m \equiv 7 \bmod 8$, then (13.12) still holds, but this time more work is needed since $H_{-m}(X)$ and $H_{-4m}(X)$ have the same degree by Corollary 7.28. We claim that

$$H_{-m}(X) = \gcd(\Phi_{m,1}(X,X), \Phi_{(m+1)/4}(X,X)). \tag{13.13}$$

To see this, first note that $H_{-m}(X)$ divides $\Phi_{(m+1)/4}(X,X)$ since in the order of discriminant $-m$, $(1+\sqrt{-m})/2$ is primitive of norm $(m+1)/4$. If we turn to the order of discriminant $-4m$, there are *no* primitive elements of norm $(m+1)/4$ (see Exercise 13.9), and (13.13) follows. Thus, to determine $H_{-m}(X)$ and $H_{-4m}(X)$, we need to factor both $\Phi_m(X,X)$ and $\Phi_{(m+1)/4}(X,X)$ into irreducibles.

Using the above process, it is now easy to compute any $H_D(X)$, assuming that we know the requisite modular equation (or equations). Some simple examples are given in Exercise 13.10.

B. Computing the Modular Equation

To complete our algorithm for finding the class equation, we need to know how to compute the modular equation $\Phi_m(X,Y) = 0$. This turns out to be the weak link in our theory, for while such an algorithm exists, it is so cumbersome that it can be implemented only for very small m.

The first step in computing $\Phi_m(X,Y)$ is to reduce to the case when m is prime. This is done by means of the following proposition:

Proposition 13.14. *Let $m > 1$ be an integer, and set $\Psi(m) = m\prod_{p|m}(1 + 1/p)$, which is the degree of $\Phi_m(X,Y)$ as a polynomial in X.*

(i) *If $m = m_1m_2$, where m_1 and m_2 are relatively prime, then*

$$\Phi_m(X,Y) = \prod_{i=1}^{\Psi(m_2)} \Phi_{m_1}(X,\xi_i),$$

where $X = \xi_i$ *are the roots of* $\Phi_{m_2}(X,Y) = 0$.

(ii) *If* $m = p^a$, *where* p *is prime and* $a > 1$, *then*

$$\Phi_m(X,Y) = \begin{cases} \dfrac{\prod_{i=1}^{\Psi(p^{a-1})} \Phi_p(X,\xi_i)}{\Phi_{p^{a-2}}(X,Y)^p}, & a > 2 \\ \dfrac{\prod_{i=1}^{p+1} \Phi_p(X,\xi_i)}{(X-Y)^{p+1}}, & a = 2, \end{cases}$$

where $X = \xi_i$ *are the roots of* $\Phi_{p^{a-1}}(X,Y) = 0$.

Proof. See Weber [102, §69]. Q.E.D.

Now let p be a prime. To compute $\Phi_p(X,Y)$, we will follow Kaltofen and Yui [66] and Yui [110]. First note by parts (iii) and (v) of Theorem 11.18, we have

$$\Phi_p(X,Y) = \Phi_p(Y,X)$$
$$\Phi_p(X,Y) \equiv (X^p - Y)(X - Y^p) \bmod p\mathbb{Z}[X,Y],$$

and we also know that $\Phi_p(X,Y)$ is monic of degree $\Psi(p) = p+1$ as a polynomial in X. Thus we can write $\Phi_p(X,Y)$ in the following form:

(13.15)
$$(X^p - Y)(X - Y^p) + p \sum_{0 \le i \le p} c_{ii} X^i Y^i + p \sum_{0 \le i < j \le p} c_{ij}(X^i Y^j + X^j Y^i),$$

where the coefficients c_{ij}'s are integers. We will use the q-expansion of the j-function to obtain a finite system of equations that can be solved uniquely for the c_{ij}'s.

By the definition of the modular equation, we have the identity

$$\Phi_p(j(p\tau), j(\tau)) = 0.$$

Substituting the q-expansions for $j(\tau)$ and $j(p\tau)$ into this equation and using (13.15), we obtain

(13.16)
$$\begin{aligned} 0 =& (j(p\tau)^p - j(\tau))(j(p\tau) - j(\tau)^p) \\ &+ p \sum_{0 \le i \le p} c_{ii} j(p\tau)^i j(\tau)^i + p \sum_{0 \le i < j \le p} c_{ij}(j(p\tau)^i j(\tau)^j + j(p\tau)^j j(\tau)^i). \end{aligned}$$

If we equate the coefficients of the different powers of q, we get an infinite number of linear equations in the variables c_{ij}. We can reduce to a finite number of equations as follows:

Proposition 13.17. *The finite system of linear equations obtained by equating coefficients of nonpositive powers of q in (13.16) has a unique solution given by the coefficients c_{ij} of the modular equation.*

Proof. Since the modular equation provides one solution, it suffices to prove uniqueness. Using (13.15), a solution of these equations gives a polynomial $F(X,Y)$ with the following three properties:

(i) $F(X,Y)$ is monic of degree $p+1$ in X.
(ii) $F(X,Y) = F(Y,X)$.
(iii) $\lim_{\mathrm{Im}\tau\to\infty} F(j(p\tau), j(\tau)) = 0$.

To explain the last property, note that the q-expansion of $F(j(p\tau), j(\tau))$ contains no nonpositive powers of q since $F(X,Y)$ comes from a solution of our finite system of equations. Since $q \to 0$ as $\mathrm{Im}\tau \to \infty$, (iii) follows.

We claim these properties force $F(X,Y) = \Phi_p(X,Y)$, which will prove uniqueness. The idea is to study $F(j(p\tau), j(\tau))$, which is a modular function for $\Gamma_0(p)$. We will first show that $F(j(p\tau), j(\tau))$ vanishes at the cusps, which means that

$$\lim_{\mathrm{Im}\tau\to\infty} F(j(p\gamma\tau), j(\gamma\tau)) = 0 \qquad \text{for all } \gamma \in \mathrm{SL}(2,\mathbb{Z}). \tag{13.18}$$

Using (11.12), this is equivalent to showing

$$\lim_{\mathrm{Im}\tau\to\infty} F(j(\sigma\tau), j(\tau)) = 0 \qquad \text{for all } \sigma \in C(p).$$

When $\sigma = \left(\begin{smallmatrix} p & 0 \\ 0 & 1 \end{smallmatrix}\right)$, we're done by (iii), and when $\sigma \neq \left(\begin{smallmatrix} p & 0 \\ 0 & 1 \end{smallmatrix}\right)$, σ must be of the form $\left(\begin{smallmatrix} 1 & i \\ 0 & p \end{smallmatrix}\right)$ since p is prime. If we set $u = \sigma\tau = (\tau + i)/p$, then $\tau = pu + i$, and

$$\begin{aligned}
\lim_{\mathrm{Im}\tau\to\infty} F(j(\sigma\tau), j(\tau)) &= \lim_{\mathrm{Im}\tau\to\infty} F(j(u), j(pu+i)) \\
&= \lim_{\mathrm{Im}u\to\infty} F(j(u), j(pu)) \\
&= \lim_{\mathrm{Im}u\to\infty} F(j(pu), j(u)) = 0,
\end{aligned}$$

where we used (ii) and (iii) above. This proves (13.18).

Thus $F(j(p\tau), j(\tau))$ is a holomorphic modular function for $\Gamma_0(p)$ which vanishes at the cusps. In the case of modular functions for $\mathrm{SL}(2,\mathbb{Z})$, we proved in Lemma 11.10 that such a function is zero, and the proof extends easily to the case of $\Gamma_0(p)$ (see Exercise 13.11). This shows that

$F(j(p\tau), j(\tau)) = 0$, so that $j(p\tau)$ is a root of $F(X, j(\tau))$ and $\Phi(X, j(\tau))$. Since the latter is irreducible over $\mathbb{C}(j(\tau))$, it must divide $F(X, j(\tau))$. Both $F(X,Y)$ and $\Phi_p(X,Y)$ are monic of the same degree, and hence they must be equal. Q.E.D.

Looking at the q-expansions for $j(\tau)$ and $j(p\tau)$, the most negative power of q in (13.16) is q^{-p^2-p}, and it follows that the system of equations described in Proposition 13.17 has p^2+p+1 equations in the $(p^2+3p+2)/2$ unknowns c_{ij}. With some cleverness, one can reduce to p^2+p equations in $(p^2+3p)/2$ unknowns (see Yui [110]). These equations have been written down explicitly by Yui [110], though the resulting expressions are *extremely* complicated. For a discussion of the computational aspects of these equations, see Kaltofen and Yui [66].

We are not quite done, for our equations for $\Phi_p(X,Y)$ involve the q-expansions of $j(\tau)$ and $j(p\tau)$. Hence we need to calculate those coefficients of the q-expansions which contribute to negative powers of q in (13.16). It suffices to do this for $j(\tau)$, and because the most negative power of q in (13.16) is q^{-p^2-p}, we need only the first p^2+p coefficients of the q-expansion of the j-function. In §21 we found some nice formulas for

$$j(\tau) = 1728\frac{g_2(\tau)^3}{\Delta(\tau)},$$

but to get the q-expansion, we need *series* expansions of the numerator and denominator. For $g_2(\tau)$, we use the classical formula

$$g_2(\tau) = \frac{(2\pi)^4}{12}\left(1 + \sum_{n=1}^{\infty} \sigma_3(n) q^n\right),$$

where $\sigma_3(n) = \sum_{d|n} d^3$ (see Lang [73, §4.1] or Serre [88, §VII.4.2]), and for $\Delta(\tau)$, we know from Theorem 12.17 that

$$\Delta(\tau) = (2\pi)^{12} q \prod_{n=1}^{\infty} (1-q^n)^{24}.$$

This is still not a series, but if we use Euler's famous identity

$$\prod_{n=1}^{\infty} (1-q^n) = \sum_{n=-\infty}^{\infty} q^{n(3n+1)/2}$$

(see Hardy and Wright [48, §19.9]), then it becomes straightforward to write a program to compute the q-expansion of $j(\tau)$. A description of how to do this is in Hermann [53] (he also gives an alternate approach to calculating the modular equation), and one finds that the first few terms of the q-expansion are

$$\begin{aligned} j(\tau) = \frac{1}{q} &+ 744 + 196884q + 21493760q^2 + 864299970q^3 \\ &+ 20245856256q^4 + 333202640600q^5 + \cdots . \end{aligned}$$

These formulas also give a second proof that the q-expansion of $j(\tau)$ has integer coefficients (see Exercise 13.12).

The conclusion of this rather long discussion is that for any integer $m > 0$, we can compute $\Phi_m(X,Y)$, which then gives us $\Phi_m(X,X)$ by setting $X = Y$. There are known algorithms for factoring $\Phi_m(X,X)$ into irreducibles, and then the discussion following Proposition 13.11 shows how to compute $H_{\mathcal{O}}(X)$. We have thus proved the following theorem:

Theorem 13.19. *Given an order $\mathcal{O}$ in an imaginary quadratic field, there is an algorithm for computing the class equation $H_{\mathcal{O}}(X)$.* Q.E.D.

The problem with this theorem is that our algorithm for computing $H_{\mathcal{O}}(X)$ requires knowing $\Phi_m(X,Y)$. Modular equations are extremely complicated polynomials and are difficult to compute. We saw in (11.22) that $\Phi_3(X,Y)$ is very large, and things get worse as m increases. For example, the printout of $\Phi_{11}(X,Y)$ takes over two single-spaced pages, and some of the coefficients have over 120 digits (see Kaltofen and Yui [66]). In general, Cohen [18] proved that the maximum of the absolute values of the coefficients of $\Phi_m(X,Y)$ is asymptotic to $\exp(6\Psi(m)\log(m))$, where $\Psi(m) = m\prod_{p|m}(1+1/p)$, so that the growth is exponential in m. Hence the above algorithm is not a practical way to compute class equations.

Recently, a more efficient approach to computing $H_D(X)$ has been developed by Kaltofen and Yui [65]. The basic idea is to compute $H_D(X)$ directly from the formula

$$H_D(X) = \prod_{i=1}^{h}(X - j(\mathfrak{a}_i)).$$

We know how to find the $h = h(D)$ reduced forms of discriminant D, and then the $\mathfrak{a}_i$'s can be taken to be the proper $\mathcal{O}$-ideals corresponding to the reduced forms via Theorem 7.7. Since $H_D(X)$ has integral coefficients, we need only compute $j(\mathfrak{a}_i)$ numerically to a sufficiently high degree of precision, and the formulas for $j(\tau)$ given in §12 are ideal for this purpose. For an example of how this works, consider the case of discriminant $D = -71$. Here, the class number is $h(-71) = 7$, and the above process shows that the minimal polynomial of $j((1+\sqrt{-71})/2)$ is

$$
\begin{aligned}
H_{-71}(X) = {} & X^7 + 5\cdot 7\cdot 31\cdot 127\cdot 233\cdot 9769\, X^6 \\
& - 2\cdot 5\cdot 7\cdot 44171287694351\, X^5 \\
& + 2\cdot 3\cdot 7\cdot 2342715209763043144031\, X^4 \\
& - 3\cdot 7\cdot 31\cdot 126502959053722086016 6039\, X^3 \\
& + 2\cdot 7\cdot 11^3\cdot 67\cdot 229\cdot 1797402619247178519 2633\, X^2 \\
& - 7\cdot 11^6\cdot 17^6\cdot 1420913330979618293\, X \\
& + (11^3\cdot 17^2\cdot 23\cdot 41\cdot 47\cdot 53)^3
\end{aligned}
\tag{13.20}
$$

(This example was taken from the preliminary version of [65]—all primes < 1000 were factored out of the coefficients.) Note that the constant term is a cube, as predicted by Theorem 12.2.

We can apply the algorithm of Theorem 13.19 to give a constructive version of Theorem 9.2, but before we do this, we need to learn about some of the recent work of Deuring, Gross and Zagier on the class equation.

C. Theorems of Deuring, Gross and Zagier

In 1946 Deuring [25] proved a remarkable result concerning prime divisors of the difference of two singular moduli. To state Deuring's theorem precisely, let $\mathcal{O}_1$ and $\mathcal{O}_2$ be orders in imaginary quadratic fields K_1 and K_2 respectively, and for $i = 1, 2$, let $\mathfrak{a}_i$ be a proper fractional $\mathcal{O}_i$-ideal. Then we have:

Theorem 13.21. *Let L be a number field containing $j(\mathfrak{a}_1)$ and $j(\mathfrak{a}_2)$, and let $\mathfrak{P}$ be a prime of L lying over the prime number p. When $K_1 = K_2$, assume in addition that p divides neither of the conductors of $\mathcal{O}_1$ and $\mathcal{O}_2$. If $j(\mathfrak{a}_1) \neq j(\mathfrak{a}_2)$, then*

$$
j(\mathfrak{a}_1) \equiv j(\mathfrak{a}_2) \bmod \mathfrak{P} \Rightarrow \begin{cases} p \text{ splits completely} \\ \text{in neither } K_1 \text{ or } K_2. \end{cases}
$$

Proof. The proof uses reduction theory of elliptic curves. See Deuring [25] or Lang [73, §13.4]. Q.E.D.

We can use this theorem to study the constant term and discriminant of the class equation:

Corollary 13.22. *Let $D < 0$ be a discriminant, and let p be prime.*

(i) *If p divides the constant term of $H_D(X)$ and $\mathbb{Q}(\sqrt{D}) \neq \mathbb{Q}(\sqrt{-3})$, then $(D/p) \neq 1$ and either $p = 3$ or $p \equiv 2 \bmod 3$.*

(ii) *If p divides the discriminant of $H_D(X)$, then $(D/p) \neq 1$.*

Proof. Let $\mathfrak{a}_1, \ldots, \mathfrak{a}_h$, $h = h(D)$, be ideal class representatives for the order of discriminant D. To prove (i), note that the constant term of the class equation is

$$C = \pm \prod_{i=1}^{h} j(\mathfrak{a}_i).$$

If $p \mid C$, then in some number field L, there is a prime $\mathfrak{P}$ containing p that divides some $j(\mathfrak{a}_i)$. Since

$$j(\mathfrak{a}_i) = j(\mathfrak{a}_i) - 0 = j(\mathfrak{a}_i) - j((1 + \sqrt{-3})/2),$$

we know by Theorem 13.21 that p splits in neither $\mathbb{Q}(\sqrt{D})$ nor $\mathbb{Q}(\sqrt{-3})$, and (i) follows immediately.

To prove (ii), note that the discriminant of $H_D(X)$ is

$$\operatorname{disc}(H_D(X)) = \prod_{i<j} (j(\mathfrak{a}_i) - j(\mathfrak{a}_j))^2.$$

Thus, if $p \mid \operatorname{disc}(H_D(X))$, then some $\mathfrak{P}$ lying over p divides some $j(\mathfrak{a}_i) - j(\mathfrak{a}_j)$. If $p \nmid D$, then Theorem 13.21 implies that p doesn't split in $\mathbb{Q}(\sqrt{D})$, and $(D/p) = -1$ follows. If $p \mid D$, then $(D/p) = 0$, so that $(D/p) \neq 1$ in either case. Q.E.D.

One of our original motivations for studying complex multiplication came from the question of when a prime can be written in the form $x^2 + ny^2$. Using the class equation, we can now prove a constructive version of our basic result, Theorem 9.2:

Theorem 13.23. *Let n be a positive integer. Then there is a monic irreducible polynomial $f_n(X)$ of degree $h(-4n)$ such that for an odd prime p not dividing n,*

$$p = x^2 + ny^2 \iff \begin{cases} (-n/p) = 1 \text{ and } f_n(X) \equiv 0 \bmod p \\ \text{has an integer solution.} \end{cases}$$

Furthermore, there is an algorithm for finding $f_n(X)$.

Proof. The order of discriminant $-4n$ is $\mathcal{O} = [1, \sqrt{-n}]$, so that by Theorem 11.1, $j(\sqrt{-n})$ is a real algebraic integer and is a primitive element of the ring class field of $\mathcal{O}$. Since $H_{-4n}(X)$ is the minimal polynomial of $j(\sqrt{-n})$, we can set $f_n(X) = H_{-4n}(X)$ in Theorem 9.2, and then the desired equivalence holds for primes dividing neither $-4n$ nor the discriminant of $H_{-4n}(X)$. But when a prime divides the discriminant, Corollary

13.22 tells us that $(-4n/p) \neq 1$. Since both sides of the desired equivalence imply $(-n/p) = 1$, the discriminant condition is superfluous. Finally, by Theorem 13.19, there is an algorithm for finding $H_{-4n}(X)$, and the theorem is proved. Q.E.D.

From a computational point of view, this result is not ideal. The polynomials $H_{-4n}(X)$ are difficult to compute, and as indicated by $H_{-56}(X)$ and $H_{-71}(X)$ (see (13.1) and (13.20)), they are excessively complicated. The real value of Theorem 13.23 is the way it links the ideas of class field theory and complex multiplication to the elementary question of when a prime can be written in the form $x^2 + ny^2$.

Deuring's study of $j(\mathfrak{a}_1) - j(\mathfrak{a}_2)$ has prompted some recent work of Gross and Zagier [46] which determines *exactly* which primes divide such a difference. Their results apply only to field discriminants, but one gets very complete information in this case. Let d_1 and d_2 be the discriminants of imaginary quadratic fields K_1 and K_2 respectively. We will assume that d_1 and d_2 are relatively prime. Then set

$$J(d_1, d_2) = \left(\prod_{i=1}^{h_1} \prod_{j=1}^{h_2} (j(\mathfrak{a}_i) - j(\mathfrak{b}_j)) \right)^{4/w_1 w_2},$$

where $\mathfrak{a}_1, \ldots, \mathfrak{a}_{h_1}$ are ideal class representatives of $\mathcal{O}_{K_1}$, $\mathfrak{b}_1, \ldots, \mathfrak{b}_{h_2}$ are ideal class representatives of $\mathcal{O}_{K_2}$, and $w_1 = |\mathcal{O}_{K_1}^*|$, $w_2 = |\mathcal{O}_{K_2}^*|$. Note that $J(d_1, d_2)$ is an integer when $d_1, d_2 < -4$, and that $J(d_1, d_2)^2$ is always an integer (see Exercise 13.13).

To state Gross and Zagier's formula for $J(d_1, d_2)^2$, we will need functions $\epsilon(n)$ and $F(m)$, which are defined as follows. First, if p is a prime, we set

$$\epsilon(p) = \begin{cases} (d_1/p) & \text{if } p \nmid d_1 \\ (d_2/p) & \text{if } p \nmid d_2. \end{cases}$$

The reader can check that this is well-defined whenever $(d_1 d_2/p) \neq -1$ (see Exercise 13.14). Then, if $n = \prod_{i=1}^r p_i^{a_i}$, we set

$$\epsilon(n) = \prod_{i=1}^{r} \epsilon(p_i)^{a_i},$$

where we assume that $(d_1 d_2/p_i) \neq -1$ for all i. Finally, $F(m)$ is defined by the formula

$$F(m) = \prod_{\substack{nn'=m \\ n,n'>0}} n^{\epsilon(n')}.$$

This is well-defined when all primes p dividing m satisfy $(d_1 d_2/p) \neq -1$.

We can now state the main theorem of Gross and Zagier [46]:

Theorem 13.24. *With the above notation,*

$$J(d_1,d_2)^2 = \pm \prod_{\substack{x^2 < d_1 d_2 \\ x^2 \equiv d_1 d_2 \bmod 4}} F\left(\frac{d_1 d_2 - x^2}{4}\right).$$

Proof. First note that $F((d_1d_2 - x^2)/4)$ is always defined since any prime p dividing $(d_1d_2 - x^2)/4$ satisfies $(d_1d_2/p) \neq -1$ (see Exercise 13.14). The paper [46] contains two proofs of this theorem, one algebraic and one analytic. The algebraic proof, which uses reduction theory of elliptic curves, is given only for the case of prime discriminants. A general version of this proof appears in Dorman [30]. Q.E.D.

This theorem gives the following corollary:

Corollary 13.25. *Let p be a prime dividing $J(d_1,d_2)^2$. Then:*
(i) $(d_1/p) \neq 1$ *and* $(d_2/p) \neq 1$.
(ii) *p divides a positive integer of the form* $(d_1d_2 - x^2)/4$.
(iii) $p \leq d_1d_2/4$.

Proof. If p divides $J(d_1,d_2)^2$, it must divide some $F((d_1d_2 - x^2)/4)$, and the formula for $F(m)$ then shows that p divides $(d_1d_2 - x^2)/4$. This easily implies parts (ii) and (iii) of the corollary.

It remains to prove part (i). We will first consider the following lemma which tells us how to compute $F(m)$:

Lemma 13.26. *Let m be a positive integer of the form $(d_1d_2 - x^2)/4$. Then $F(m) = 1$ unless m can be written in the form*

$$m = p^{2a+1} p_1^{2a_1} \cdots p_r^{2a_r} q_1^{b_1} \cdots q_s^{b_s},$$

where $\epsilon(p) = \epsilon(p_1) = \cdots = \epsilon(p_r) = -1$ and $\epsilon(q_1) = \cdots = \epsilon(q_s) = 1$. In this case,

$$F(m) = p^{(a+1)(b_1+1)\cdots(b_s+1)}.$$

In particular, $p \mid F(m)$ means that p is the only prime dividing m with an odd exponent and $\epsilon(p) = -1$.

Proof. See Exercises 13.15 and 13.16. Q.E.D.

We can now complete the proof of Corollary 13.25. The above lemma shows that $\epsilon(p) = -1$ for any prime p dividing $F(m)$. It is easy to see that $\epsilon(p) = -1$ implies $(d_1/p) \neq 1$ and $(d_2/p) \neq 1$ (see Exercise 13.14), and the corollary is proved. Q.E.D.

Note that this corollary implies Deuring's theorem in the case of relatively prime field discriminants. We should also mention that when $d_1 d_2 \equiv 1 \bmod 8$, one gets better upper bounds on p (see Exercise 13.17).

If we apply Corollary 13.25 when $d_2 = -3$, then we can strengthen Deuring's result about the constant term of the class equation:

Corollary 13.27. *Let d_K be the discriminant of an imaginary quadratic field K, and assume that $3 \nmid d_K$. If p is a prime dividing the constant term of $H_{d_K}(X)$, then $(d_K/p) \neq 1$ and either $p = 3$ or $p \equiv 2 \bmod p$. Furthermore, $p \leq 3|d_K|/4$.*

Proof. If $\mathfrak{a}_1, \ldots, \mathfrak{a}_h$ are ideal class representatives of $\mathcal{O}_K$, then

$$J(d_K, -3)^2 = \left(\prod_{i=1}^{h} j(\mathfrak{a}_i)\right)^{4/3w},$$

where $w = |\mathcal{O}_K^*|$. Thus the primes dividing $J(d_K, -3)^2$ are the same as the primes dividing the constant term of $H_{d_K}(X)$, and we are done by the previous corollary. Q.E.D.

For an example of how good these estimates are, consider $H_{-56}(X)$. We know from (13.1) that the constant term is

$$(2^8 \cdot 11^2 \cdot 17 \cdot 41)^3.$$

Corollary 13.27 gives us the estimate $p \leq 3|-56|/4 = 42$, which is as good as one can get. The reader should also check the constant term of $H_{-71}(X)$ given in (13.20)—the estimate is again as good as possible. Of course, one could use Theorem 13.24 to compute these constant terms directly (see Exercise 13.18).

Gross and Zagier also have similar theorems for primes dividing the discriminant of the class equation. Rather than give the formula for the multiplicities of the primes, we will just state the following corollary of their result:

Theorem 13.28. *Let d_K be the discriminant of an imaginary quadratic field K, and let p be a prime dividing the discriminant of $H_{d_K}(X)$. Then $(d_K/p) \neq 1$ and $p \leq |d_K|$.*

Proof. In the case of prime discriminants, this is proved by Gross and Zagier in [46], and the general case is in Dorman [29]. Q.E.D.

This theorem strengthens Deuring's result about the discriminant of the class equation. For an example of the bound $p \leq |d_K|$, consider $H_{-56}(X)$.

One computes that its discriminant is

$$-2^{116}\cdot 7^{13}\cdot 11^{10}\cdot 17^{6}\cdot 29^{4}\cdot 31^{2}\cdot 37^{2}\cdot 41^{2}\cdot 43^{2}\cdot 47^{2}\cdot 53^{2}.$$

Theorem 13.28 gives the bound $p \le 56$ on the primes that can appear, which again is the best possible.

D. Exercises

13.1. Use Corollary 11.37 to prove Proposition 13.2.

13.2. If $\mathcal{O}$ is an order in an imaginary quadratic field and m is a positive integer, then we define $r(\mathcal{O},m) = |\{\alpha \in \mathcal{O} : \alpha \text{ is primitive and } N(\alpha) = m\}/\mathcal{O}^*|$, where $\mathcal{O}^*$ acts by multiplication.

(a) Prove that $r(\mathcal{O},m)$ is finite.

(b) For fixed m, prove that there are only finitely many orders $\mathcal{O}$ such that $r(\mathcal{O},m) > 0$.

13.3. Let $F(X,Y) \in \mathbf{C}[X,Y]$, and suppose that $F(X_0,X_0) = 0$. Then X_0 is a root of both $F(X,X_0)$ and $F(X,X)$.

(a) If $F(X,Y) = X^3 + Y^3 + XY$, then show that 0 is a root of $F(X,0)$ and $F(X,X)$ of different multiplicities. Note that the polynomial $F(X,Y)$ is symmetric.

(b) If $F(X,Y)$ and X_0 satisfy the additional condition that

$$\lim_{X\to X_0} \frac{F(X,X)}{F(X,X_0)}$$

exists and is nonzero, then show that X_0 is a root of $F(X,X_0)$ and $F(X,X)$ of the same multiplicity.

13.4. This exercise is concerned with the proof of (13.7). Recall that $\tilde{\sigma}(\tau_0) = \tau_0$, where $\tilde{\sigma} = \left(\begin{smallmatrix} a & b \\ c & d\end{smallmatrix}\right)$ has relatively prime entries and determinant $m > 1$.

(a) Prove that $c \neq 0$.

(b) When $j(\tau_0) = 1728$, we can assume $\tau_0 = i$. Show that $m^2 = (ci + d)^4$ implies $c = \pm\sqrt{m}$ and $d = 0$. Since $\tilde{\sigma}(i) = i$, conclude that $a = 0$ and $b = \pm\sqrt{m}$, and derive a contradiction.

(c) When $j(\tau_0) = 0$, argue as in (b) to complete the proof of (13.7).

13.5. Let $m > 1$, and let $\mathcal{O} = [1,\tau_0]$ be an order in an imaginary quadratic field. Consider the sets

$$\mathcal{A} = \{\alpha \in \mathcal{O} : \alpha \text{ is primitive and } N(\alpha) = m\}/\mathcal{O}^*$$

$$\mathcal{B} = \{\sigma \in C(m) : j(\sigma\tau_0) = j(\tau_0)\}.$$

In the proof of Theorem 13.4, we showed how an element $[\alpha] \in \mathcal{A}$ determines a unique $\sigma \in \mathcal{B}$. Prove that the map $[\alpha] \mapsto \sigma$ defines a bijection $\mathcal{A} \xrightarrow{\sim} \mathcal{B}$.

13.6. The goal of this exercise is to prove the formula for the degree N of $\Phi_m(X,X)$ given in Proposition 13.8.

(a) Prove that q^{-N} is the most negative power of q in the q-expansion of $\Phi_m(j(\tau), j(\tau))$.

(b) If $\sigma = \left(\begin{smallmatrix} a & b \\ 0 & d \end{smallmatrix}\right) \in C(m)$, then use (11.19) to show that the q-expansion of $j(\tau) - j(\sigma\tau)$ is

$$\begin{aligned} q^{-1} - \zeta_m^{ab} q^{-a/d} + \cdots & \qquad \text{when } a < d \\ -\zeta_m^{ab} q^{-a/d} + q^{-1} + \cdots & \qquad \text{when } a > d \\ (1 - \zeta_m^{ab}) q^{-1} + \cdots & \qquad \text{when } a = d, \end{aligned}$$

where $\zeta_m = e^{2\pi i/m}$. The last possibility can occur only when m is a perfect square, and in this case, $\zeta_m^{ab} \neq 1$ since $\sigma \in C(m)$.

(c) Given a, we know that $d = m/a$. In part (a) of Exercise 11.9 we showed that the number of possible $\sigma \in C(m)$ with this a and d was

$$\frac{d}{e}\phi(e),$$

where $e = \gcd(a,d)$. Use this formula and (b) to show that the degree N of $\Phi_m(X,X)$ equals

$$\sum_{\substack{a|m \\ a<\sqrt{m}}} \frac{d}{e}\phi(e) + \sum_{\substack{a|m \\ a>\sqrt{m}}} \frac{a}{d} \cdot \frac{d}{e}\phi(e) + \phi(\sqrt{m}).$$

(d) Show that the first two sums in the above expression are equal. This proves the formula for N given in Proposition 13.8.

13.7. This exercise is concerned with some examples of Theorem 13.4.

(a) Verify that $r(-3,3) = r(-12) = 1$, $r(-8,3) = r(-11,3) = 2$, and also show that $r(D,3) = 0$ for all other discriminants. This proves that

$$\Phi_3(X,X) = \pm H_{-3}(X) H_{-12}(X) H_{-8}(X)^2 H_{-11}(X)^2.$$

(b) Use the method of (a) to write down the factorization of $\Phi_5(X,X)$.

13.8. The proof of Proposition 13.11 requires the following facts about the orders of discriminant -3 and -4 ($\mathbf{Z}[\omega]$ and $\mathbf{Z}[i]$ respectively).

(a) If $m > 1$, show that $r(-3,m) = 1$ if and only if $m = 3$.

(b) If $m > 1$, show that $r(-4, m) = 1$ if and only if $m = 2$.

13.9. Let $m \equiv 3 \bmod 4$ be an integer > 3. Show that the order $\mathbf{Z}[\sqrt{-m}]$ of discriminant $-4m$ has no primitive elements of norm $(m+1)/4$.

13.10. In this exercise we will illustrate the algorithm given in the text for computing $H_D(X)$.

(a) Show that $H_{-56}(X)$ is determined by knowing $\Phi_{14}(X, X)$.

(b) Show that $H_{-11}(X)$ and $H_{-44}(X)$ are determined by knowing $\Phi_{11}(X, X)$.

(c) Show that $H_{-7}(X)$ and $H_{-28}(X)$ are determined by knowing $\Phi_7(X, X)$ and $\Phi_2(X, X)$.

13.11. Let $f(\tau)$ be a modular function for $\Gamma_0(m)$ which vanishes at the cusps.

(a) If γ_i, $i = 1, \ldots, |C(m)|$ are coset representatives for $\Gamma_0(m) \subset \mathrm{SL}(2, \mathbf{Z})$, then show that

$$\prod_{i=1}^{|C(m)|} f(\gamma_i \tau)$$

is a modular function for $\mathrm{SL}(2, \mathbf{Z})$ which vanishes at infinity.

(b) If in addition $f(\tau)$ is holomorphic on $\mathfrak{h}$, then show that $f(\tau)$ is identically zero. Hint: use (a) and Lemma 11.10.

13.12. Use the formulas

$$g_2(\tau) = \frac{(2\pi)^4}{12}\left(1 + \sum_{n=1}^{\infty} \sigma_3(n) q^n\right)$$

$$\Delta(\tau) = (2\pi)^{12} q \prod_{n=1}^{\infty} (1 - q^n)$$

to show that the coefficients of the q-expansion of $j(\tau)$ are integral. This is the classical method used to prove Theorem 11.8.

13.13. Let $J(d_1, d_2)$ be as defined in the text.

(a) If d_1, $d_2 < -4$, then show that $J(d_1, d_2)$ is an integer. Hint: use Galois theory.

(b) Show that $J(d_1, d_2)^2$ is always an integer. Hint: when d_1 or d_2 is -3, recall that $j((1 + \sqrt{-3})/2) = 0$. Theorem 12.2 will be useful.

13.14. Let $\epsilon(m)$ and $F(n)$ be as defined in the text, and let p be a prime number.

(a) Show that $\epsilon(p)$ is defined whenever $(d_1 d_2 / p) \neq -1$.

(b) If p divides a number of the form $(d_1d_2 - x^2)/4$, then show that $(d_1d_2/p) \neq -1$.

(c) Show that $\epsilon(p) = -1$ implies that $(d_1/p) \neq 1$ and $(d_2/p) \neq 1$.

13.15. Exercises 13.15 and 13.16 will prove Lemma 13.26. In this exercise we will show that any positive integer of the form $m = (d_1d_2 - x^2)/4$ satisfies $\epsilon(m) = -1$. We will need the following extension of the Legendre symbol. Let $D \equiv 0, 1 \bmod 4$, and let $\chi : (\mathbb{Z}/D\mathbb{Z})^* \to \{\pm 1\}$ be the homomorphism from Lemma 1.14 (so that $\chi([p]) = (D/p)$ when p is a prime not dividing D). Then for any integer m relatively prime to D, set

$$\left(\frac{D}{m}\right) = \chi([m]).$$

(a) Show that (D/m) is multiplicative in D and m and depends only on the congruence class of m modulo D. Also, when $m = p_1^{a_1} \cdots p_r^{a_r}$ is positive, show that

$$\left(\frac{D}{m}\right) = \prod_{i=1}^{r} \left(\frac{D}{p_i}\right)^{a_i},$$

where (D/p_i) is the usual Kronecker symbol. Thus, when m is odd and positive, (D/m) is just the Jacobi symbol. Finally, show that $(D/-1) = \operatorname{sgn}(D)$. Hint: see Lemma 1.14.

(b) We will need the following limited version of quadratic reciprocity for (D/m). Namely, if $D \equiv 1 \bmod 4$ is relatively prime to $m \equiv 0, 1 \bmod 4$, then prove that $(D/m) = (m/|D|)$. Furthermore, if D and m have opposite signs, then prove that $(D/m) = (m/D)$.

(c) Let m be a positive integer such that $\epsilon(m)$ is defined. If m is relatively prime to d_1, then show that $\epsilon(m) = (d_1/m)$.

(d) Now we can prove that $\epsilon(m) = -1$ when $m = (d_1d_2 - x^2)/4$. We can assume $d_1 \equiv 1 \bmod 4$, and write $m = ab$, where $a \mid d_1$, $a \equiv 1 \bmod 4$ and $\gcd(d_1, b) = 1$. Then $d_1 = ad$, where $d \equiv 1 \bmod 4$.

(i) Show that $\epsilon(m) = (d_2/a)(d_1/b)$.

(ii) Show that $(d_1/b) = (a/d_2)(d/-1)$. Hint: $(d_1/b) = (d_1/4b) = (a/4b)(d/4b)$. Then use $4ab = d_1d_2 - x^2$ and quadratic reciprocity. Remember that a and d have opposite signs and that a has no square factors.

(iii) Use quadratic reciprocity to prove that $\epsilon(m) = -1$. Hint: remember that $d_2 < 0$.

13.16. Let m be a positive integer such that $\epsilon(m) = -1$. The goal of this exercise is to compute $F(m)$. We will use the function $s(m)$ defined by

$$s(m) = \sum_{\substack{n|m \\ n>0}} \epsilon(m).$$

Note that $s(m)$ is defined whenever $\epsilon(m)$ is. Given a prime p, let $\nu_p(m)$ be the highest power of p dividing m.

(a) If m_1 and m_2 are relatively prime integers such that $\epsilon(m_1)$ and $\epsilon(m_2)$ are defined, then prove that

$$F(m_1 m_2) = F(m_1)^{s(m_2)} F(m_2)^{s(m_1)}.$$

(b) Suppose that $m = p_1^{a_1} \cdots p_r^{a_r} q_1^{b_1} \cdots q_s^{b_s}$, where $\epsilon(p_i) = -1$ and $\epsilon(q_i) = 1$ for all i. Prove that

$$s(m) = \begin{cases} 0 & \text{some } a_i \text{ is odd} \\ \prod_{i=1}^{s}(b_i + 1) & \text{all } a_i\text{'s are even.} \end{cases}$$

(c) If $\epsilon(m) = -1$, show that there is at least one prime p with $\epsilon(p) = -1$ and $\nu_p(m)$ odd. Conclude that $s(m) = 0$.

(d) Suppose that $\epsilon(m) = -1$, and that m is divisible by two primes p and q with $\epsilon(p) = \epsilon(q) = -1$ and $\nu_p(m)$ and $\nu_q(m)$ odd. Prove that $F(m) = 1$. Hint: write $m = p^{2a+1}q^{2b+1}m'$, and use (a)–(c).

(e) Finally, suppose that m is divisible by a unique prime p with $\epsilon(p) = 1$ and $\nu_p(m)$ odd. Then m can be written $m = p^{2a+1}p_1^{a_1} \cdots p_r^{a_r} q_1^{b_1} \cdots q_s^{b_s}$, where $\epsilon(p) = \epsilon(p_i) = -1$ and $\epsilon(q_i) = 1$ for all i. Prove that

$$F(m) = p^{(a+1)(b_1+1)\cdots(b_s+1)}.$$

Hint: show that $F(p^{2a+1}) = p^{a+1}$, and use (a)–(c).

By (d) and (e), we see that when $\epsilon(m) = -1$, $F(m)$ is computed by the formulas given in Lemma 13.26. Thus Lemma 13.26 is an immediate corollary of this exercise and the previous one.

13.17. Let p be a prime dividing $J(d_1, d_2)^2$. In Corollary 13.25, we showed that $p \leq d_1 d_2/4$. In some cases, this estimate can be improved.

(a) If $d_1 d_2 \equiv 1 \bmod 8$, then prove that $p < d_1 d_2/8$. Hint: use $p \mid (d_1 d_2 - x^2)/4$. When $p = 2$, note that $d_1 d_2 \equiv 1 \bmod 8$ implies $d_1 d_2 \geq 33$.

(b) If $d_1 \equiv d_2 \equiv 5 \bmod 8$, then prove that $p < d_1 d_2/16$. Hint: when p is odd, we have $p \mid (d_1 d_2 - x^2)/8$. To rule out the case $2p =$

$(d_1d_2 - x^2)/4$, use Exercise 13.15 and Lemma 13.26. When $p = 2$, see (a).

13.18. Use Theorem 13.24 to compute the constant terms of $H_{-56}(X)$ and $H_{-71}(X)$, and compare your results with (13.1) and (13.20). Hint: use Lemma 13.26 to compute $F(m)$.

§14. ELLIPTIC CURVES

The theory of complex multiplication has enabled us to prove some wonderful results, but our treatment is still far from complete. In particular, we need to acquaint the reader with the more modern form of the theory, where elliptic functions are replaced by elliptic curves. Thus, in this last section of the book, we will give some of the basic definitions and theorems concerning elliptic curves, and we will discuss complex multiplication and elliptic curves over finite fields. Then, to illustrate the power of what we've done, we will examine two recent primality tests that involve elliptic curves, one of which makes use of the class equation. Our treatment of these topics will not be self-contained, for our purpose is mostly to entice the reader into learning more about this lovely subject. Excellent introductions to elliptic curves are available, notably the books by Husemöller [58], Koblitz [67] and Silverman [93], and more advanced topics are discussed in the books by Lang [73] and Shimura [90].

A. Elliptic Curves and Weierstrass Equations

Given a field K of characteristic different from 2 or 3, an *elliptic curve E over K* is an equation of the form

$$y^2 = 4x^3 - g_2x - g_3, \tag{14.1}$$

where

$$g_2, g_3 \in K \quad \text{and} \quad \Delta = g_2^3 - 27g_3^2 \neq 0.$$

For reasons that will soon become clear, this equation is called the *Weierstrass equation* of E. When K has characteristic 2 or 3, a more complicated defining equation is needed (see Silverman [93, Appendix A]).

Given an elliptic curve E over K, we define $E(K)$ to be the set of solutions

$$E(K) = \{(x,y) \in K \times K : y^2 = 4x^3 - g_2x - g_3\} \cup \{\infty\}.$$

The symbol ∞ appears because in algebraic geometry, it is best to work with homogeneous equations in projective space. Equation (14.1) defines a curve in the affine space K^2, but in the projective space $\mathbb{P}^2(K)$ there is an extra "point at infinity" (see Exercise 14.1 for the details). Given a field extension $K \subset L$, we can also define $E(K) \subset E(L)$ in an obvious way.

Over the complex numbers $\mathbf{C}$, the Weierstrass $\wp$-function gives us elliptic curves as follows. Let $L \subset \mathbf{C}$ be a lattice, and let $\wp(z) = \wp(z; L)$ be the corresponding $\wp$-function. Then we have the differential equation

$$\wp'(z)^2 = 4\wp(z)^3 - g_2(L)\wp(z) - g_3(L)$$

of Theorem 10.1, which gives us the elliptic curve

$$y^2 = 4x^3 - g_2(L)x - g_3(L).$$

If $z \notin L$, then $\wp(z)$ and $\wp'(z)$ are defined, and the differential equation shows that $(\wp(z), \wp'(z))$ is in $E(\mathbf{C})$. Since $\wp(z)$ and $\wp'(z)$ are also periodic for L, we get a well-defined mapping

$$(\mathbf{C} - L)/L \longrightarrow E(\mathbf{C}) - \{\infty\}.$$

It is easy to show that this map is a bijection (see Exercise 14.2), and consequently we get a bijection

$$\mathbf{C}/L \simeq E(\mathbf{C}) \tag{14.2}$$

by sending $0 \in \mathbf{C}$ to $\infty \in E(\mathbf{C})$. Both $\mathbf{C}/L$ and $E(\mathbf{C})$ have natural structures as Riemann surfaces, and it can be shown that the above map is biholomorphic.

The unexpected fact is that *every* elliptic curve over $\mathbf{C}$ arises from a *unique* Weierstrass $\wp$-function. More precisely, we have the following result:

Proposition 14.3. *Let E be an elliptic curve over $\mathbf{C}$ given by the Weierstrass equation*

$$y^2 = 4x^3 - g_2x - g_3, \qquad g_2, g_3 \in \mathbf{C}, \quad g_2^3 - 27g_3^2 \neq 0,$$

then there is a unique lattice $L \subset \mathbf{C}$ such that

$$g_2 = g_2(L)$$
$$g_3 = g_3(L).$$

Proof. The existence of L was proved in Corollary 11.7, and the uniqueness follows from the from the proof of Theorem 10.9 (see Exercise 14.3). Q.E.D.

Proposition 14.3 is often called uniformization theorem for elliptic curves. Note that it is a consequence of the properties of the j-function.

The mention of the j-function prompts our next definition: if an elliptic curve E over a field K is defined by the Weierstrass equation (14.1), then the *j-invariant* $j(E)$ is defined to be the number

$$j(E) = 1728\frac{g_2^3}{g_2^3 - 27g_3^2} = 1728\frac{g_2^3}{\Delta} \in K.$$

Note that $j(E)$ is well-defined since $\Delta \neq 0$, and the factor of 1728 doesn't cause trouble since K has characteristic different from 2 and 3 (the definition of the j-invariant is more complicated in the latter case—see Silverman [93, Appendix A]). Over the complex numbers, notice that

$$j(L) = j(E)$$

whenever E is the elliptic curve determined by the lattice $L \subset \mathbf{C}$.

To define isomorphisms of elliptic curves, let E and E' be elliptic curves over K, defined by Weierstrass equations $y^2 = 4x^3 - g_2x - g_3$ and $y^2 = 4x^3 - g_2'x - g_3'$ respectively. Then E and E' are *isomorphic over* K if there is a nonzero $c \in K$ such that

$$g_2' = c^4 g_2$$
$$g_3' = c^6 g_3.$$

In this case, note that the map sending (x,y) to (c^2x, c^3y) induces a bijection

$$E(K) \simeq E'(K).$$

It is trivial to check that isomorphic elliptic curves have the same j-invariant.

Over the complex numbers, isomorphisms of elliptic curves are related to lattices and j-invariants as follows:

Proposition 14.4. *Let E and E' be elliptic curves corresponding to lattices L and L' respectively. Then the following statements are equivalent:*

(i) *E and E' are isomorphic over* $\mathbf{C}$.

(ii) *L and L' are homothetic.*

(iii) $j(E) = j(E')$.

Proof. This follows easily from Theorem 10.9. We leave the details to the reader (see Exercise 14.4). Q.E.D.

What is more interesting is that part of this proposition generalizes to any algebraically closed field:

Proposition 14.5. *Let E and E' be elliptic curves over a field K.*

(i) *E and E' have the same j-invariant if and only if they are isomorphic over a finite extension of K.*

(ii) *If K is algebraically closed, then E and E' have the same j-invariant if and only if they are isomorphic over K.*

Proof. The proof is basically a transcription of the algebraic part of the proof of Theorem 10.9—see Exercise 14.4. Q.E.D.

Over nonalgebraically closed fields, nonisomorphic elliptic curves may have the same j-invariant (see Exercise 14.4 for an example over $\mathbb{Q}$). Later, we will discuss the isomorphism classes of elliptic curves over a finite field.

Finally, we need to discuss the group structure on an elliptic curve. The basic idea is to translate the addition law for the Weierstrass $\wp$-function into algebraic terms. To see how this works, let E be an elliptic curve over K, and let P_1 and P_2 be two points in $E(K)$. Our goal is to define $P_1 + P_2 \in E(K)$. If $P_1 = \infty$, we define

$$P_1 + P_2 = \infty + P_2 = P_2,$$

and the case $P_2 = \infty$ is treated similarly. Thus ∞ will be the identity element of $E(K)$. For the remaining cases, we may write $P_1 = (x_1, y_1)$ and $P_2 = (x_2, y_2)$. If $x_1 \neq x_2$, then we define

$$P_1 + P_2 = (x_3, y_3),$$

where x_3 and y_3 are given by

$$\begin{aligned} x_3 &= -x_1 - x_2 - \frac{1}{4}\left(\frac{y_1 - y_2}{x_1 - x_2}\right)^2 \\ y_3 &= -y_1 - (x_3 - x_1)\left(\frac{y_1 - y_2}{x_1 - x_2}\right). \end{aligned} \tag{14.6}$$

These formulas come from the addition laws for $\wp(z+w)$ and $\wp'(z+w)$ (see Theorem 10.1 and Exercise 14.5).

We still need to consider what happens when $x_1 = x_2$. In this case, the Weierstrass equation implies that $y_1 = \pm y_2$, so that there are two cases to consider. When $y_1 = -y_2$, we define

$$P_1 + P_2 = \infty.$$

This formula tells us that the inverse of $(x, y) \in E(K)$ is $(x, -y)$. Finally, suppose that $P_1 = P_2$, where $y_1 = y_2 \neq 0$. Here, we define

$$P_1 + P_2 = 2P_1 = (x_3, y_3),$$

where x_3 and y_3 are given by

$$(14.7)\qquad \begin{aligned} x_3 &= -x_1 - x_2 - \frac{1}{16}\left(\frac{12x_1^2 - g_2}{y_1}\right)^2 \\ y_3 &= -y_1 - (x_3 - x_1)\left(\frac{12x_1 - g_2}{2y_1}\right). \end{aligned}$$

These formulas come from the duplication laws for $\wp(2z)$ and $\wp'(2z)$ (see (10.13) and Exercise 14.5). The major fact is that we get a group:

Theorem 14.8. *If E is an elliptic curve over a field K, then $E(K)$ is a group (with ∞ as identity) under the binary operation defined above.*

Proof. See Husemöller [58], Koblitz [67] or Silverman [93] for a proof. These references also explain a lovely geometric interpretation of the above formulas. Q.E.D.

If E is an elliptic curve over K and $K \subset L$ is a field extension, then it is easy to show that $E(K)$ is a subgroup of $E(L)$.

Over the complex numbers, we saw in (14.2) that there is a bijection $\mathbf{C}/L \simeq E(\mathbf{C})$. Notice that both of these objects are groups: $\mathbf{C}/L$ has a natural group structure induced by addition of complex numbers, and $E(\mathbf{C})$ has the group structure defined in Theorem 14.8. It is immediate that the map $\mathbf{C}/L \simeq E(\mathbf{C})$ is a group isomorphism.

B. Complex Multiplication and Elliptic Curves

The next topic to discuss is the complex multiplication of elliptic curves. The idea is to take the theory developed in §§10 and 11 and translate lattices into elliptic curves. The crucial step is to get an algebraic description of complex multiplication, which can then be used over arbitrary fields.

Let's start by describing the endomorphism ring of an elliptic curve E over $\mathbf{C}$. Namely, if E corresponds to the lattice L, we define

$$\mathrm{End}_{\mathbf{C}}(E) = \{\alpha \in \mathbf{C} : \alpha L \subset L\}.$$

This is clearly a subring of $\mathbf{C}$, and note that $\mathbf{Z} \subset \mathrm{End}_{\mathbf{C}}(E)$. Then we say that E has *complex multiplication* if $\mathbf{Z} \neq \mathrm{End}_{\mathbf{C}}(E)$. From Theorem 10.14, it follows that E has complex multiplication if and only if L does, and in this case, $\mathrm{End}_{\mathbf{C}}(E)$ is an order $\mathcal{O}$ in an imaginary quadratic field.

Given $\alpha \in \mathcal{O}$, the inclusion $\alpha L \subset L$ gives us a group homomorphism $\alpha : \mathbf{C}/L \to \mathbf{C}/L$. Combined with (14.2), we see that $\alpha \in \mathrm{End}_{\mathbf{C}}(E)$ induces induces a group homomorphism

$$\alpha : E(\mathbf{C}) \to E(\mathbf{C}).$$

In terms of the x and y coordinates of a point in $E(\mathbb{C})$, this map can be described as follows:

Proposition 14.9. *Given $\alpha \neq 0 \in \operatorname{End}_{\mathbb{C}}(E)$, there is a rational function $R(x) \in \mathbb{C}(x)$ such that for $(x,y) \in E(\mathbb{C})$, we have*

$$\alpha(x,y) = (R(x), \frac{1}{\alpha}R'(x)y),$$

where $R'(x) = (d/dx)R(x)$.

Proof. Given $\alpha L \subset L$, we saw in Theorem 10.14 that there is a rational function $R(x)$ such that $\wp(\alpha z) = R(\wp(z))$. Differentiating with respect to z gives $\wp'(\alpha z)\alpha = R'(\wp(z))\wp'(z)$, and thus $\wp'(\alpha z) = (1/\alpha)R'(\wp(z))\wp'(z)$. Since $\alpha : E(\mathbb{C}) \to E(\mathbb{C})$ comes from $\alpha : \mathbb{C}/L \to \mathbb{C}/L$ via the map $z \mapsto (\wp(z), \wp'(z))$, the proposition follows. Q.E.D.

Because of the algebraic nature of $\alpha \in \operatorname{End}_{\mathbb{C}}(E)$, we write $\alpha : E \to E$ instead of $\alpha : E(\mathbb{C}) \to E(\mathbb{C})$. When $\alpha \neq 0$, we say that α is an *isogeny* from E to itself. The most important invariant of an isogeny is its *degree* $\deg(\alpha)$, which is defined to the the order of its kernel. More precisely, if E corresponds to the lattice L, then it is easy to see that the kernel of $\alpha : E(\mathbb{C}) \to E(\mathbb{C})$ is isomorphic to $L/\alpha L$ (see Exercise 14.6). Thus, by Theorem 10.14, it follows that

$$\deg(\alpha) = |L/\alpha L| = N(\alpha),$$

where $N(\alpha)$ is the norm of $\alpha \in \mathcal{O} = \operatorname{End}_{\mathbb{C}}(E)$.

For an example of complex multiplication, consider the elliptic curve E defined by

$$y^2 = 4x^3 - 30x - 28.$$

We claim that $\operatorname{End}_{\mathbb{C}}(E) = \mathbb{Z}[\sqrt{-2}]$, and that for $(x,y) \in E(\mathbb{C})$, complex multiplication by $\sqrt{-2}$ is an isogeny of degree 2 given by the formula

$$\sqrt{-2}(x,y) = \left(-\frac{2x^2+4x+9}{4(x+2)}, -\frac{1}{\sqrt{-2}}\frac{2x^2+8x-1}{4(x+2)^2}y\right). \tag{14.10}$$

It turns out that the major work of this claim was proved in §10 when we considered the lattice $L = [1, \sqrt{-2}]$. Namely, in the discussion surrounding (10.21) and (10.22), we showed that for some λ,

$$g_2(\lambda L) = \frac{5 \cdot 27}{2}$$
$$g_3(\lambda L) = \frac{7 \cdot 27}{2}.$$

If we set $\lambda' = \sqrt{3/2}\lambda$, then it follows that

$$g_2(\lambda' L) = 30$$
$$g_3(\lambda' L) = 28,$$

which implies that E has complex multiplication by $\sqrt{-2}$. Furthermore, the formula for $\wp(\sqrt{-2}z)$ given in (10.21) and (10.22) easily combines with Proposition 14.9 to prove (14.10) (see Exercise 14.7).

For an elliptic curve E over an arbitrary field K, we can't use lattices to define complex multiplication. But as indicated by Proposition 14.9, there is a purely algebraic definition of the endomorphism ring $\mathrm{End}_K(E)$ that depends only on the defining equation of E (see Silverman [93, Chapter III]). Because of the group structure of E, $\mathrm{End}_K(E)$ always contains $\mathbb{Z}$, and if K has characteristic zero, we say that E has *complex multiplication* if $\mathrm{End}_{\overline{K}}(E) \neq \mathbb{Z}$, where $\overline{K}$ is the algebraic closure of K (thus the complex multiplications may only be defined over finite extensions of K). When K is a finite field, we will see below that End_K is always bigger than $\mathbb{Z}$. For this reason, the term "complex multiplication" is rarely used when K has positive characteristic.

When $K \subset \mathbb{C}$, we can describe the endomorphism ring $\mathrm{End}_K(E)$ as follows. Namely, let $\alpha \in \mathrm{End}_{\mathbb{C}}(E)$, and use Proposition 14.9 to write $\alpha(x,y) = (R(x), (1/\alpha)R'(x)y)$ for $(x,y) \in E(\mathbb{C})$. Then

$$\alpha \in \mathrm{End}_K(E) \iff R(x), \frac{1}{\alpha}R'(x) \in K(x).$$

Another interesting case is when $K = \mathbb{F}_q$ is a finite field. Here, the map sending (x,y) to (x^q, y^q) clearly defines a group homomorphism $E(L) \to E(L)$ for any field L containing K (see Exercise 14.8). This gives an element $Frob_q \in \mathrm{End}_K(E)$, which is called the *Frobenius endomorphism* of E. It will play an important role later on. Notice that this map is *not* of the form $(R(x), (1/\alpha)R'(x)y)$.

In this abstract setting, one can still define the degree of an isogeny $\alpha \neq 0 \in \mathrm{End}_K(E)$. When $K \subset \mathbb{C}$, the degree of α is again the order of $\ker(\alpha): E(\mathbb{C}) \to E(\mathbb{C})$, while over a finite field, the degree is more subtle to define. For example, the Frobenius isogeny $Frob_q$ always has degree q even though $Frob_q : E(L) \to E(L)$ is injective for any field $K \subset L$. See Silverman [93, §III.4] for a precise definition of the degree of an isogeny.

Besides isogenies from E to itself (which are recorded by $\mathrm{End}_K(E)$), one can also define the notion of an isogeny α between different elliptic curves E and E' over the same field K. For simplicity, we will confine our remarks to the case $K = \mathbb{C}$. In this situation, E and E' correspond to lattices L and L'. If $\alpha \neq 0$ is a complex number such that $\alpha L \subset L'$, then multiplication by α induces a map

$$\alpha : E(\mathbb{C}) \longrightarrow E'(\mathbb{C})$$

with kernel $L'/\alpha L$, and we say that α is an *isogeny* from E to E'. As in Proposition 14.9, one can show that α is essentially algebraic in nature (see Exercise 14.9), so that we can write α as $\alpha : E \to E'$, and we say that α is an *isogeny* from E to E'.

The notion of isogeny has a close relation to the modular equation. We define an isogeny $\alpha : E \to E'$ to be *cyclic* if its kernel $L'/\alpha L$ is cyclic. Then we have:

Proposition 14.11. *Let E and E' be elliptic curves over* $\mathbf{C}$. *Then there is a cyclic isogeny α from E to E' of degree m if and only if $\Phi_m(j(E), j(E'))$* $= 0$.

Proof. This follows easily from the analysis of $\Phi_m(u,v) = 0$ given in Theorem 11.23 (see Exercise 14.10). Q.E.D.

For a more complete treatment of these topics, see Lang [73, Chapters 2 and 5] and Silverman [93, Chapter III].

C. Elliptic Curves over Finite Fields

So far, we've translated concepts about lattices into concepts about elliptic curves. If this were all that happened, there would be no special reason to study elliptic curves. The important point is that the algebraic formulation allows us to state some fundamentally new results, the most interesting of which involve elliptic curves over a finite field $\mathbf{F}_q$. As usual, we will assume that $\mathbf{F}_q$ has characteristic greater than 3, i.e., $q = p^a$, $p > 3$.

When E is an elliptic curve over $\mathbf{F}_q$, the group of solutions $E(\mathbf{F}_q)$ is a finite Abelian group, and it is easy to see that its order $|E(\mathbf{F}_q)|$ is at most $2q + 1$ (see Exercise 14.11). In 1934, Hasse proved the following stronger bound conjectured by Artin:

Theorem 14.12. *If E is an elliptic curve over* $\mathbf{F}_q$, *then*

$$q + 1 - 2\sqrt{q} \leq |E(\mathbf{F}_q| \leq q + 1 + 2\sqrt{q}.$$

Proof. We will discuss some of the ideas used in the proof. The key ingredient is the isogeny $Frob_q \in \mathrm{End}_{\mathbf{F}_q}(E)$ defined by $Frob_q(x,y) = (x^q, y^q)$.

We can form the isogeny $1 - Frob_q$, and it follows easily that if $\overline{\mathbf{F}}_q$ is the algebraic closure of $\mathbf{F}_q$, then

$$E(\mathbf{F}_q) = \ker(1 - Frob_q : E(\overline{\mathbf{F}}_q) \to E(\overline{\mathbf{F}}_q))$$

(see Exercise 14.12). The next step is to show that $1 - Frob_q$ is a separable isogeny, which implies that

$$|E(\mathbf{F}_q)| = \deg(1 - Frob_q). \tag{14.13}$$

From here, the proof is a straightforward consequence of the basic properties of isogenies (see Silverman [93, Chapter V, Theorem 1.1]). Q.E.D.

In 1946, Weil proved a similar result for algebraic curves over finite fields, and in 1974, Deligne proved a vast generalization (conjectured by Weil) to higher dimensional algebraic varieties. For further discussion and references, see Ireland and Rosen [59, Chapter 11] and Silverman [93, §V.2].

Elliptic curves over finite fields come in two types, *ordinary* and *supersingular*, as determined by their endomorphism rings:

Theorem 14.14. *If E is an elliptic curve over $\mathbf{F}_q$, then the endomorphism ring $\mathrm{End}_{\overline{\mathbf{F}}_q}(E)$ is either an order in an imaginary quadratic field or an order in a quaternion algebra.*

Remarks

(i) We say that E is *ordinary* in the former case and *supersingular* in the latter.

(ii) Notice that for elliptic curves over a finite field K, $\mathrm{End}_{\overline{K}}(E)$ is *always* larger than $\mathbf{Z}$.

Proof. See Silverman [93, Chapter V, Theorem 3.1]. Q.E.D.

There are many known criteria for E to be supersingular (see Husemöller [58, p. 258] for an exhaustive list). Over a prime field $\mathbf{F}_p$, there is a special criterion which will be useful later on:

Proposition 14.15. *Let E be an elliptic curve over $\mathbf{F}_p$. Then E is supersingular if and only if*

$$|E(\mathbf{F}_p)| = p+1.$$

Proof. See Silverman [93, Chapter V, Exercise 5.10]. Q.E.D.

It is interesting to note that $|E(\mathbf{F}_p)| = p+1$ is the center of the range $p+1-2\sqrt{p} \le |E(\mathbf{F}_p)| \le p+1+2\sqrt{p}$ allowed by Hasse's theorem.

From the point of view of endomorphisms, ordinary elliptic curves over finite fields behave like elliptic curves over $\mathbf{C}$ with complex multiplication, since in each case, the endomorphism ring is an order in an imaginary quadratic field. This suggests a deeper relation between these two classes, which leads to our next topic, reduction of elliptic curves.

The basic idea of reduction is the following. Let K be a number field, and let E be an elliptic curve defined by

$$y^2 = 4x^3 - g_2x - g_3, \qquad g_2, g_3 \in K.$$

If $\mathfrak{p}$ is prime in $\mathcal{O}_K$, we want to "reduce" E modulo $\mathfrak{p}$. This can't be done in general, but suppose that g_2 and g_3 can be written in the form α/β, where $\alpha, \beta \in \mathcal{O}_K$ and $\beta \notin \mathfrak{p}$. Then we can define $[g_2]$ and $[g_3]$ in $\mathcal{O}_K/\mathfrak{p}$. If, in addition, we have

$$\Delta = [g_2]^3 - 27[g_3]^2 \neq 0 \in \mathcal{O}_K/\mathfrak{p},$$

then

$$y^2 = 4x^3 - [g_2]x - [g_3]$$

is an elliptic curve $\overline{E}$ over the finite field $\mathcal{O}_K/\mathfrak{p}$. In this case we call $\overline{E}$ the *reduction* of E modulo $\mathfrak{p}$, and we say that E has *good reduction* modulo $\mathfrak{p}$.

When E has complex multiplication and good reduction, Deuring, drawing on examples of Gauss, discovered an astonishing relation between the complex multiplication of E and the number of points in $\overline{E}(\mathcal{O}_K/\mathfrak{p})$. Rather than state his result in its full generality, we will present a version that concerns only elliptic curves over the prime field $\mathbf{F}_p$.

To set up the situation, let $\mathcal{O}$ be an order in an imaginary quadratic field K, and let L be the ring class field of $\mathcal{O}$. Let p be a prime in $\mathbf{Z}$ which splits completely in L, and we will fix a prime $\mathfrak{P}$ of L lying above p, so that $\mathcal{O}_L/\mathfrak{P} \simeq \mathbf{F}_p$. Finally, let E be an elliptic curve over L which has good reduction at $\mathfrak{P}$. With these hypotheses, the reduction $\overline{E}$ is an elliptic curve over $\mathbf{F}_p$. Then we have the following theorem:

Theorem 14.16. *Let $\mathcal{O}$, L, p and $\mathfrak{P}$ be as above, and let E be an elliptic curve over L with* $\mathrm{End}_{\mathbf{C}}(E) = \mathcal{O}$. *If E has good reduction modulo $\mathfrak{P}$, then there is $\pi \in \mathcal{O}$ such that $p = \pi\overline{\pi}$ and*

$$|\overline{E}(\mathbf{F}_p)| = p + 1 - (\pi + \overline{\pi}).$$

Furthermore, $\mathrm{End}_{\overline{\mathbf{F}}_p}(\overline{E}) = \mathcal{O}$, *and every elliptic curve over $\mathbf{F}_p$ with endomorphism ring (over $\overline{\mathbf{F}}_p$) equal to $\mathcal{O}$ arises in this way.*

Proof. The basic idea is that when the above hypotheses are fulfilled, reduction induces an isomorphism

$$\mathrm{End}_{\mathbf{C}}(E) \xrightarrow{\sim} \mathrm{End}_{\overline{\mathbf{F}}_p}(\overline{E})$$

that preserves degrees. The proof of this fact is well beyond the scope of this book (see Lang [73, Chapter 13, Theorem 12]).

From the above isomorphism, it follows that there is some $\pi \in \mathrm{End}_{\mathbf{C}}(E)$ which reduces to $Frob_p \in \mathrm{End}_{\overline{\mathbf{F}}_p}(\overline{E})$. Since $Frob_p$ has degree p, so does π. Over the complex numbers, we know that the degree of $\pi \in \mathcal{O} = \mathrm{End}_{\mathbf{C}}(E)$ is just its norm, so that $N(\pi) = p$. Thus we can write $p = \pi\overline{\pi}$ in $\mathcal{O}$.

It is now trivial to compute the number of points on $\overline{E}$. As we noted in (14.13),

$$|\overline{E}(\mathsf{F}_p)| = \deg(1 - Frob_p).$$

Since the reduction map preserves degrees, it follows that

$$\begin{aligned}\deg(1 - Frob_p) &= \deg(1 - \pi) = N(1 - \pi) = (1 - \pi)(1 - \overline{\pi}) \\ &= p + 1 - (\pi + \overline{\pi})\end{aligned}$$

since $p = \pi\overline{\pi}$. This proves the desired formula for $|\overline{E}(\mathsf{F}_p)|$.

For a proof of the final part of the theorem, see Lang [73, Chapter 13, Theorems 13 and 14]. Q.E.D.

The remarkable fact is that we've already seen two examples of this theorem. First, in (4.24), we stated the following result of Gauss: if $p \equiv 1 \bmod 3$ is prime, then

(14.17) If $4p = a^2 + 27b^2$ and $a \equiv 1 \bmod 3$, then $N = p + a - 2$, where N is the number of solutions modulo p of $x^3 - y^3 \equiv 1 \bmod p$.

We can relate this to Deuring's theorem as follows. The coordinate change $(x, y) \mapsto (3x/(1 + y), 9(1 - y)/(1 + y))$ transforms the curve $x^3 = y^3 + 1$ into the elliptic curve E defined by $y^2 = 4x^3 - 27$ (see Exercise 14.13). Gauss didn't count the three points at infinity that lie on $x^3 = y^3 + 1$, and when these are taken into account, then (14.17) asserts that $|E(\mathsf{F}_p)| = p + 1 + a$. Since $p \equiv 1 \bmod 3$, we can write $p = \pi\overline{\pi}$ in $\mathsf{Z}[\omega]$, $\omega = e^{2\pi i/3}$. In §4, we saw that π may be chosen to be primary, which means $\pi \equiv \pm 1 \bmod 3$. Thus we may assume $\pi \equiv 1 \bmod 3$, so that $\pi = A + 3B\omega$, $A \equiv 1 \bmod p$. Then an easy calculation shows that

$$4p = (-(2A - 3B))^2 + 27B^2.$$

Since $2A - 3B = \pi + \overline{\pi}$ and $-(2A - 3B) \equiv 1 \bmod 3$, it follows that (14.17) may be stated as follows:

If $p = \pi\overline{\pi}$ in $\mathsf{Z}[\omega]$ and $\pi \equiv 1 \bmod 3$, then $|E(\mathsf{F}_p)| = p + 1 - (\pi + \overline{\pi})$.

Since E is the reduction of $y^2 = 4x^3 - 27$, which has complex multiplication by $\mathsf{Z}[\omega]$ (see Exercise 14.13), Gauss's observation (14.17) really is a special case of Deuring's theorem.

Similarly, one can check that Gauss's last diary entry, which concerned the number of solutions of $x^2 + y^2 + x^2y^2 \equiv 1 \bmod p$, is also a special case of Deuring's theorem. See the discussion following (4.24) and Exercise 14.14.

As a application of Deuring's theorem, we can give a formula for the number of elliptic curves over F_p which have a preassigned number of points. We first need some notation. Given an order $\mathcal{O}$ in an imaginary quadratic field K, we define the *Hurwitz class number* $H(\mathcal{O})$ to be the weighted sum of class numbers

$$H(\mathcal{O}) = \sum_{\mathcal{O} \subset \mathcal{O}' \subset \mathcal{O}_K} \frac{2}{|\mathcal{O}'^*|} h(\mathcal{O}').$$

We also write $H(\mathcal{O})$ as $H(D)$, where D is the discriminant of $\mathcal{O}$. Then we have the following theorem of Deuring:

Theorem 14.18. *Let $p > 3$ be prime, and let $N = p + 1 - a$ be an integer, where $-2\sqrt{p} \leq a \leq 2\sqrt{p}$. Then the number of elliptic curves E over $\mathbf{F}_p$ which have $|E(\mathbf{F}_p)| = N = p + 1 - a$ is*

$$\frac{p-1}{2} H(a^2 - 4p).$$

Proof. Let π be a root of $x^2 - ax + p$. Since $-2\sqrt{p} \leq a \leq 2\sqrt{p}$, the quadratic formula shows that $\mathcal{O}_a = \mathbf{Z}[\pi]$ is an order in an imaginary quadratic field K. One can also check that p doesn't divide the conductor of $\mathcal{O}_a$ (in fact, it doesn't divide the discriminant), and hence the same is true for any order $\mathcal{O}'$ containing $\mathcal{O}_a$ (see Exercise 14.15).

We will start with the case $a \neq 0$, which by Proposition 14.15 means that all of the elliptic curves involved are ordinary. Given an order $\mathcal{O}'$ containing $\mathcal{O}_a$ and a proper $\mathcal{O}'$-ideal $\mathfrak{a}$, we will produce a collection of elliptic curves E_c with good reduction modulo p. Namely, let L' be the ring class field of $\mathcal{O}'$. Since $p = \pi\bar{\pi}$ in $\mathcal{O}_a \subset \mathcal{O}'$, it follows from Theorem 9.4 that p splits completely in L'. Thus, if $\mathfrak{P}$ is any prime of L' containing p, then $\mathcal{O}_{L'}/\mathfrak{P} \simeq \mathbf{F}_p$.

First, assume that $\mathcal{O}' \neq \mathbf{Z}(i)$ or $\mathbf{Z}[\omega]$, $\omega = e^{2\pi i/3}$, and which implies that $j(\mathfrak{a}) \neq 0, 1728$. If we let

$$k = \frac{27 j(\mathfrak{a})}{j(\mathfrak{a}) - 1728},$$

then we define the collection of elliptic curves E_c over L' by the Weierstrass equations

$$y^2 = 4x^3 - kc^2 x - kc^3,$$

where $c \in \mathcal{O}_L - \mathfrak{P}$ is arbitrary. A computation shows that $j(E) = j(\mathfrak{a})$. We can reduce k modulo L provided that $j(\mathfrak{a}) - 1728 \notin \mathfrak{P}$. Since $1728 = j(i)$, Theorem 13.21 implies that

$$j(\mathfrak{a}) \equiv 1728 \bmod \mathfrak{P} \Rightarrow p \text{ does not split in } K \text{ or } \mathbf{Q}(i)$$

(when $K = \mathbf{Q}(i)$, note that the conductor condition of Theorem 13.21 is satisfied). However, p splits in K, and thus $j(\mathfrak{a}) - 1728 \notin \mathfrak{P}$, as desired.

Then one computes that in $\mathcal{O}_{L'}/\mathfrak{P} \simeq \mathbf{F}_p$,

$$\Delta = [kc^2]^3 - 27[kc^3]^2 = 1728[c^6]\left[\frac{27^3 j(\mathfrak{a})^2}{(j(\mathfrak{a}) - 1728)^3}\right].$$

By the argument used to prove $j(\mathfrak{a}) - 1728 \notin \mathfrak{P}$, Theorem 13.21 and $j(\omega) = 0$ show that $j(\mathfrak{a}) \notin \mathfrak{P}$. It follows that E_c has good reduction modulo $\mathfrak{P}$ since $c \notin \mathfrak{P}$.

If $\mathcal{O}' = \mathbf{Z}[i]$ or $\mathbf{Z}[\omega]$, then $L' = K$. Here, we will use the collection of elliptic curves E_c defined by

$$y^2 = 4x^3 - cx, \qquad c \notin \pi\mathbf{Z}[i]$$
$$y^2 = 4x^3 - c, \qquad c \notin \pi\mathbf{Z}[\omega].$$

One easily checks that these curves have good reduction modulo π and $\mathcal{O}'$ as their endomorphism ring.

Theorem 14.16 assures us that every ordinary elliptic curve $\overline{E}$ over $\mathbf{F}_p$ arises from reduction of some elliptic curve with complex multiplication. Given this, it follows without difficulty that $\overline{E}$ is in fact the reduction of one of the E_c's constructed above (see Exercise 14.16).

Given $\mathcal{O}'$, there are $h(\mathcal{O}')$ distinct j-invariants $j(\mathfrak{a})$, and hence for a fixed a, we have

$$\sum_{\mathcal{O}_a \subset \mathcal{O}'} h(\mathcal{O}')$$

distinct collections of elliptic curves E_c. Furthermore, another application of Theorem 13.21 shows that different collections reduce to curves with different j-invariants. Since each collection E_c gives us $p-1$ curves over $\mathbf{F}_p$, we get

$$(p-1)\sum_{\mathcal{O}_a \subset \mathcal{O}'} h(\mathcal{O}')$$

elliptic curves over $\mathbf{F}_p$. But which of these have $p + 1 - a$ points on them? The problem is that Theorem 14.16 implies that $|\overline{E}_c(\mathbf{F}_p)|$ is determined by *some* element of $\mathcal{O}'$ of norm p, but it need not be π! All curves in a given collection have the same j-invariant, but they need not be isomorphic over $\mathbf{F}_p$, and hence they may have different numbers of points. In fact, this is always the case:

Proposition 14.19. *Let E and E' be elliptic curves over $\mathbf{F}_p$. If E is ordinary, then E and E' are isomorphic over $\mathbf{F}_p$ if and only if $j(E) = j(E')$ and $|E(\mathbf{F}_p)| = |E'(\mathbf{F}_p)|$.*

Proof. One direction of the proof is obvious, but the other requires some more advanced concepts. We will give the details since this result doesn't appear in standard references. The key ingredient is a theorem of Tate, which asserts that curves with the same number of points over a finite field K are isogenous over K (see Husemöller [58, §13.8]). Applying this to $|E(\mathsf{F}_p)| = |E'(\mathsf{F}_p)|$, we get an isogeny $\lambda : E \to E'$ defined over F_p. Replacing λ by $1-\lambda$ if necessary, we may assume that λ is separable. Since E and E' have the same j-invariant, we can also find an isomorphism $\phi : E' \to E$ defined over some extension F_{p^a} (see Proposition 14.5). Thus $\phi \circ \lambda \in \mathrm{End}_{\overline{\mathsf{F}}_p}(E)$. Since E is ordinary, the endomorphism ring is an order in an imaginary quadratic field, and it follows that $\mathbb{Z}[Frob_p]$ has finite index in $\mathrm{End}_{\overline{\mathsf{F}}_p}(E)$. In Exercise 14.15, we saw that p does not divide the conductor of $\mathbb{Z}[\pi] \simeq \mathbb{Z}[Frob_p]$, and it follows that p does not divide the index m of $\mathbb{Z}[Frob_p] \subset \mathrm{End}_{\overline{\mathsf{F}}_p}(E)$. Thus $m\phi \circ \lambda \in \mathbb{Z}[Frob_p]$, which implies that $m\phi \circ \lambda = \phi \circ m\lambda$ is defined over F_p. Since $m\lambda$ is separable, the standard properties of isogenies imply that ϕ is defined over F_p (see Silverman [93, Chapter III, Corollary 4.11]). Q.E.D.

We claim that the collection $\overline{E}_c$ contains exactly $(p-1)/|\mathcal{O}'^*|$ curves with $p+1-a$ points. This will immediately imply our desired formula. Let's first consider the case when $\overline{E}_c$ corresponds to a j-invariant $j(\mathfrak{a}) \neq 0$ or 1728. Here, the only solutions of $N(\alpha) = p$ in $\mathcal{O}'$ are $\alpha = \pm\pi$ and $\pm\overline{\pi}$ (see Exercise 14.17). Thus, for each c, Deuring's theorem tells us that

$$|\overline{E}_c(\mathsf{F}_p)| = p + 1 \pm a.$$

However, the curves $\overline{E}_c$ fall into two isomorphism classes, each consisting of $(p-1)/2$ curves, corresponding to whether $[c] \in \mathsf{F}_p^{*2}$ or not (see Exercise 14.18). By the above proposition, nonisomorphic curves have a different number of elements, and hence we see that exactly half of the E_c's have $p + 1 - a$ elements. Since $\mathcal{O}'^* = \{\pm 1\}$, we get $(p-1)/2 = (p-1)/|\mathcal{O}'^*|$ curves with $p+1-a$ points.

When $j(\mathfrak{a}) = 1728$, things are a bit more complicated. Here, $\mathcal{O}' = \mathbb{Z}[i]$, and $p = \pi\overline{\pi}$ implies that $p \equiv 1 \bmod 4$. The only solutions of $N(\alpha) = p$ are $\alpha = \pm\pi, \pm\overline{\pi}, \pm i\pi$, and $\pm i\overline{\pi}$ (see Exercise 14.17), and thus there are at most four possibilities for $|\overline{E}_c(\mathsf{F}_p)|$. But there are four isomorphism classes of curves with $j = 1728$ in this case, each consisting of $(p-1)/4$ curves (see Exercise 14.18). It follows that there are exactly $(p-1)/|\mathcal{O}'^*|$ curves with $p+1-a$ points. The case $j = 0$ is similar and is left to the reader.

It remains to study the case $a = 0$, which concerns the number of supersingular curves over F_p. Since Theorem 14.16 doesn't apply to this case, we will take a more indirect approach. Given any a in the range $2\sqrt{p} \leq a \leq 2\sqrt{p}$, we just proved that when $a \neq 0$, there are $(p-1)/2H(a^2-4p)$ elliptic curves over F_p with $p+1-a$ points. Let SS denote the number of supersingular curves. Since there are $p(p-1)$ elliptic curves over F_p (see

Exercise 14.19), it follows that

$$p(p-1) = SS + \sum_{0<|a|\leq 2\sqrt{p}} \frac{p-1}{2} H(a^2-4p). \tag{14.20}$$

However, we claim that there is a class number formula

$$2p = \sum_{0\leq |a|\leq 2\sqrt{p}} H(a^2-4p). \tag{14.21}$$

Since (14.20) and (14.21) imply that $SS = (p-1)/2H(-4p)$, we need only prove (14.21).

To prove this, note that $H(a^2-4p) = H(\mathcal{O}_a)$, so that by definition, the right-hand side of (14.21) equals

$$\sum_{0\leq |a|\leq 2\sqrt{p}} \sum_{\mathcal{O}_a\subset \mathcal{O}'} \frac{2}{|\mathcal{O}'^*|} h(\mathcal{O}').$$

If we define the function $\chi(a)$ by

$$\chi(a) = \begin{cases} 1 & \text{if } \mathcal{O}_a \subset \mathcal{O}' \\ 0 & \text{otherwise,} \end{cases}$$

then the above sum can be written as

$$\sum_{\mathcal{O}'} \left(\frac{2}{|\mathcal{O}'^*|} \sum_{0\leq |a|\leq 2\sqrt{p}} \chi(a) \right) h(\mathcal{O}').$$

It is easy to prove that the quantity in parentheses is $r(\mathcal{O}', p)$, which we defined in §13 to be $|\{\pi \in \mathcal{O}' : N(\pi) = p\}/\mathcal{O}'^*|$ (see Exercise 14.20). Thus the right-hand side of (14.21) becomes

$$\sum_{\mathcal{O}'} r(\mathcal{O}', p) h(\mathcal{O}').$$

In Corollary 13.9 we proved that this quantity equals $2p$, and (14.21) is proved. This completes the proof of Theorem 14.18. Q.E.D.

Recall that Corollary 13.9 was part of our study of the polynomial $\Phi_n(X,X)$. It is rather unexpected that the modular equation has a connection with supersingular curves over F_p. This is just more evidence of the amazing richness of the study of elliptic curves. To pursue these topics further, the reader should consult Lang [73] and Shimura [90]. Also, see the monographs by Cassou-Noguès and Taylor [15] and by Gross [45] for an introduction to some of the current research concerning elliptic curves and complex multiplication.

D. Elliptic Curve Primality Tests

In the past few years, there have been some surprising applications of elliptic curves to problems involving factoring and primality. In 1985, Lenstra announced an elliptic curve factoring method [76], and a year later, Goldwasser and Kilian adapted Lenstra's method to obtain an elliptic curve primality test [43]. Both methods use the properties of elliptic curves over finite fields. We will concentrate on the Goldwasser–Kilian Test and its recent variation, the Goldwasser–Kilian–Atkin Test. This last test is especially interesting, for it uses the class equations studied in §13. Thus, the polynomial $H_{-4n}(X)$, which appears in our critierion for when p is of the form x^2+ny^2, can actually be used to prove that p is prime! Our treatment of these tests will not be complete, and for further details, we refer the reader to the articles by Goldwasser and Kilian [43], Lenstra [76] and Morain [79].

Given a potential prime l, the goal of these tests is to prove the primality of l by considering elliptic curves over the field $\mathbf{Z}/l\mathbf{Z}$. Since we don't know that l is prime, we must treat $\mathbf{Z}/l\mathbf{Z}$ as a ring, and thus we need a theory of elliptic curves over *rings*. Fortunately, the basic ideas carry over quite easily. Let R be any commutative ring with identity where 2 and 3 are units. Then an *elliptic curve E over R* is a Weierstrass equation of the usual form

$$y^2 = 4x^3 - g_2x - g_3, \qquad g_2, g_3 \in R,$$

where we now require that

$$\Delta = g_2^3 - 27g_3^2 \in R^*. \tag{14.22}$$

Note that since Δ is a unit in R, the j-invariant

$$j(E) = 1728\frac{g_2^3}{\Delta} \in R$$

is defined.

Given an elliptic curve E over R, we set

$$E_0(R) = \{(x,y) \in R \times R : y^2 = 4x^3 - g_2x - g_3\} \cup \{\infty\}.$$

The reason for the new notation is that $E_0(R)$ may *fail* to be a group! To see this, consider $P_1 = (x_1, y_2)$ and $P_2 = (x_2, y_2)$ in $E_0(R)$. If $x_1 \neq x_2$, then we would like to define

$$P_1 + P_2 = (x_3, y_3),$$

where x_3 and y_3 are given by the formulas (14.6). The problem comes from the denominator $x_1 - x_2$: it is nonzero in R, but it need *not* be invertible! For this reason, the binary operation is only partially defined on $E_0(R)$. Using tools from algebraic geometry, one can define a superset $E(R)$ of $E_0(R)$ which is a group, but we prefer to use $E_0(R)$ because it is easier to work with in practice.

If E is an elliptic curve over $\mathbb{Z}/l\mathbb{Z}$, the potentially incomplete group structure on $E_0(\mathbb{Z}/l\mathbb{Z})$ is not a problem. Namely, if we ever found P_1 and P_2 in $E_0(\mathbb{Z}/l\mathbb{Z})$ such that $P_1 + P_2$ wasn't defined, then it would follow automatically that l must be composite, and the noninvertible denominator would give us factor of l (just compute the appropriate gcd). This observation is the driving force of Lenstra's elliptic curve factoring algorithm (see [76]).

Before discussing the Goldwasser–Kilian Test, let's review some basic ideas concerning primality testing. We regard l as an input of length $[\log_{10} l]$, where $[\ \ ]$ is the greatest integer function. The length is thus bounded by a constant times $\ln l$, which we express by writing $[\log_{10} l] = O(\ln l)$. The most interesting question concerning a primality test is its running time: given an input l, how long, as a function of $\ln l$, does it take a given algorithm to prove that l is (or is not) prime? The simplest algorithm (divide by all numbers $\leq \sqrt{l}$) requires

$$\sqrt{l} = e^{(1/2)\ln l}$$

divisions, and hence runs in *exponential* time. What we really want is a algorithm that runs in *polynomial* time, i.e., where is running time is $O((\ln l)^d)$ for some fixed d. Right now, no polynomial time algorithm is known, although there is a candidate for one—see Wagon [99] for further details.

Another sort of algorithm commonly used is what is called a *probabilistic primality test*. Such a test has two outputs, "prime" and "composite or unluckily prime." In the former case, the program proves the primeness of l, while in the latter case, it says either that l is composite or that l is prime and we were unlucky. A nice discussion of probabilistic primality tests may be found in Wagon's article [99]. For our purposes, we will explain this concept by considering the following very special probabilistic primality test.

Let l be our potential prime, relatively prime to 6, and suppose that we have an elliptic curve E over $\mathbb{Z}/l\mathbb{Z}$ with the following two properties:
(i) $l + 1 - 2\sqrt{l} \leq |E_0(\mathbb{Z}/l\mathbb{Z})| \leq l + 1 + 2\sqrt{l}$.
(ii) $|E_0(\mathbb{Z}/l\mathbb{Z})| = 2q$, where q is an odd prime.
In certain situations, this setup can be used to prove primality:

Lemma 14.23. *Let l and E be as above, and assume $l > 13$. Let $P \neq \infty$ be in $E_0(\mathbb{Z}/l\mathbb{Z})$. If qP is defined and equal to ∞ in $E_0(\mathbb{Z}/l\mathbb{Z})$, then l is prime.*

Proof. Assume that l is not prime, and let $p \leq \sqrt{l}$ be a prime divisor of l. Using the natural map $\mathbb{Z}/l\mathbb{Z} \to \mathbb{Z}/p\mathbb{Z} = \mathsf{F}_p$, we can reduce the equation of E modulo p, and by (14.22), we get an elliptic curve $\overline{E}$ over F_p. Furthermore, we get a natural map

$$E_0(\mathbb{Z}/l\mathbb{Z}) \longrightarrow \overline{E}(\mathsf{F}_p)$$

which takes $P = (x,y) \neq \infty$ in $E_0(\mathbb{Z}/l\mathbb{Z})$ to $\overline{P} = (\overline{x},\overline{y}) \neq \infty$ in $\overline{E}(\mathsf{F}_p)$. Since this map is also clearly a homomorphism (wherever defined), it follows that $q\overline{P} = \infty$ in $\overline{E}(\mathsf{F}_p)$. But q is prime, so that $\overline{P}$ is a point of order q, and hence

$$q \leq |\overline{E}(\mathsf{F}_p)| \leq p + 1 + 2\sqrt{p},$$

where the second inequality comes from Hasse's theorem (Theorem 14.12). Since $p \leq \sqrt{l}$, this implies that

$$q \leq \sqrt{l} + 1 + 2\sqrt[4]{l} = (\sqrt[4]{l} + 1)^2.$$

However, by assumption, we have

$$2q = |E_0(\mathbb{Z}/l\mathbb{Z})| \geq l + 1 - 2\sqrt{l} = (\sqrt{l} - 1)^2.$$

Combining these two inequalities, we obtain

$$\sqrt{l} - 1 \leq \sqrt{2}(\sqrt[4]{l} + 1),$$

which is easily seen to be impossible for $l > 13$. This contradiction proves the lemma. Q.E.D.

To convert this lemma into a probabilistic primality test, we need one more observation. Namely, if l is prime and $|E_0(\mathbb{Z}/l\mathbb{Z})| = 2q$, q an odd prime, then $E_0(\mathbb{Z}/l\mathbb{Z})$ must be a cyclic group, and hence exactly $q-1$ of the $2q-1$ nonidentity elements have order q. Thus, the probability that a randomly chosen $P \neq \infty$ doesn't prove primality (i.e., has order $\neq q$) is $q/(2q-1) \simeq 1/2$, assuming that q is large.

Now we can state the test. Given E and l be as above, pick k randomly chosen points $P_1,\ldots,P_k$ from $E_0(\mathbb{Z}/l\mathbb{Z})$, and then compute $qP_1,\ldots,qP_k$. If any one of these is defined and equals ∞, then by the above lemma, we have a proof of primality. If none of $qP_1,\ldots,qP_k$ satisfy this condition, then either l is composite, or l is prime and we were unlucky. To see how unlucky, suppose that l were prime. Then our test fails only if all of $P_1,\ldots,P_k$ have order $\neq q$. By the above paragraph, the probability of this happening is

$$\left(\frac{q}{2q-1}\right)^k \simeq \frac{1}{2^k}.$$

So we can't guarantee a proof of primality, but we have to be mighty unlucky not to find one.

This test depended on the assumptions (i) and (ii) above. The first assumption is quite reasonable, since by Hasse's theorem it holds if l is prime. So if (i) fails, we have a proof of compositeness. But the second assumption, that $|E_0(\mathbb{Z}/l\mathbb{Z})|$ is twice a prime, is a very special, and certainly fails for most elliptic curves. An added difficulty is that $|E_0(\mathbb{Z}/l\mathbb{Z})|$ is a very large number (by (i), it has the same order of magnitude as l). Thus, even if $|E_0(\mathbb{Z}/l\mathbb{Z})| = 2q$ were twice a prime, we'd be unlikely to know it, since we'd have to prove that q, a number roughly the size of $l/2$, is also prime.

To overcome these problems, Goldwasser and Kilian used two ideas. The first idea is quite simple:

(14.24)
Choose *lots* of elliptic curves E over $\mathbb{Z}/l\mathbb{Z}$ at random.
If we get one where $|E_0(\mathbb{Z}/l\mathbb{Z})| = 2q$, q a probable prime,
then use the above special test to check for primality.

Notice the word "probable." Using known probabilistic compositeness tests (described in Wagon [99]), one can efficiently reduce to the case where $|E_0(\mathbb{Z}/l\mathbb{Z})|$ is of the form $2q$, where q is *probably* prime. If the special test succeeds, we have proved that l is prime, provided that q is prime. Then the second idea is

(14.25) Make the above process recursive.

This means proving q is prime by applying the special test to an elliptic curve over $\mathbb{Z}/q\mathbb{Z}$ of order $2q'$, q' a probable prime. In this way the primality of q' implies the primality of q. Since each iteration reduces the size by a factor of 2 (i.e., q is about the size of $l/2$, q' is about the size of $q/2$, etc.), it follows that in $O(\ln l)$ steps the numbers will get small enough that primality can be verified easily.

The algorithm contained in (14.24) and (14.25) is the heart of the Goldwasser–Kilian primality test (see their article [43] for a fuller discussion). The key unanswered question concerns (14.24): when l is prime, how many elliptic curves do we have to choose before finding one where $|E(\mathbb{Z}/l\mathbb{Z})|$ is twice a prime? The following result of Lenstra plays a crucial role:

Theorem 14.26. *Let l be a prime, and let*

$$S = \{2q : q \text{ prime}, \ l+1-\sqrt{l} \leq 2q \leq l+1+\sqrt{l}\}.$$

Then there is a constant $c_1 > 0$, independent of l and S, such that the number of elliptic curves E over $\mathbb{F}_l$ satisfying $|E(\mathbb{F}_l)| \in S$ is at least

$$c_1 \cdot \frac{(|S|-2)\sqrt{l}(l-1)}{\ln l}.$$

Remark. Notice that the elliptic curves described in this theorem satisfy $l+1-\sqrt{l} \leq |E(\mathbb{F}_l)| \leq l+1+\sqrt{l}$, which is more restrictive than the bound given by Hasse's theorem. The proof below will explain the reason for this.

Proof. Given $2q \in S$, write $2q = l+1-a$. Then we proved in Theorem 14.18 that the number of curves with $2q = l+1-a$ points is $(l-1)/2$

$\times H(a^2 - 4l)$, where $H(a^2 - 4l)$ is the Hurwitz class number defined earlier. Using classically known bounds on class numbers, Lenstra proved in [76] that for $2q \in S$, with at most two exceptions, there is the estimate

$$H(a^2 - 4l) \geq c \cdot \frac{\sqrt{|a^2 - 4l|}}{\ln l}$$

where c is a constant independent of the discriminant (see [76, Proposition 1.8]). We are assuming that $|a| \leq \sqrt{l}$, which implies $\sqrt{|a^2 - 4l|} \geq \sqrt{3l}$, and consequently

$$\frac{l-1}{2} H(a^2 - 4l) \geq c_1 \cdot \frac{\sqrt{l}(l-1)}{\ln l},$$

where $c_1 = \sqrt{3}c/2$. The theorem follows immediately. Q.E.D.

By this theorem, we are reduced to knowing the number of primes in the interval $[(l+1)/2 - \sqrt{l}/2, (l+1)/2 + \sqrt{l}/2]$. By the Prime Number Theorem, the probability that a number in the interval $[0, N]$ is prime is $1/\ln N$. It is conjectured that this holds for intervals of shorter length. Applied to the above, we get the following conjecture:

Conjecture 14.27. *There is a constant $c_2 > 0$ such that, for all sufficiently large primes l, the number of primes in the interval $[(l+1)/2 - \sqrt{l}/2, (l+1)/2 + \sqrt{l}/2]$ is at least*

$$c_2 \cdot \frac{\sqrt{l}}{\ln l}.$$

If this conjecture were true, then Theorem 14.26 would imply that when l is large, there is a constant c_3, independent of l, such that at least

$$c_3 \cdot \frac{l(l-1)}{(\ln l)^2}$$

elliptic curves E over $\mathbb{Z}/l\mathbb{Z}$ have order $|E(\mathbb{F}_l)| = 2q$ for some prime q (see Exercise 14.21). Since there are $l(l-1)$ elliptic curves over $\mathbb{F}_l$, it follows that there is a probability of at least

$$c_3/(\ln l)^2 \tag{14.28}$$

that $|E(\mathbb{F}_l)|$ has the desired order.

Now we can explain how many curves need to be chosen in (14.24). Namely, pick an integer k, and pick $k(\ln l)^2/c_3$ randomly chosen elliptic curves over $\mathbb{Z}/l\mathbb{Z}$. If l were prime, could all these curves fail to have order twice a prime? By (14.28), the probability of this happening is less than

$$\left(1 - \frac{c_3}{(\ln l)^2}\right)^{k(\ln l)^2/c_3} \simeq \frac{1}{e^k}.$$

It remains to give a run time analysis of the Goldwasser–Kilian Test. For (14.24), we need to pick $O((\ln l)^2)$ curves and count the number of points on each one. By an algorithm of Schoof (see Morain [79, §5.5]), it takes $O((\ln l)^8)$ to count the points on each curve. Once a curve with $|E_0(\mathbf{Z}/l\mathbf{Z})| = 2q$ is found, we then need to pick points $P \in E_0(\mathbf{Z}/l\mathbf{Z})$ and compute qP. These operations are bounded by $O((\ln l)^8)$ (see Goldwasser and Kilian [43, §4.3]), and thus the run time of (14.24) is $O((\ln l)^{10})$. By (14.25), we have to iterate this $O(\ln l)$ times, so that the run time of the whole algorithm is $O((\ln l)^{11})$.

The above analysis is predicated on Conjecture 14.27, which may be very difficult to prove (or even false!). But now comes the final ingredient: using known results about the distribution of primes, Goldwasser and Kilian were able to prove that their algorithm terminates with a run time of $O(k^{11})$ for at least

$$1 - O(2^{-k^{1/\ln\ln k}})$$

percent of the prime inputs of length k (see [43, Theorem 3]). Thus the Goldwasser–Kilian Test is *almost* a polynomial time probabilistic primality test!

In practice, the implementation of the Goldwasser–Kilian Test is more complicated than the algorithm sketched above. The main difference is that the order $|E_0(\mathbf{Z}/l\mathbf{Z})|$ is allowed to be of the form mq, where m may be bigger than 2 but is still small compared to q. This means that fewer elliptic curves must be tried before finding a suitable one, and thus the algorithm runs faster. For the details of how this is done, see Goldwasser and Kilian [43, §4.4] or Morain [79, §§2.2.2 and 7.7].

The most "expensive" part of the Goldwasser–Kilian Test is the $O((\ln l)^8)$ spent counting the points on a given elliptic curve. So rather than starting with E and then computing $|E_0(\mathbf{Z}/l\mathbf{Z})|$ the hard way, why not use the theory developed earlier to *predict* the order? This is the basis of the Goldwasser–Kilian–Atkin Test, which we will discuss next.

The wonderful thing about this test is that it brings us back to our topic of primes of the form $x^2 + ny^2$. To see why, let l be a prime, and let n be a positive integer such that l can be written as

$$l = a^2 + nb^2, \qquad a, b \in \mathbf{Z}.$$

We will use this information to produce an elliptic curve over F_l with $l + 1 - 2a$ points on it. The basic idea is to use the characterization of primes of the form $x^2 + ny^2$ proved in §13:

$$l = x^2 + ny^2 \Rightarrow \begin{cases} (-n/l) = 1 \text{ and } H_{-4n}(X) \equiv 0 \bmod l \\ \text{has an integer solution,} \end{cases}$$

where $H_{-4n}(X)$ is the class equation for discriminant $-4n$ (see the proof of Theorem 13.23). Thus $l = a^2 + nb^2$ gives us a solution j of the congruence $H_{-4n}(X) \equiv 0 \bmod p$, and for simplicity, we will suppose that $j \not\equiv 0, 1728 \bmod l$. Define $k \in \mathsf{F}_l$ to be the congruence class

$$k = \left[\frac{27j}{j-1728}\right],$$

and then consider the two elliptic curves

$$\begin{aligned} y^2 &= 4x^3 - kx - k \\ y^2 &= 4x^3 - c^2kx - c^3k, \end{aligned} \tag{14.29}$$

where $c \in \mathsf{F}_l$ is a nonsquare. We have the following result:

Proposition 14.30. *Of the two elliptic curves over F_l defined in (14.29), one has order $l+1-2a$, and the other has order $l+1+2a$, where $l = a^2 + nb^2$.*

Proof. Let L be the ring class field of the order $\mathcal{O} = \mathbb{Z}[\sqrt{-n}]$, and let $H_{-4n}(X) = \prod_{i=1}^{h}(X - j(\mathfrak{a}_i))$ be the class equation. If $\mathfrak{P}$ is prime in L, then the isomorphism $\mathcal{O}_L/\mathfrak{P} \simeq \mathbb{Z}/l\mathbb{Z} = \mathsf{F}_l$ shows that our solution j of $H_{-4n}(X) \equiv 0 \bmod l$ satisfies $j \equiv j(\mathfrak{a}_i) \bmod \mathfrak{P}$ for some i. It follows that the curves (14.29) are members of the corresponding collection $\overline{E}_c$ constructed in the proof of Theorem 14.18, and our proposition then follows immediately since $l = \pi\overline{\pi}$ in $\mathcal{O}$, where $\pi = a + b\sqrt{-n}$. Q.E.D.

The curves (14.29) don't make sense when $j \equiv 0, 1728 \bmod l$, but the proof of Theorem 14.18 makes it clear how to proceed in these cases.

We can now sketch the Goldwasser–Kilian–Atkin Test. Given a potential prime l, one searches for the smallest n with l of the form $a^2 + nb^2$. Once we succeed, we check if either $l + 1 \pm 2a$ is twice a probable prime q. If not, we look for the next n with $l = a^2 + nb^2$. We continue this until $l + 1 \pm 2a$ has the right form, and then we apply the special primality test embodied in Lemma 14.23, using the two curves given in (14.29). In this way, we can prove that l is prime, provided that q is prime. Then, as in the regular Goldwasser–Kilian Test, we make the whole process recursive.

In practice, the implementation of the Goldwasser–Kilian–Atkin Test improves the run time by allowing the order $l + 1 \pm 2a$ to be more complicated than just twice a prime. The complete description of an implementation can be found in Morain's article [79].

For our purposes, this test is wonderful because it relates so nicely to our problem of when a prime is of the form $x^2 + ny^2$. But from a practical point of view, the situation is less than ideal, for the test requires knowing $H_{-4n}(X)$, a polynomial with notoriously large coefficients. So in implementing the Goldwasser–Kilian–Atkin Test, one of the main goals is to avoid computing the full class equation. Different authors have taken different approaches to this problem, but the basic idea in each case is to use the Weber functions $\mathfrak{f}(\tau)$, $\mathfrak{f}_1(\tau)$ and $\mathfrak{f}_2(\tau)$ from §12. In [79, §6.2], Morain uses formulas of Weber, such as the one quoted in §12

$$\mathfrak{f}(\sqrt{-105})^6 = \sqrt{2}^{-13}(1+\sqrt{3})^3(1+\sqrt{5})^3(\sqrt{3}+\sqrt{7})^3(\sqrt{5}+\sqrt{7}),$$

to determine a root of $H_{-4n}(X)$ modulo l when n is one of Euler's convenient numbers (as defined in §3). Another approach, suggested by Kaltofen, Valente and Yui [64], is to use the methods of Kaltofen and Yui [65] to compute the minimal polynomials of the Weber functions. To see the potential savings, consider the case $n = 14$. We proved in §12 that

$$\mathfrak{f}_1(\sqrt{-14})^2 = \frac{\sqrt{2}+1+\sqrt{2\sqrt{2}-1}}{\sqrt{2}}$$

$$j(\sqrt{-14}) = \left(\mathfrak{f}_1(\sqrt{-14})^{16} + \frac{16}{\mathfrak{f}_1(\sqrt{-14})^8}\right)^3.$$

It is clear which one has the simpler minimal polynomial! The papers by Kaltofen, Valente and Yui [64] and Morain [79] give more details on the various implementions of the Goldwasser–Kilian–Atkin Test.

Primality testing is a good place to end this book, for primes are the basis of all number theory. We started in §1 with concrete questions concerning $p = x^2 + y^2$, $x^2 + 2y^2$ and $x^2 + 3y^2$, and followed the general question of $x^2 + ny^2$ through various wonderful areas of number theory. The theory of §8 was rather abstract, and even the ring class fields of §9 were not very intuitive. Complex multiplication helped bring these ideas down to earth, and now elliptic curves bring us back to the question of proving that a given number is prime. Fermat and Euler would have loved it.

E. Exercises

14.1. Let K be a field, and let $\mathbf{P}^2(K)$ be the projective plane over K, which is the set $K^3 - \{0\}/\sim$, where we set $(\lambda x, \lambda y, \lambda z) \sim (x, y, z,)$ for all $\lambda \in K^*$.

(a) Show that the map $(x, y) \mapsto (x, y, 1)$ defines an injection $K^2 \to \mathbf{P}^2(K)$ and that the complement $\mathbf{P}^2(K) - K^2$ consists of those points with $z = 0$ (this is called the *line at infinity*).

(b) Given an elliptic curve E over K defined by the Weierstrass equation $y^2 = 4x^3 - g_2x - g_3$, we form the equation

$$y^2z = 4x^3 - g_2xz^2 - g_3z^3,$$

which is a homogeneous equation of degree 3. Then we define

$$\tilde{E}(K) = \{(x,y,z) \in \mathbf{P}^2(K) : y^2z = 4x^3 - g_2xz^2 - g_3z^3\}.$$

To relate this to $E(K)$, show that

$$\tilde{E}(K) = \{(x,y,1) \in \mathbf{P}^2(K) : y^2 = 4x^3 - g_2x - g_3\} \cup \{0,1,0\}.$$

Thus the projective solutions consist of the solutions of the affine equation together with one point at infinity, $(0,1,0)$. This is the point denoted ∞ in the text.

14.2. Let $L \subset \mathbf{C}$ be a lattice, and let $y^2 = 4x^3 - g_2(L)x - g_3(L)$ be the corresponding elliptic curve. Then show that the map $z \mapsto (\wp(z), \wp'(z))$ induces a bijection

$$(\mathbf{C} - L)/L \xrightarrow{\sim} E(\mathbf{C}) - \{\infty\}.$$

Hint: use Lemma 10.4 and part (b) of Exercise 10.14. Note also that $\wp'(z)$ is an odd function.

14.3. Prove Proposition 14.3.

14.4. In this exercise we will study elliptic curves with the same j-invariant.

(a) Prove Propositions 14.4 and 14.5.

(b) Consider the elliptic curves $y^2 = 4x^3 - g_3$, where g_3 is any nonzero integer. These curves all have j-invariant 0, so that they are all isomorphic over $\mathbf{C}$. Show that over $\mathbf{Q}$, these curves break up into infinitely many isomorphism classes.

14.5. In this exercise we will study the addition and duplication laws of $\wp'(z)$.

(a) Use formula (14.6) and the addition law for $\wp(z+w)$ (see Theorem 10.1) to conjecture and prove an addition law for $\wp'(z+w)$.

(b) Use formula (14.7) and the duplication law for $\wp(2z)$ (see 10.13) to conjecture and prove a duplication law for $\wp'(2z)$.

14.6. If L and L' are lattices and $\alpha L \subset L'$, where $\alpha \neq 0$, show that the kernel of the map $\alpha : \mathbf{C}/L \to \mathbf{C}/L'$ is isomorphic to $L'/\alpha L$.

14.7. Complete the proof (begun in (14.10)) that

$$(x,y) \mapsto \left(-\frac{2x^2+4x+9}{4(x+2)}, -\frac{1}{\sqrt{-2}}\frac{2x^2+8x-1}{4(x+2)^2}y \right)$$

defines the isogeny of $y^2 = 4x^3 - 30x - 28$ given by complex multiplication by $\sqrt{-2}$. Hint: use the discussion surrounding (10.21) and (10.22).

14.8. Let E be an elliptic curve the finite field F_q, and for any extension $\mathsf{F}_q \subset L$, define $Frob_q : E(L) \to E(L)$ by $Frob_q(x,y) = (x^q, y^q)$.

(a) Show that $Frob_q$ is a group homomorphism.

(b) Show that $Frob_q$ is not of the form $(R(x), (1/\alpha)R'(x)y)$ for any rational function $R(x)$.

14.9. Formulate and prove a version of Proposition 14.9 that applies to lattices L and L' such that $\alpha L \subset L'$ for some $\alpha \in \mathsf{C}^*$.

14.10. Use Theorem 11.23 to prove Proposition 14.11.

14.11. If E is an elliptic curve over F_q, then prove that $|E(\mathsf{F}_q)| \leq 2q+1$. Hint: given x, how many y's can satisfy $y^2 = 4x^3 - g_2 x - g_3$?

14.12. If E is an elliptic curve over F_q, then show that

$$E(\mathsf{F}_q) = \ker(Frob_q : E(\overline{\mathsf{F}}_q) \to E(\overline{\mathsf{F}}_q)),$$

where $\overline{\mathsf{F}}_q$ is the algebraic closure of F_q. Hint: for $x \in \overline{\mathsf{F}}_q$, recall that $x \in \mathsf{F}_q$ if and only if $x^q = x$.

14.13. This exercise is concerned with the relation between Gauss's claim (14.17) and Theorem 14.16.

(a) Verify that the transformation $(x,y) \mapsto (3x/(1+y), 9(1-y)/(1+y))$ takes the curve $x^3 = y^3 + 1$ into the elliptic curve E defined by $y^2 = 4x^3 - 27$.

(b) The projective version of (a) is given by $(x,y,z) \mapsto (3x, 9(1-y), 1+y)$. Check that $(0,-1)$ on $x^3 = y^3 + 1$ is the only point that maps to $\infty = (0,1,0)$ on E.

(c) Check that $x^3 = y^3 + 1$ has three points at infinity. Hint: remember that $p \equiv 1 \bmod 3$.

(d) Show that E has complex multiplication by $\mathsf{Z}[\omega]$, $\omega = e^{2\pi i/3}$. Hint: see Exercise 10.17.

14.14. The last entry in Gauss' mathematical diary states that for a prime $p \equiv 1 \bmod p$,

> If $p = a^2 + b^2$ and $a + bi$ is primary, then $N = p - 2a - 3$,
>
> where N is the number of solutions modulo p of
>
> $x^2 + y^2 + x^2y^2 = 1 \bmod p$.

Show that this is a special case of Theorem 14.16. Hint: use the change of variables $(x,y) \mapsto ((1+x)/2(1-x), (1+x^2)y/(1-x)^2)$ to transform the curve $x^2 + y^2 + x^2y^2 = 1$ into the elliptic curve $y^2 = 4x^3 + x$. See the discussion surrounding (4.24) for more details and references.

14.15. Prove that p does not divide the discriminant of the order $\mathcal{O}_a$ defined in the proof of Theorem 14.18.

14.16. Let E be an elliptic curve over a field K, and assume that its j-invariant j is different from 0 and 1728. Then define $k \in K$ to be the number

$$k = \frac{27j}{j-1728}.$$

Then show that the Weierstrass equation for E can be written in the form

$$y^2 = 4x^3 - c^2kx - c^3k$$

for a unique $c \in K^*$. Hint: $c = g_3/g_2$.

14.17. Let $\mathcal{O}$ be an order in an imaginary quadratic field, and let p be a prime not dividing the conductor of $\mathcal{O}$. If $\pi \in \mathcal{O}$ satisfies $N(\pi) = p$, then prove that all solutions $\alpha \in \mathcal{O}$ of $N(\alpha) = p$ are given by $\alpha = \epsilon\pi$ or $\epsilon\overline{\pi}$ for $\epsilon \in \mathcal{O}^*$. Hint: this can be proved using unique factorization of ideals prime to the conductor (see Exercise 7.26).

14.18. Let $\overline{E}_c$ be one of the collections of elliptic curve over $\mathbf{F}_p$ which appear in the proof of Theorem 14.18, and let j denote their common j-invariant. By Exercise 14.16, note that $\overline{E}_c$ consists of *all* elliptic curves over $\mathbf{F}_p$ with this j-invariant.

(a) If $j \neq 0, 1728$, show that the curves break up into two isomorphism classes, each consisting of $(p-1)/2$ curves. Hint: consider the subgroup of squares in $\mathbf{F}_p^*$.

(b) If $j = 1728$ and $p \equiv 1 \bmod 4$, then show that there are four isomorphism classes, each consisting of $(p-1)/4$ curves.

(c) If $j = 0$ and $p \equiv 1 \bmod 3$, then show that there are six isomorphism classes, each consisting of $(p-1)/6$ curves.

14.19. In this exercise, we will sketch two proofs that there are $q(q-1)$ elliptic curves over the finite field $\mathbf{F}_q$. As ususal, $q = p^a$, $p > 3$.

(a) Adapt the proof of Exercise 14.16 to show that there are q possible j-invariants for elliptic curves over $\mathbf{F}_q$, and show that there are $q-1$ curves with a given j-invariant. This gives $q(q-1)$ elliptic curves.

(b) A second way to prove the formula is to show that there are exactly q solutions $(g_2, g_3) \in \mathbf{F}_q^2$ of the equation $g_2^3 - 27g_3^2 = 0$. We can write this as $(g_2/3)^3 = g_3^2$, and once we exclude the trivial solution $(0,0)$ we need to study solutions of $u^3 = v^2$ in the group $\mathbf{F}_q^*$. So prove the following general fact: if G is a finite Abelian group and a and b are relatively prime integers, then the equation $u^a = v^b$ has exactly $|G|$ solutions in $G \times G$.

14.20. Let $\mathcal{O}'$ be an order in an imaginary quadratic field. Given a integer m which isn't a perfect square, show that

$$|\{\alpha \in \mathcal{O}' : N(\alpha) = m\}| = 2 \sum_{0 \leq |a| \leq 2\sqrt{m}} \chi(a),$$

where $\chi(a)$ is defined by

$$\chi(a) = \begin{cases} 1 & \text{if } \mathcal{O}' \text{ contains a root of } x^2 - ax + m \\ 0 & \text{otherwise.} \end{cases}$$

14.21. Use Theorem 14.26 to show when Conjecture 14.27 is true, there is a constant $c_3 > 0$ such that for all sufficiently large primes l, there are at least

$$c_3 \cdot \frac{l(l-1)}{(\ln l)^2}$$

elliptic curves E over F_l with $|E(\mathsf{F}_l)|$ twice a prime.

REFERENCES

1. T. M. Apostol, *Modular Functions and Dirichlet Series in Number Theory*, Springer-Verlag, Berlin, Heidelberg, and New York, 1976.
2. E. Artin, *Galois Theory*, University of Notre Dame, Notre Dame, Indiana, 1959.
3. A. Baker, *Linear forms in the logarithms of algebraic numbers I*, Mathematika **13** (1966), pp. 204–216.
4. W. E. H. Berwick, *Modular invariants expressible in terms of quadratic and cubic irrationalities*, Proc. Lon. Math. Soc. **28** (1927), pp. 53–69.
5. K. R. Biermann, E. Schuhmann, H. Wussing and O. Neumann, *Mathematisches Tagebuch 1796–1814 von Carl Friedrich Gauss*, 3rd edition, Ostwalds Klassiker 256, Leipzig, 1981.
6. B. J. Birch, *Diophantine analysis and modular functions*, in *Algebraic Geometry, Papers Presented at the Bombay Colloquium, 1968*, Oxford University Press, London, 1969, pp. 35–42.
7. B. J. Birch, *Weber's class invariants*, Mathematika **16** (1969), pp. 283–294.
8. Z. I. Borevich and I. R. Shafarevich, *Number Theory*, Academic Press, New York, 1966.
9. J. M. Borwein and P. B. Borwein, *Pi and the AGM*, Wiley, New York, 1987.
10. W. E. Briggs, *An elementary proof of a theorem about the representations of primes by quadratic forms*, Canadian J. Math. **6** (1954), pp. 353–363.

11. G. Bruckner, *Charakterisierung der galoisschen Zahlkörper, deren zerlegte Primzahlen durch binäre quadratische Formen gegeben sind*, Math. Nachr. **32** (1966), pp. 317–326.

12. D. A. Buell, *Class Groups of Quadratic Fields I, II*, Math. Comp. **30** (1976), pp. 610–623 and **48** (1987), pp. 85–93.

13. W. K. Bühler, *Gauss: A Biographical Study*, Springer-Verlag, Berlin, Heidelberg, and New York, 1981.

14. J. J. Burkhardt, *Euler's work on number theory: a concordance for A. Weil's Number Theory*, Historia Math. **13** (1986), pp. 28–35.

15. P. Cassou–Noguès and M. J. Taylor, *Elliptic Functions and Rings of Integers*, in Progress in Math. **66**, Birkhäuser, Boston, Basel, and Stuttgart, 1987.

16. K. Chandrasekharan, *Elliptic Functions*, Springer-Verlag, Berlin, Heidelberg, and New York, 1985.

17. S. Chowla, *An extension of Heilbronn's class number theorem*, Quarterly J. Math. **5** (1934), pp. 304–307.

18. P. Cohen, *On the coefficients of the transformation polynomials for the elliptic modular function*, Math. Proc. Camb. Phil. Soc. **95** (1984), pp. 389–402.

19. H. Cohn, *A Classical Invitation to Algebraic Numbers and Class Fields*, Springer-Verlag, Berlin, Heidelberg, and New York, 1978.

20. H. Cohn, *A Second Course in Number Theory*, Wiley, New York, 1962. (Reprinted as *Advanced Number Theory*, Dover, New York, 1980.)

21. H. Cohn, *Introduction to the Construction of Class Fields*, Cambridge University Press, Cambridge, 1985.

22. M. J. Collinson, *The origins of the cubic and biquadratic reciprocity laws*, Arch. Hist. Exact Sci. **17** (1977), pp. 63–69.

23. D. Cox, *The arithmetic-geometric mean of Gauss*, L'Ens. Math. **30** (1984), pp. 275–330.

24. M. Deuring, *Die Klassenkörper der Komplexen Multiplikation*, in *Enzyklopädie der Mathematischen Wissenschaften*, Band I 2, Heft 10, Teil II, Teubner, Stuttgart, 1958.

25. M. Deuring, *Teilbarkeitseigenschaften der singulären Moduln der elliptischen Funktionen und die Diskriminante der Klassengleichung*, Commentarii Math. Helv. **19** (1946), pp. 74–82.

26. L. E. Dickson, *History of the Theory of Numbers*, Carnegie Institute, Washington D.C., 1919–1923. (Reprint by Chelsea, New York, 1971.)

27. P. G. L. Dirichlet, *Werke*, Berlin, 1889–1897. (Reprint by Chelsea, New York, 1969.)

28. P. G. L. Dirichlet, *Zahlentheorie*, 4th edition, Vieweg Brunswick, 1894.

29. D. Dorman, *Singular moduli, modular polynomials, and the index of the closure of* $\mathbb{Z}[j(\tau)]$ *in* $\mathbb{Q}(j(\tau))$, Math. Annalen **283** (1989), pp. 177–191.

30. D. Dorman, *Special values of the elliptic modular function and factorization formulae*, J. reine angew. Math. **383** (1988), pp. 207–220.

31. H. Edwards, *Fermat's Last Theorem*, Springer-Verlag, Berlin, Heidelberg, and New York, 1977.

32. W. Ellison and F. Ellison, *Théorie des nombres*, in *Abrégé d'Histoire des Mathématiques 1700–1900*, Vol. I, ed. by J. Dieudonné, Hermann, Paris, 1978, pp. 165–334.

33. L. Euler, *Opera Omnia*, Series prima, Vols. I–V, Teubner, Leipzig and Berlin, 1911–1944.

34. P. Eymard and J. P. Lafon, *Le journal mathématique de Gauss*, Revue d'Histoire des Sciences **9** (1956), pp. 21–51.

35. P. de Fermat, *Oeuvres*, Gauthier-Villars, Paris, 1891–1896.

36. D. Flath, *Introduction to Number Theory*, Wiley, New York, 1988.

37. W. Franz, *Die Teilwert der Weberschen Tau-Funktion*, J. reine angew. Math. **173** (1935), pp. 60–64.

38. G. Frei, *Leonhard Euler's convenient numbers*, Math. Intelligencer **3** (1985), pp. 55–58 and 64.

39. G. Frei, *On the development of the genus of quadratic forms*, Ann. Sc. Math. Québec **3** (1979), pp. 5–62.

40. P.–H. Fuss, *Correspondance Mathématique et Physique*, St. Petersburg, 1843. (Reprint by Johnson Reprint Corporation, New York and London, 1968.)

41. C. F. Gauss, *Disquisitiones Arithmeticae*, Leipzig, 1801. Republished in 1863 as Volume I of *Werke* (see [42]). French translation, *Recherches Arithmétiques*, Paris, 1807. (Reprint by Hermann, Paris, 1910.) German translation, *Untersuchungen über Höhere Arithmetik*, Berlin, 1889. (Reprint by Chelsea, New York, 1965.) English Translation, Yale, New Haven, 1966. (Reprint by Springer-Verlag, Berlin, Heidelberg, and New York, 1986.)

42. C. F. Gauss, *Werke*, Gottingen and Leipzig, 1863–1927.

43. S. Goldwasser and J. Kilian, *Almost all primes can be quickly certified*, Proc. 18th Annual ACM Symp. on Theory of Computing (STOC), Berkeley, 1986, pp. 316–329.

44. J. J. Gray, *A commentary on Gauss's mathematical diary, 1796–1814, with an English translation*, Expo. Math. **2** (1984), pp. 97–130. (Gauss' diary has also been translated into French [34] and German [5].)

45. B. Gross, *Arithmetic on Elliptic Curves with Complex Multiplication*, Lecture Notes in Math. **776**, Springer-Verlag, Berlin, Heidelberg, and New York, 1980.

46. B. Gross and D. Zagier, *On singular moduli*, J. reine angew. Math. **355** (1985), pp. 191–220.

47. E. Grosswald, *Representations of Integers as Sums of Squares*, Springer-Verlag, Berlin, Heidelberg, and New York, 1985.

48. G. H. Hardy and E. M. Wright, *An Introduction to the Theory of Numbers*, 5th edition, Clarendon Press, Oxford, 1979.

49. H. Hasse, *Bericht über neuere Untersuchungen und Probleme as der Theorie def algebraischen Zahlkörper, I, Ia and II*, Jahresber. Deutch. Math. Verein **35** (1926), pp. 1–55, **36** (1927), pp. 233–311, and Erg. Bd. **6** (1930), pp. 1–201. (Reprint by Physica-Verlag, Würzburg Vienna, 1965.)

50. H. Hasse, *Number Theory*, Springer-Verlag, Berlin, Heidelberg, and New York, 1980.

51. H. Hasse, *Zur Geschlechtertheorie in quadratischen Zahlkörpern*, J. Math. Soc. Japan **3** (1951), pp. 45–51.

52. K. Heegner, *Diophantische Analysis und Modulfunktionen*, Math. Zeit. **56** (1952), pp. 227–253.

53. O. Hermann, *Über die Berechnung der Fourierkoeffizienten der Funktion* $j(\tau)$, J. reine angew. Math. **274/275** (1974), pp. 187–195.

54. I. N. Herstein, *Topics in Algebra*, 2nd edition, Wiley, New York, 1975.

55. C. S. Herz, *Computation of singular* j*-invariants*, in *Seminar on Complex Multiplication*, Lecture Notes in Math. **21**, Springer-Verlag, Berlin, Heidelberg, and New York, 1966, pp. VIII-1 to VIII-11.

56. C. S. Herz, *Construction of class fields*, in *Seminar on Complex Multiplication*, Lecture Notes in Math. **21**, Springer-Verlag, Berlin, Heidelberg, and New York, 1966, pp. VII-1 to VII-21.

57. L.-K. Hua, *Introduction to Number Theory*, Springer-Verlag, Berlin, Heidelberg, and New York, 1982.

58. D. Husemöller, *Elliptic Curves*, Springer-Verlag, Berlin, Heidelberg, and New York, 1987.

59. K. Ireland and M. Rosen, *A Classical Introduction to Modern Number Theory*, Springer-Verlag, Berlin, Heidelberg, and New York, 1982.

60. M. Ishida, *The Genus Fields of Algebraic Number Fields*, Lecture Notes in Math. **555**, Springer-Verlag, Berlin, Heidelberg, and New York, 1976.

61. C. G. J. Jacobi, *Gesammelte Werke*, Vol. 6, Berlin, 1891. (Reprint by Chelsea, New York, 1969.)

62. G. Janusz, *Algebraic Number Fields*, Academic Press, New York, 1973.

63. B. W. Jones, *The Arithmetic Theory of Quadratic Forms*, Carus Monographs 10, MAA, Washington D. C., 1950.

64. E. Kaltofen, T. Valente and N. Yui, *An improved Las Vegas primality test*, preliminary report, 1989.

65. E. Kaltofen and N. Yui, *Explicit construction of the Hilbert class fields of imaginary quadratic fields by integer lattice reduction*, preliminary version, 1989.

66. E. Kaltofen and N.Yui, *On the modular equation of order 11*, in *Third MACSYMA User's Conference, Proceedings*, General Electric, 1984, pp. 472–485.

67. N. Koblitz, *Introduction to Elliptic Curves and Modular Forms*, Springer-Verlag, Berlin, Heidelberg, and New York, 1984.

68. L. Kronecker, *Werke*, Leipzig, 1895–1931. (Reprint by Chelsea, New York, 1968.)

69. J. L. Lagrange, *Oeuvres*, Vol. 3, Gauthier-Villars, Paris, 1869.

70. E. Landau, *Über die Klassenzahl der binären quadratischen Formen von negativer Discriminante*, Math. Annalen **56** (1903), pp. 671–676.

71. E. Landau, *Vorlesungen über Zahlentheorie*, Hirzel, Leipzig, 1927.

72. S. Lang, *Algebraic Number Theory*, Springer-Verlag, Berlin, Heidelberg, and New York, 1986.

73. S. Lang, *Elliptic Functions*, 2nd edition, Springer-Verlag, Berlin, Heidelberg, and New York, 1987.

74. A. M. Legendre, *Essai sur la Théorie des Nombres*, Paris, 1798. Third edition retitled *Théorie des Nombres*, Paris, 1830. (Reprint by Blanchard, Paris, 1955.)

75. A. M. Legendre, *Recherches d'analyse indéterminée*, in *Histoire de l'Académie Royale des Sciences, 1785*, Paris, 1788, pp. 465–559.

76. H. W. Lenstra, Jr., *Factoring integers with elliptic curves*, Annals of Math. **126** (1987), pp. 649–673.

77. D. Marcus, *Number Fields*, Springer-Verlag, Berlin, Heidelberg, and New York, 1977.

78. G. B. Mathews, *Theory of Numbers*, Deighton Bell, Cambridge, 1892. (Reprint by Chelsea, New York, 1961.)

79. F. Morain, *Implementation of the Goldwasser–Kilian–Atkin primality testing algorithm*, draft, 1988.

80. J. Neukirch, *Class Field Theory*, Springer-Verlag, Berlin, Heidelberg, and New York, 1986.

81. J. Oesterlé, *Nombres des classes des corps quadratiques imaginaires*, Asterisque **121–122** (1985), pp. 309–323.

82. T. Ono, *Arithmetic of Algebraic Groups and its Applications*, St. Paul's International Exchange Series, Occasional Papers VI, St. Paul's University, 1986.

83. H. Orde, *On Dirichlet's class number formula*, J. London Math. Soc. **18** (1978), pp. 409–420.

84. G. J. Reiger, *Die Zahlentheorie bei C. F. Gauss*, in *C. F. Gauss, Gedenkband Anlässlich des 100. Todestages, am 23. Februar 1955*, Teubner, Leipzig, 1957, pp. 38–77.

85. P. Roquette, *On class field towers*, in *Algebraic Number Theory*, ed. by J. W. S. Cassels and A. Fröhlich, Academic Press, New York, 1967, pp. 231–249.

86. W. Scharlau and H. Opolka, *From Fermat to Minkowski*, Springer-Verlag, Berlin, Heidelberg, and New York, 1985.

87. R. Schertz, *Die singulären Werte der Weberschen Funktionen* $\mathfrak{f}$, $\mathfrak{f}_1$, $\mathfrak{f}_2$, γ_2, γ_3, J. reine angew. Math. **286/287** (1976), pp. 46–74.

88. J.-P. Serre, *A Course in Arithmetic*, Springer-Verlag, Berlin, Heidelberg, and New York, 1973.

89. D. Shanks, *Class number, a theory of factorization, and genera*, in *1969 Number Theory Institute*, Proc. Symp. Pure Math. **20**, AMS, Providence, Rhode Island, 1971, pp. 415–440.

90. G. Shimura, *Arithmetic Theory of Automorphic Functions*, Princeton University Press, Princeton, New Jersey, 1971.

91. C. L. Siegel, *Equivalence of Quadratic Forms*, Am. J. Math. **63** (1941), pp. 658–680.

92. C. L. Siegel, *Über die Classenzahl quadratischer Zahlkörper*, Acta Arithmetica **1** (1935), pp. 83–86.

93. J. H. Silverman, *The Arithmetic of Elliptic Curves*, Springer-Verlag, Berlin, Heidelberg, and New York, 1986.

94. H. J. S. Smith, *Note on the modular equation for the transformation of the third order*, Proc. London Math. Soc. **10** (1878), pp. 87–91.

95. H. J. S. Smith, *Report on the Theory of Numbers*, Reports of the British Association, 1859–1865. (Reprint by Chelsea, New York, 1965.)

96. H. M. Stark, *A complete determination of the complex quadratic fields of class number one*, Michigan Math. J. **14** (1967), pp. 1–27.

97. H. M. Stark, *Class numbers of complex quadratic fields*, in *Modular Functions of One Variable I*, Lecture Notes in Math. **320**, Springer-Verlag, Berlin, Heidelberg, and New York, 1973, pp. 153–174.

98. H. M. Stark, *On the "gap" in a theorem of Heegner*, J. Number Theory **1** (1969), pp. 16–27.

99. S. Wagon, *The evidence: primality testing*, Math. Intelligencer **8** (1986), pp. 58–61.

100. J. Wallis, *Opera Mathematica*, Oxford, 1695–1699. (Reprint by G. Olms, Hindesheim, New York, 1972.)

101. H. Weber, *Beweis des Satzes, daß jede eigentlich primitive quadratische Form unendliche viele Primzahlen darzustellen fähig ist*, Math. Annalen **20** (1882), pp. 301–329.

102. H. Weber, *Lehrbuch der Algebra*, Vol. III, 2nd edition, Braunschweig, 1908. (Reprint by Chelsea, New York, 1961.)

103. H. Weber, *Zur Komplexen Multiplikation elliptischer Funktionen*, Math. Annalen **33** (1889), pp. 390–410.

104. A. Weil, *Basic Number Theory*, 3rd edition, Springer-Verlag, Berlin, Heidelberg, and New York, 1974.

105. A. Weil, *La cyclotomie jadis et naguère*, L'Ens. Math. **20** (1974), pp. 247–263. Reprinted in *Essais Historiques sur la Théorie des Nombres*, Monograph **22**, L'Ens. Math., Geneva, 1975, pp. 39–55, and in Vol. III of *André Weil: Collected Papers*, Springer-Verlag, Berlin, Heidelberg, and New York, 1980, pp. 311–327.

106. A. Weil, *Number Theory: An Approach Through History*, Birkhäuser, Boston, Basel, and Stuttgart, 1984.

107. A. Weil, *Two lectures on number theory: past and present*, L'Ens. Math. **20** (1974), pp. 87–110. Reprinted in *Essais Historiques sur la Théorie des Nombres*, Monograph **22**, L'Ens. Math., Geneva, 1975, pp. 7–30, and in Vol. III of *André Weil: Collected Papers*, Springer-Verlag, Berlin, Heidelberg, and New York, 1980, pp. 279–302.

108. P. J. Weinberger, *Exponents of the class groups of complex quadratic fields*, Acta Arith. **22** (1973), pp. 117–124.

109. E. T. Whittaker and G. N. Watson, *A Course of Modern Analysis*, 4th edition, Cambridge University Press, Cambridge, 1963.

110. N. Yui, *Explicit form of the modular equation*, J. reine angew. Math. **299/300** (1978), pp. 185–200.

111. D. Zagier, *Zetafunktionen und Quadratiche Korper*, Springer-Verlag, Berlin, Heidelberg, and New York, 1981.

112. D. Zagier, *L-Series of elliptic curves, the Birch–Swinnerton-Dyer conjecture, and the class number problem of Gauss*, Notices of the AMS, **31** (1984), pp. 739–743.

INDEX

D

Q

QA 246 .C69 1989

Cox, David A.

Primes of the form x2 + ny2